SHORT HISTORY OF THE ART OF DISTILLATION FROM THE BEGINNINGS UP TO THE DEATH OF CELLIER BLUMENTHAL

SHORT HISTORY

OF THE

ART OF DISTILLATION

FROM THE BEGINNINGS UP TO THE DEATH
OF CELLIER BLUMENTHAL

BY

R. J. FORBES

WITH 203 ILLUSTRATIONS

LEIDEN

E. J. BRILL

1948

Printed in the Netherlands

CONTENTS

NOTE TO THE READER

Books full of notes are rightly disliked by the general reader who wants a continuous story and does not bother about the evidence for every scrap of opinion stated in the text. Therefore, the author has appended a full bibliography to this book for those who wish to study the subject further, to which bibliography the numbers in parentheses in the text refer.

As far as the sources of the biographies recounted in this story are not stated, these "lives" have been compiled from the data in J. CH. POGGENDORFF's *Handwörterbuch* (488) and HABERLING, HÜBOTTER & VIERORDT, *Biographisches Lexikon der hervorragenden Aerzte aller Zeiten und Völker* (2nd edition, Berlin, 1929 . . . .), which have been carefully compared and corrected with those in the *Encyclopedia Britannica, Larousse* and other reliable works in different languages. As to the data on the Arabian chemists they have been compiled from MIELI's *Science Arabe* (427) and the references in the *Enzyklopaedie des Islam,* unless other sources are indicated.

On other points as few notes as possible refer to the pages of the works of the authors discussed. In general these indications are of little use, as one has to study far more than these few lines in the original work to understand the correct meaning of the passage or lines cited in this text. In looking up these references one should never forget to study the whole scheme and meaning of the book cited as these are generally of the utmost importance for the correct interpretation of the lines quoted.

Amsterdam 1944   R. J. FORBES

CHAPTER ONE

INTRODUCTION

"Je fais profession de ne rien dire que ce que je scay
et de n'escrire rien que ce que j'ay fait"
                                    CHRISTOPHE GLASER (261)

Distillation is an art and even an ancient one. It is strange to find that the history of this oldest and still most important method of producing chemically pure substances has never been written. The reader looking at the bibliography appended to this book might object that many data existed. This may be true but a proper history of the art from its origin upto the present time was lacking. Three small works that attempted to give this history were failures. There is the booklet by SCHREINER (566), that by DUJARDIN (197) and that by SCHELENZ (559). SCHELENZ as an apothecary looks at distillation from this narrow point of view and gives hardly any facts on the chemical technology side of that matter. But his data, titles and quotations, like those of DUJARDIN are often faulty and cause much trouble to the reader who desires to look up the original literature. Though DUJARDIN gives many interesting facts on the industrial sides of distillation his book makes the impression of a notebook, giving small paragraphs of facts but no continuous history. Thus apart from two excellent but far too short essays by UNDERWOOD (635) (636) there is no historical survey of this major operation in laboratory and technology.

Now the history of distillation forms part of that of chemistry and chemical technology, a fact which most of the above mentioned writers have overlooked or stressed too little. But the history of chemistry can not be understood without looking at the cultural background, that is the history of civilisation. If therefore we often refer to these traits of civilisation in the course of our discussion this is because we must suppose a strong influence of these aspects of civilisation on the history of distillation. In our opinion this is too often forgotten in the books on historical-technical subjects but it is eminently important to draw our sketch on the proper background which makes the lines stand out as they should.

The history of distillation is closely connected with that of alcoholic beverages and this has often led authors to wrong conclusions as to the origin of distillation. For the ethnologist who studies the civilisation of many primitive tribes will be struck by the many more or less primitive forms of distillation which he meets all over the world. Thus SPEIGHT (592) tells us that among African tribes the favourite native intoxicating liquour is made by fermenting oatmeal gruel with water. This is the drink called "manawa", kaffir beer or skokiaon. Races along the shores of the Victoria Nyanza are more addicted to banana wine, a comparatively mild and inoccuous beverage, but the natives make a distilled liquour from it which is highly intoxicating. The distilling apparatus consists of calabashes and bamboo reeds, the distillate being collected in calabashes. The resultant liquour

Fig. 1. Native still for cajapoetih-oil (Indian Institute Amsterdam)

is too punguent even for the African native and has to be diluted before drinking.

VON HUMBOLDT, the great German explorer (1769-1859), who travelled several years in South America (1799-1804) and Siberia (1829-1830) when speaking of the preparation of alcoholic beverages by the Kalmuks discusses the history of distillation in general (311). He points to the fact that even the earliest travellers in Asia found the tribes in the steppes preparing distilled drinks. RUBROUCK (1250 A.D.) for instance mentions a rice-wine and another alcoholic beverage from milk called ceracina (terracina) or caracosus. The widespread word arak is first found in PIGAFETTI's description of the voyage of MAGELLAN and here it is expressly said that arak is "vino di riso distillato, vino fatto lambicco e chiaro come acqua". All such

terms as raki, arak, etc. point to rice-wine though we should remember with VON HUMBOLDT that such terms do not only denote distilled
beverages but also others made by interrupting fermentation at some
point during the process. Only too often do modern authors make
this mistake and state that every drink of this kind is distilled thus
creating false data which cause much confusion. VON HUMBOLDT
quite correctly states that for instance the Vedas never mention
anything like distillation, therefore the famous ancient drink called
"soma" is certainly not a distillate. On many other points too he
brings the proofs to refute the wrong opinions revived at regular
intervals that distillation was known long before Christ in the East,
in India or in Egypt, opinions which we find in the works of FAIRLEY
(216) and others.

It would not be difficult to quote many scores of examples of primitive distilling apparatus from the works of VON HUMBOLDT and
modern ethnologists. BUCH (104) describes an apparatus used by the
Wotjaks, the cooling tube of which passes a trough of water, and
in which a "kumyška" with 5-20 % of alcohol is prepared from fermented milk. A similar apparatus is used by the Tartars of Astrachan.
KLEMM (329) describes such an apparatus of the Kalmuks in which
they distill their araki or raka from fermented milk. Rich Kalmuks
do not distill their raki three times but six times or more, as VON
HUMBOLDT remarks. This distilled Kalmuk beverage is already mentioned by MANGGON (483) in 1250 A. D. TARAJANZ tells us that
the Armenians have very good distilling apparatus and even seem to
use the principle of the Liebig cooler (613). WEULE quotes other
examples, e.g. a distilling apparatus of Amboina with long bamboo
cooling tubes and an apparatus of the Kazaks the receiver of which
is placed in a trough with running water. Any amount of further
examples can be collected from HARTWICH's *Die menschlichen Ge-
nussmittel* (Leipzig, 1911) or handbooks like those of HOFMANN
and GILDEMEISTER (254), GILG and SCHÜRHOFF (255), SCHELENZ
(558), and TSCHIRSCH (631).

But again, it is not superfluous to state that too many alcoholic
beverages of primitive peoples have been stated, without proof, to be
distilled. MANN (407) has proved that many of such "distilled"
drinks of primitive Indian tribes are characterised by a high percentage
of fusel oil and a low percentage of alkaloids (like many of the
imported modern substitutes) and it is highly doubtful whether these
drinks were ever distilled. It is rather naive to believe with WEULE

that distillation must have been known very early because the boiling point of alcohol is lower than that of water! But the idea that distillation was discovered long ago by primitive tribes is a very tenacious cne. One finds it in the works of GESNER (246) who attributes the discovery to the "Barbarians, Carthaginians or Arabs", but also in those of CRAWLEY (146). The latter says on this point: "Apart from fermented drinks primitive peoples knew distilled drinks and infusions. Distillation has been known in the East especially in China from the remotest antiquity. It is an invention difficult to trace to its sources but it seems attested for a few peoples at the stage of lower barbarism and in the higher stage of barbarism it is very generally known. Some of the more primitive American Indians seem to have been acquainted with the process, a primitive form of distillation was found by COOK in the Pacific islands in the eighteenth century. It is known too, but little used in the ancient mediterranean civilisation."

MAURIZIO (421) has even stronger opinions. She maintains that distillation is very old and that it was used in preparing drinks from the juices of trees (palms, etc.), mead, kumiss, beer, etc. There are still primitive distilleries in which they are prepared like camphor and essential oils. Only seldom are vapours of distillate cooled with water. Generally such an apparatus has a bent tube leading from the alembic into the receiver. Often the distilling or "burning" is preceded by a "strengthening", that is a primary concentration by distillation. The peoples of the Asian steppes use both, as the earliest tales of the Kalmuks, Bashkirs and Tartars tell us. MAURIZIO is convinced that the primitive peoples invented distillation on their own and that the art was lost in the Middle Ages. To prove her thesis she cites a long list compiled by FAIRLY in which 15 kinds of brandy are mentioned which were distilled long before Christ. Her argument ends with the words: "It is impossible to find one's way in the knot of contradictions and the belief in the "holy" classical antiquity. If the primitive and prehistorical peoples had known how to write there would have been a rustling in the forest of books. But now they belong to the silent distance. Even Menghin who undertook to use ethnological methods in prehistory did not hear a sound."

With due respect for the important material gathered by Dr. MAURIZIO in her book and the masterly discussion of many aspects of her subject we must point to a serious error as soon as she discusses the history of distillation. In no place has she proved that the primitive peoples which she mentions knew distillation earlier than the civilisations (China, Persia, the Arabs) with which they were in contact. Neither MAURIZIO nor those who are of the same opinion have ever tried to prove the originality of the primitive tribes with proper data (however relative they might be). And if we state that in no civilised country is there any trace of the distillation of alco-

Fig. 2. Distilling kanaga-oil at Serang (Java) (Indian Institute Amsterdam)

holic beverages before this operation was discovered in the north of Italy, that is in the eleventh century A.D., their contentions loose much ground. It is true that the principles of distillation were known earlier, as we shall see, and therefore they may have reached primitive peoples living on the fringe of the civilised world, but even this is not certain. The story of the development of distillation by the primitive peoples all over the world is a subject which merits special study. But this study will only bear fruit when we know better the story of distillation in the civilised regions of the world and when the contact of these with their more primitive neighbours is better established by proper historical data. The typology of the distilling apparatus will be the mainstay of this special study. But in meantime we are sorry that MAURIZIO and so many other authors on the subject have not taken to heart the wise words which SAYCE wrote in his *Primitive Arts and Crafts* on the diffusion of discoveries and inventions. Again HARRISON has written some remarkable lines on the subject in his *Evolution in material culture* (*Report of the British Association for the Advancement of Science* 1930, pp. 137-159). He warns against the underestimation of inventions which seem very simple to us. One forgets too often that at the back of a simple distilling apparatus there are a mass of experiences and experiments and that it represents the combination of several principles of natural science with the ability to make the proper apparatus to execute the operation. One is too often apt to overlook this and to assume the invention of such an apparatus at different places in the world and various dates. But it is better to say with HARRISON: "Simple primary discoveries such as the plasticity or maleability of metals may be repeated but any further step in metallurgy makes diffusion more likely". And applying this to distillation of alcoholic beverages we should say that it is correcter to suppose a priori the diffusion of the invention of distillation and distilling apparatus unless and until we find proofs to the contrary. And taking this as our guiding principle we can not believe at present that distillation was discovered by several primitive tribes at different dates and places. We must adopt the general opinion that distillation was first discovered by the Alexandrian Chemists in the first century A.D. until we have further proof.

Another strong opinion which crops up regularly is that which states that the Celts invented the distillation of alcoholic beverages. URE (638) puts it in this way: "This art of evoking the fiery demon of drunkenness from its attempered state in wine and beer was unknown to the ancient Greeks and Romans. It seems to have been invented by the Barbarians of the north of Europe as a solace of their cold and humid clime". MAURIZIO repeats the statement of I. A. DAVIDSOHN from his article in the *Int. Monatschrift z. Erforschung des Alkoholismus* of 1912 (Heft 8), that the Irish distilled very early in history and ascribes the invention to St. Patrick. Anyway the "mead-song" of the bard TALIESIN is said to refer to distilled mead. Old Irish Celtic ruins near Cashel were said to contain the remains of bronze distilling apparatus. DAVIDSOHN only repeats the opinion of SCARISBRICK that the English army invading Ireland (1172-1174) found the natives using whisky. This word is not connected with Viscaya, but seems derived from a Celtic „uisge-beatha", that is water of life, aqua vitae as VON LIPPMANN proved (373).

The latter writer proved the unhistoricity of DAVIDSOHN's opinions, but the theory of the Celtic origin of distillation seems to crop up regularly in modern times as it did many years ago. At the beginning of the nineteenth century we find DONOVAN maintaining that "Sir James Ware supposed that ardent spirit was distilled in Ireland earlier than in Engeland" (189). But all these contentions remain unproved, the evidence is as weak as the philological thesis which

SCHRADER put forward to prove that the Celts were the inventors of the distillation of whisky from beer and possibly from mead too.

We shall take one more of these many loose statements, that which says that the Chinese have known distillation since times immemorial, supported by PIQUE (483), CRAWLEY (146) and so many others, and examine it in detail. This story is an old one, for a hundred years ago we find that DEMACHY (168) says that the Chinese have discovered alcohol.

Fig. 3. Cassia-oil distillery in China (Indian Institute Amsterdam)

In China too distillation first appears in connection with alchemy. It is therefore important to trace the connection between Alexandrian and Chinese alchemy. Unfortunately this is still a tangled question, but though the battlecries have not yet subsided on the field, it is already possible to trace the bare outline of the resultant scheme.

Many writers like PARTINGTON believed that Chinese alchemy was a child of Arabian chemistry and therefore dated after 1000 A.D. This meant indirect dependence on the Alexandrian school (468). But this opinion is untenable because READ (503) correctly pointed out that authentic remarks on alchemical subjects can be found in the writings of SSU MA CH'IEN (116 B.C.), in documents of 92 B.C. and in the books of WEI PO YANG (142 A.D.) and others before the first half of the fourth century when the famous KO HUNG flourished.

However, READ is of the opinion too that more translations of Chinese alchemical works are necessary to prove an independant evolution in that country.

On the other hand it is important to remember with DAVIS (159) that Chinese and Alexandrian chemistry are alike in being erected on the nature theory of the doctrine of the contraries and are very much alike in materials and apparatus with which they worked. This is also pointed out by CHIKASHIGE(131), who mentions such methods as digestion, heating in manure, etc., though he does not specify distillation methods. Quite probably dualistic philosophy came to China from Iran as the Near Eastern dualistic theories among which Alexandrian chemistry grew up were certainly of Iranian origin. Iranian influences must have penetrated China by the great inland travel routes along which a considerable amount of chemical knowledge may have been carried in the course of many centuries.

The dualistic theory entailed the belief that all substances were born by the interaction of the Male Principle (the positive, active fiery, hot, light principle) and the Female Principle (the negative, passive, cold, heavy, dark principle). This distortion of the older Yin-Yang theory of the Chinese philosophers attracted the Taoists, who by their alchemical practices hoped to gain the immortality which their religion did not promise them. In the West the dualistic tendencies can be found in the writings of ZOSIMOS and his followers (third century A.D. and later) but they did not for a long the time change the essentials of Alexandrian chemical science, which was a true chemistry, dealing with the colouring of metals and the interaction of different substances such as sulphur, arsenic etc., and not an alchemy in the sense in which this term is usually understood.

There seems little reason to believe that the chemistry of Alexandria and Hellenistic Egypt had any direct contact with Chinese alchemy (616). There is some slight likeness of materials and methods, which is only natural seeing the simple reactions and materials studied by both. But to the early chemistry of the West were unknown the two central notions of Chinese alchemy, the prolongation of life and the philosopher's stone, a minute portion of which would transmute a large quantity of base metal. Since the use of cinnabar in the magical Kansu oracle bones many centuries before Christ, the Chinese seem to have had an idea that a Potion of Life could be prepared, a Potion of Immortality which was mostly a concoction containing colloidal gold and the preparation of which usually started with the treatment of the magical cinnabar. The manufacture of gold from baser metals with the help of the philosopher's stone was not attempted to satisfy the greed of the Chinese alchemist, he made gold because of its magical efficacy and because it was one of the base materials for the Potion of Immortality.

It seems now certain that both the notions of the Philosopher's Stone and the Potion of Immortality came back to the West by the way of Iran (159) (616) where they are first found in the alchemical texts of the Arabs. There they mingled with Alexandrian chemical tradition to form one body that was transmitted to the Latins. In the alchemy of the Middle Ages the two traditions are still discernible.

These ideas even penetrated southwards to Burma, where the zaw-gyee (a word derived from the term "yogi") who is more a magician than a fakir is concerned with the preparation of the "stone of live metal", which does not seem to involve distillation (420).

BARNES (46) in reviewing TSAO's paper (630) mentions that the Chinese alchemists employed their distilling apparatus chiefly in the preparation of mercury, roasting cinnabar and decomposing the oxides of mercury. A very simple form of still consisted of a porcelain container ressembling a short-necked round-bottom flask in appearance. Substances like cinnabar and gold were put in the flask and its neck closed with a suitable piece of porcelain. The flask was then inverted and put in a jar neck-downwards, the mouth of the jar was sealed, and jar and flask were heated in an oven. More efficient designs of still-heads and still for sublimation and distillation with still-heads projecting from the oven and long air-cooled condensor tubes leading to cooled receivers appear as early as the twelfth century, though the same type is already used in the third century in the West!

But the Chinese were not only interested in alchemy but also in practical chemistry and they certainly knew how to distill alcoholic drinks before the Jesuits became scientific advisers of the court of Cathay. Now the grape vine was introduced into China from Bactria by the general CHAN K'IEN as early as 128 B.C.. But the Chinese acquired the art of wine-making from a Turkish tribe of Turkistan as late as the T'ang dynasty about the middle of the seventh century A.D. A work of 1331 contains the word a-la-ki which is not derived from "alcohol" but from the Arabic " 'araq". The first author who offers a coherent notice and intelligent discussion of the subject of grape-wine is LI-SHI-CHEN at the end of the sixteenth century. He is well acquainted with the fact that this kind of wine was anciently only

made in the Western Countries, and that the method of manufacturing
it was but introduced under the T'ang after the subjugation of KAO-
CHAN. He discriminates between two types of grape-wine, the fermented and the distilled. In the latter case "ten catties of grapes are
taken and an equal quantity of great leaven (distiller's grains), which
are subjected to a process of fermentation. The whole is then placed
in an earthen kettle and steamed. The drops are collected in a vessel
and this liquid is of red colour(!) and very pleasing". In a preceeding
notice on distillation he states that this is not an ancient method, but
was practised only from the Yüan (Mongol) period, he then describes it in its application to rice-wine in the same manner as to
grape-wine.

After reviewing the Chinese evidence LAUFER (360) states that it
is certain that distillation was a Western invention and was unknown
to the Chinese of the earliest centuries A.D. "What we know of Arabic
knowledge of distillation makes it certain that this method, as stated
by Li Shi-Chen, was first practised under the Mongols, that is in the
fourteenth century. It is hence reasonable to hold (at least for the
present) also that distilled grape-wine was not made earlier in China
than the epoch of the Yüan". The same view is held by J. DUDGEON
(*The beverages of the Chinese*) and GRUPPY (271). And even in the
days of the Jesuit DE PAUW so many centuries later distillation was
hardly a general technological method in China.

On the other hand we must not forget that we have only very little
knowledge of Chinese technological literature. Chinese literary tradition is essentially a literary and philological one with strong ethical
and moralistic tendencies. It was written down by the leading intel-
lectuals, the "literati", a class whose only education consisted in the
extemporisation and imitation of the classics. Technology was not
written down nor read by the leaders of Chinese civilisation who, like
the Greeks and Romans, paid little attention to such subjects. It is,
however, quite possible that LAUFER's view will prove to be too pessi-
mistic and that the Chinese did distil at earlier periods. At present we
can cite only two authors mentioned by WYLIE, which might give us
new and interesting data on the problem, if their works were translated and edited (133) (628).

The earliest of these two is TOU P'ING (11435, 99314), a chemist
who flourished in the first half of the eleventh century. He was the
author of a treatise on alcoholic liquors called *Chiu P'u* (2260,
9515) which consists of brief notices regarding different kinds of

liquors and celebrated distillers. The second author is CHU I-CHUNG (2544, 5507, 2875), a chemist who wrote about 1120 A.D. a standard treatise called *Pei-Shan Chiu Ching* (8871, 9663, 2260, 2122) on the distillation of spirits. The first part contains a general discussion of spirituous liquors, the remainder gives ample details on the compendium of ferments and the various methods of distillation. As in the West the distillation of spirituous liquors cannot be dated earlier than 1100 and Chinese literary tradition does not mention alcohol before the Yüan period (that is two hundred years later), it would be of the utmost importance to translate these books as soon as possible. But even if it were proved that the distillation of spirituous liquors was practised earlier in China than in Europa and Western Asia, there remains the fact that the distillation apparatus very probably came from the West. Anyway it is certain that the fables of the hoary age of distillation in China can be dismissed. We must look to the West for the origin of distillation.

Neither is there any doubt that distillation was not practised by the ancient Assyrians or Egyptians as is sometimes stated in technological literature. None of the Sumerian, Assyrian or Babylonian texts from ancient Mesopotamia mention any compound among the numerous drugs (622) that can be definitely stated to be a distillate, nor is there any verb that can be interpreted to mean "to distil". Of course several of these substances were prepared from solids by processes involving dry distillation or other destructive handlung of the prime material, or from liquids by inspissation.

Again no Egyptian text mentions any distillation of a liquid combined with the condensing of the vapours and collecting the distillate. Of course in this country too drugs were concocted by dry distillation of several substances. Sometimes wrong or doubtful translations suggest the use of some distillation process. Thus EBBELL (204) in his translation of the Ebers papyrus suggests that ᵓ*d* · *Š* means pine-tar and *šft* turpentine (prepared by distilling pine-tar). But the former word is not pine-tar but cedaroil (ERMAN-GRAPOW's *Dictionary* I, 239.14) and the latter word was proved by the author to mean cedar (oil) pitch (*Ambix*, vol. II, p. 72).

Though PARTINGTON was right in stating that the Ebers papyrus contains no reference to distillation, he mentions that the Berlin medical papyrus does. The Berlin papyrus he means is the Berlin papyrus No. 3038, interpreted and translated by ERMAN and KREBS (211) and later by WRESZINSKY (669). Recipe No. 46 of this papyrus reads: "A remedy for the *srjit*-illness: *jbnw-nhš* (an unknown drug) pulverize with the leaves of the aâm-plant, put it in seven ... Put it in a pot, the cover of which is pierced. Stick a tube through this hole, half of which the man shall hold in his mouth. And the man shall drink beer and spit it out again. Watch closely the result." Obviously this is a recipe involving an inhalation process and it has nothing whatever to do with distillation.

Evidently in the Ancient Near East distillation was unknown, though some substances were prepared by dry distillation, melting out or cracking other substances by some heating process and others were obtained as the residues of some process of evaporation. But in no case is a distillate collected from any type of apparatus which we could in any way call a distillation apparatus. We will have to wait until the period of the Alexandrian chemists to read about such things as stills, alembics, etc.

But the problem of the origin of distillation may not yet be solved by these few pages. It may be, as RAZI says in his Introduction to his *Secret of Secrets* that "in the book there are chapters which the learned and the searchers have not noticed".

CHAPTER TWO

THE ALEXANDRIAN CHEMISTS

(100-900 A.D.)

"We will begin to gather the fairest fruits of the
work in hand and trust to track down the truth. Now
from a true theory of nature our problem must be
set out."   (STEPHANOS, *First Letter*)

Modern authors like DARMSTAEDTER (157), VON LIPPMANN
(372), DIELS (177), NEUGEBAUER (453), DUJARDIN (197), SARTON
(550) and so many others agree on the point that distillation of
liquids was unknown in classical Antiquity and that distilling apparatus was unknown too. But the classical authors do mention instruments which are subsequently employed in distilling or counted among
distilling apparatus by later generations and on the other hand the
ancient natural scientists came quite close to the proper understanding
of the principle of distillation, especially in their discussions on the
cycle of water in nature.

The ancient scientists were always intrigued by the conversion of
salt water into fresh and the reverse in connection with their experience
with rain-, well-, river- and sea-water. The old theory of THALES
and HIPPON (253) which PLATO also defends and which was destined
to rule the Middle Ages believed that the earth swam on the Ocean
(the ancient Apsû or fresh-water-deep of the Sumerians and Baby-
lonians). From this Ocean the earth sucks in water like a sponge and
drives it again to the sea by the way of wells and rivers. According
to this theory only the light, fresh parts rose and fell again on the
earth to feed the rivers. ANAXIMANDER and EMPEDOCLES agreed with
this theory in general.

But ARISTOTLE looked deeper into the problem and declared that
the sun drew up the "atmis" or "humor" by its heat, to drop it afterwards in the form of rain. As against PLATO he believed that all
the water on earth was of meteorological origin. Older physicists had
stated that rain was a condensation of air without explaining this
theory more closely. But ARISTOTLE attacked this theory and that of
PLATO (given in *Phaidon* cap. 60) in his *Meteorologica* (I. 9. 11).
PLATO had written: "The origin of rivers is this: The water, rising

by the sun's heat, falls back in the form of rain and collects in a large
cavern...". ARISTOTLE, however, believes in a materia prima, which
can be rarified more of less by heat (382) and condensed by cold.
The properties of this prime matter, therefore, continue to exist however
the qualities may change. Thus air is transformed into water if it is
condensed into rain, etc. This opinion had great influence in later
periods. LUCIAN for instance mentions in his *True Histories* that air
compressed in vessels will change into water by this compression and
this water is drunk by the inhabitants of the stars among whom the
scene of his romance is laid.

These speculations came quite close to a correct theory of the
natural cycle of water and so close to a proper understanding of the
principles of distillation. Thence many older and even some modern
authors have believed that ARISTOTLE and his pupils practised distillation for which theory they thought to find a proof in another
passage of the *Meteorologica* (II. 3) reading: "Salt water when it
turns into vapour becomes sweet and the vapour does not form salt
water again, when it condenses. This I know by experiment. The
same thing is true in every case of this kind: wine and all fluids that
evaporate and condense into a liquid state become water. They are
all water modified by a certain admixture, the nature of which deter-
mines the flavour". And he goes on to say: "If one plunges a water-
tight vessel of wax into the ocean, it will hold, after 24 hours, a
certain quantity of water, that filtered into it through the waxen
walls, and this water will be found to be potable, because the earthy
and salty components have been sieved off".

The latter passage has aroused much discussion. ARISTOTLE repeats
the same story in his *Hist. Anim.* IX. 2, so that there is no doubt
that the words "waxen vessel" (angeion kerinon) are correct. VON
LIPPMANN and DIELS have tried to extract some meaning from this
passage by substituting "angeion keraminon" (earthenware vessels)
for "angeion kerinon". Now a fifth century commentator, OLYMPIO-
DOR, says that several people were able to repeat the experiment and
find ARISTOTLE's assertion true. DIELS therefore reasoned that possibly
the original experiment (which DEMOCRITUS must have made) was
conducted with an earthenware vessel, but that somehow the word
"keraminon" was changed into "kerinon" in ARISTOTLE's version of
the phenomenon and remained in the versions of his works since.
STEPHANIDES (603) however doubts this and points out that there
were similar legends on a waxen vessel in Antiquity and that we have

no right to change the word "kerinon", however mysterious the passage may remain to us.

In none of these passages, however, is there any reference to the distillation of sea-water (372). But when five centuries later ALEXANDER of Aphrodisias (18) comments on the evaporation and condensation theories of ARISTOTLE, he says: "Per hunc quidem modum maris aquam potabilem nonulli reddunt: lebetes enim hujusmodi aqua plenos multi ignis imponentes et vaporem in operculis superimpositis colligentes et recipientes in aquam permutato utuntur potu". This might be read as a sign that ALEXANDER knew the distillation of seawater, but by then it was about 200 A.D!

He also mentions that "sailors at sea boil sea water and suspend large sponges from the mouth of a bronzen vessel to imbibe what is evaporated. In drawing this off the sponges, they find it to be sweet water". This is a method which was well-known to PLINY and DIOSCORIDES, and which they mention in connection with the preparation of distillates from tar. Also "distillare" (also written destillare) means "to drip off, to drop off" in classical writings and in this sense we find it used by CELSUS, SENECA and others when they speak of patients that have a cold. PLINY uses the word in connection with snow. But "distillare" is never used in our sense.

Nor does chemical technology in Antiquity show any signs of the application of distillation. It is not used in the manufacture of alcoholic drinks, nor in the oil- and fat-industry or in the manufacture of perfumes and essential oils or cosmetics. We do hear a lot about the water-bath, which DIOSCORIDES uses preferably for the rendering of fats (II. 95) and which will play a great part in our story.

Agricultural or medical works do not mention distillation of essential oils. CATO, VARRO, COLUMELLA or GALEN mention only the preparation of perfumes by digestion or maceration of flowers and spices with olive- or sesame-oil, never by distillation with water and steam.

Nor do the ancients speak of the distillation of alcholic beverages. HOEFER had drawn the inference from the passage of ARISTOTLE which we have cited (303) but modern authors have quite rightly rejected this opinion. Nor are statements of older authors such as SACHS (543) correct, which say that the ancients made brandy by "philosophical distillation" that is by freezing out wine. It is only true that some wines were rendered more viscous by evaporation (558) (77). Such a boiled down must was a very popular drink of which the

Romans knew two or three qualities differing in the grade of inspissation. VARRO (Lib. I of his *de vita populi Romani,* see Non. 551, 9) mentions that must boiled down to half the volume was called "sapa", if boiled down to one third "defrutum" (PLINY XIV, 80). Spiced wine filtered through linen was called "vinum saccatum" (373), in the sixteenth century this operation would have been called "destillatio per filtrum".

Very frequent are the references to the inflammability of special qualities of wine. ARISTOTLE ascribes them to exhalations as we saw, and THEOPHRAST (*de Igne* 67) quotes them, but this does not mean that the manufacture of alcohol by distillation was known, as some believe.

An ancestor of the distillation process is described by DIOSCORIDES when he deals with the preparation of distillates from tar, and it was to have a very long life, being still mentioned by EVONYMUS in the sixteenth century. The former remarks: "Of Pitch is made oleum picinum, the watry matter of it which swims on ye top as whey doth of milke being separated. This is taken away in ye seething of the Pitch by laying cleen wool ouer it which when it is made moyst by the steam thereof ascending upon it is squeesed out into a vessel and this is donne as long as the Pitch is seething". The tar water which is separated is called "serum picis" or "pisselaion", the tar distillate "piccinum" or "pissinum". PLINY mentions that the "best hails from Bruttium", it is "most fatty and resinous". Probably both these ancient authors derived their knowledge of the process from SEXTIUS NIGER. It might be described as the first crude attempt to isolate a chemical product by "distillation". SCHELENZ's theory that the distilling apparatus was derived from the charcoal pit can be dismissed without further commentary.

We do meet different forms of dry distillation in the manufacture of charcoal, a very old process. THEOPHRAST in his *Enquiry into plants* (IX. 3. 1-3) (621) describes the construction and operation of charcoal pits in detail. Less wasteful methods are discussed by PLINY (*Nat. Hist.* XVI, 52-60, sect. 21 & 22), who gives the dry distillation of pitchpine in billets and the production of tar as a by-product. He uses a kind of furnace with fires on both sides. The first liquid that flows from the furnace "like water" is called "cedrium" and is used for embalming in Syria and other places.

Another very old process that combines dry distillation, distillation, melting down and sublimation is carried out in two pots placed upon each other and separated by a perforated plate. One pot (usually the upper one) is filled with the substance to be treated (an ore, herbs, etc.) and a fire kindled around it. The "distillate" then drips down into the lower pot. The ancients have no special name for it, but later it is referred to as "destillatio per descensum".

Sublimation was used for the manufacture of mercury and is described by DIOSCORIDES (V. 110), who says: "Putting an iron Spoon having Cinnabaris in an earthen pot, they cover the Cup dawbing it with clay, then they make a fire under it with coals; and ye soot that sticks to ye pot being scraped off and collected becomes Hydrargyrum".

All these operations were also used in the very old bitumen industry, most of them seem to hail from Mesopotamia (see the author's *Bitumen and Petroleum in Antiquity*, Leyden, 1936 and the article in *Ambix* vol. II, 1938/1939, pp. 68-92).

These methods would bear fruit when they were used by the Alexandrian chemists as instruments for their new science, chemistry.

The clash of Greek and pre-classical science in Hellenistic Alexandria was destined to bear rich fruits and even after four centuries Alexandrian science was far from exhausted and produced this new art of enquiry into the composition of matter. The rationalistic element, child of Greek science was to protect this new chemistry for several centuries from the inroads of gnosis and all those other cultural streams that kept the minds busy in the first centuries of our era and which hindered so many branches of science in their growth by the introduction of mystical, quasi-philosophical and very speculative elements.

If we want to draw a proper sketch of the development of distillation methods it is necessary to pay closer attention to this oldest phase of chemistry. It was long believed that two papyri, e.g. the papyrus Reuvens No. X of the Leyden Museum of Antiquities and the papyrus Holmiensis at Stockholm (which both originally formed part of the Anastasi collection) of the third century A.D., were the oldest chemical books, as they certainly go back to far older originals of which they are copies. But a closer study by VON LIPPMANN, DIELS and LAGERCRANTZ has shown that the original opinion of BERTHELOT is not correct. For these papyri contain recipes for the imitation of precious and semiprecious stones and the colouring of metals to imitate gold and silver and they represent a notebook of recipes for artisans rather than a chemist's handbook (314). The first papyrus probably hails from Thebes in Upper Egypt, the latter from Alexandria. They date from a period when there was much demand for imitation jewelry.

But JENSEN quite correctly attacks the opinion that not only these papyri but also the important contributions of the Egyptian craftsmen and even these in the Ancient East in general are to be attributed to a "secret science of the priests" (314) (315). It is true that the different crafts were often exercised under the supervision of certain priests but this means no more than that these guilds, like every other aspect of social life in Antiquity, had a strong religious character. We may compare them with the medieval guilds adoring their patron saint together or devoting themselves to other religious duties. There are no Egyptian texts which speak of priests exercising a craft, but some mention their leadership in certain crafts such as building activities, sculpturing, etc. The members of the Egyptian Academy of Science (the Royal Society of those days called the "House of Life") are mostly priests but do not exercise the craft themselves though they are the leaders of groups of sculptors, artists and builders. All these works had a strong religious aspect and therefore priestly control could not be dispensed with if the new creation were to have its proper magical effect. The control by the priest was therefore not a tyranny but a religious necessity in the eyes of the ancients. What is often forgotten in these discussions is the actual skill of the craftsmen of Hellenistic Egypt which in many branches formed the apex for many centuries to come. Their material conquests transmitted by Byzantium and the Arabs were to exercise strong influence on medieval handicraft.

It is very important for the youthful science of chemistry to read in the two above-mentioned papyri, that the late-Egyptian craftsmen had achieved great skill in the handling of fire and the construction of furnaces, the manufacture of all kinds of earthenware and fireproof material, the production of glass and glazes, while the metallurgy and especially the art of the jeweller was very well developed. In the instruments of these craftsmen and those which were used in contemporary cooking the young science was to find inspiration for its specific apparatus.

The Hellenistic era was one of those lucky periods in which craftsmanship and science met and stimulated each other. The young chemistry had the typical rationalistic traits of the older Greek science. Therefore though these chemists were generally Egyptian Christians, e.g. Copts, this chemistry long kept the characteristics of the older classical natural science. Only relatively late in Antiquity can we speak

of a degeneration when alchemical elements (in our sense) are introduced. The texts make it clear that numerous chemical phenomena have
been discovered which include multifarious fusion, sublimation and
distillation processes though we can not trace any interest in the general
properties of matter. This early, unmodern phase of chemistry is of
course steeped in a deeply religious atmosphere as were all aspects of
life in those days.

JENSEN (315) characterizes this early phase of chemistry thus:
"This alchemy is a late branch of Hellenistic science. In a deeply
religious society which believed in achieving mastery over nature, this
new art became the mystery that would lead towards redemption. When
in the course of time the early chemists were disappointed in their
hopes, some delivered the art into the hands of the masses. Then it
deteriorated into the search for gold by becoming strongly affected
by Neoplatonic beliefs".

In fact there is nothing of the gold-making alchemist about those
early scientists. Neither the word "chemistry" nor "alchemy" was used
in their texts, but they always refer to The Work or The Divine and
Sacred Art (673).

The most important manuscripts from which we draw our details
about their achievements are the codex Marcianus 299 of St. Marc's,
Venice copied about the XIth or XIIth century, the manuscripts of the
Bibliothèque Nationale, Paris Nos. 2325, 2327, 2249, 2250, 2251,
2252 and 2329. The first two from Paris, copied in the XIIIth and
XVth century respectively are the most important ones, the second of
these being a more complete copy of the first, though many of the
illustrations have been altered considerably. These texts deal mainly
with original authors before ZOSIMOS (who lived at the end of the
third century) and commentators of later date. Other important manuscripts are the Codex Holkamicus No. 20 and the Codex Vatican.
graec. No. 1134.

The available evidence allows us to divide the ancient chemists of
the first and second century A. D. into three schools (616):

a) The *followers of* DEMOCRITUS including ISIS, IAMBLICHOS, MOSES,
OSTANES, EUGENIUS and others. They concentrate on the superficial colouring of metals and the preparation of alloys.

b) The *school of* MARIA *the Jewess and* COMARIUS including HER-
MES, CLEOPATRA and possibly AGATHODAEMON. The members of
this school invented and employed very frequently apparatus for
distillation and sublimation, designed the water-bath or bain-marie

and used the kerotakis-process, hot ashbath and dung-bath. The texts belonging to this school are partly practical, partly mystical. To the latter class belongs the single page of CLEOPATRA's *Chrysopoeia* which survived (Fig. 4) and which shows both drawings of apparatus and symbols, though the latter may be of later date.

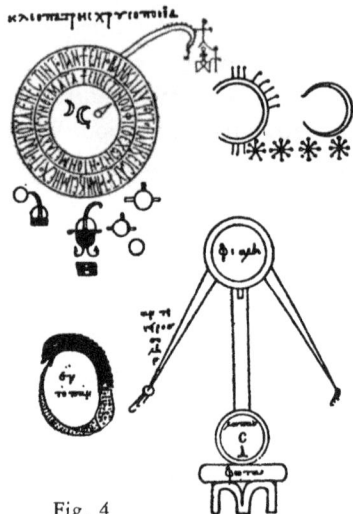

Fig. 4

c) *Fragments* of the writings of PAMNENES, CHYMES, PIBECHIOS and PETASIOS are too small to allow us to classify their contents.

Their heirs are the *authors of the third and fourth century* including ZOSIMOS, AFRICANUS, HELIODORUS, PELAGIUS and others. AFRICANUS may be identical with the well-known late classical author SEXTUS JULIUS AFRICANUS. ZOSIMOS is heir to the ideas of MARIA and CLEOPATRA. Perhaps SYNESIOS, the well-known author of the fifth century was an original chemist too.

After ZOSIMOS a series of *commentators* up to the tenth century make very little advance on the knowledge displayed by ZOSIMOS. Among them we find PHILOSOPHUS CHRISTIANUS, HERACLIUS, JUSTINIANUS, PAPPUS, THEOPHRASTUS, HIEROTHEUS, ARCHELAUS (the latter three belonging to the eight and ninth century) and SAMANAS and COSMAS, both of the tenth century. Then the Arabian chemists take the lead in the development of the art.

Before discussing further details of the distillation and other apparatus it will be well to summarize the main achievements and defects of this early phase of chemistry. Among the achievements we must reckon:

a) The discovery of the distillation, sublimation and other primary processes.

b) The design of the proto-types of our present chemical apparatus.

c) The collection of new facts on the properties of materials.

Among the defects and limitations of these chemists we count:

a) The fact, that all reactions were carried out in molten condition and never in solution.

b) This was partly due to the fact that the ordinary acids (sulphuric-, muriatic-, and nitric acid) were unknown, the strongest acid available to them being acetic acid.

c) The use of distillation apparatus with very insufficient cooling, so that only liquids with boiling points higher than that of water could be recovered somewhat efficiently. Still the distilling apparatus as used by ZOSIMOS and SYNESIOS was a marked progress when compared with the primitive "distilling" apparatus of DIOSCORIDES and other classical writers. Even in the days of ALEXANDER of Aphrodisias the vapours evolved in the still were

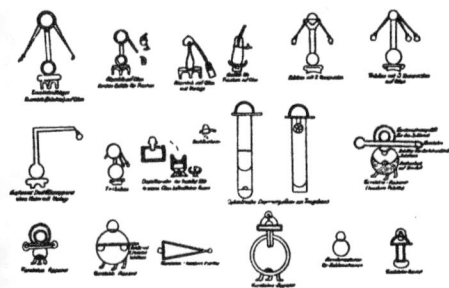

Fig. 5

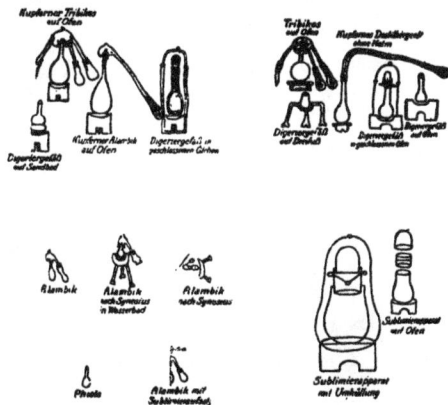

Fig. 6

Illustrations of different apparatus from Hellenistic manuscripts
according to SÜSSENGUTH and FERCHL

condensed and transported into a separate receiving flask and good earthenware or glass apparatus had been evolved sufficient to purify higher boiling liquids (636).

In Figs. 5 & 6 illustrations are given from the above-mentioned manuscripts showing some of the apparatus used. Figs. 7 & 8 give a few of these instruments more in detail with the original legends. The class of apparatus that interests us most is the *distilling apparatus*. This is already very far advanced in the writings of MARIA the Jewess who is generally considered to have invented it. It already consists of the three necessary elements, the cucurbit and alembic, a tube for transporting the distillate and vapours and the receiving flask. There is no trace of such an apparatus in the papyrus Holmiensis or the papyrus Leidensis (66) (371).

The *cucurbit* proper is still separate from the alembic, which carries a spout or tube leading to the receiver. The cucurbit is usually

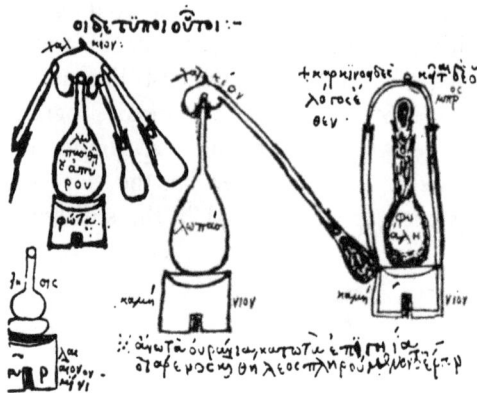

Fig. 7. Detailed reproductions of chemical apparatus used by the Alexandrian chemists (acc. to SHERWOOD TAYLOR)

called "bikos" (βίκος) (bekos, bikion), a word that in HERODOTUS still means "jar" but that is also used in the sense of "vessel" and "cup". Apart from this word we also meet the term "lopas" (λωπάς) (lepás) e.g. in the writings of CLEOPATRA. A few times it is called "lebes" (λέβης) which meant originally a "kettle in which water was boiled". The word "sikuá" (σίκυά) (calabash) is also used, but this becomes the general term only much later and then in its Latin form "cucurbit". SYNESIOS uses both the words "lopas" and "botarion".

The *alembic* is a separate and important element of the distilling

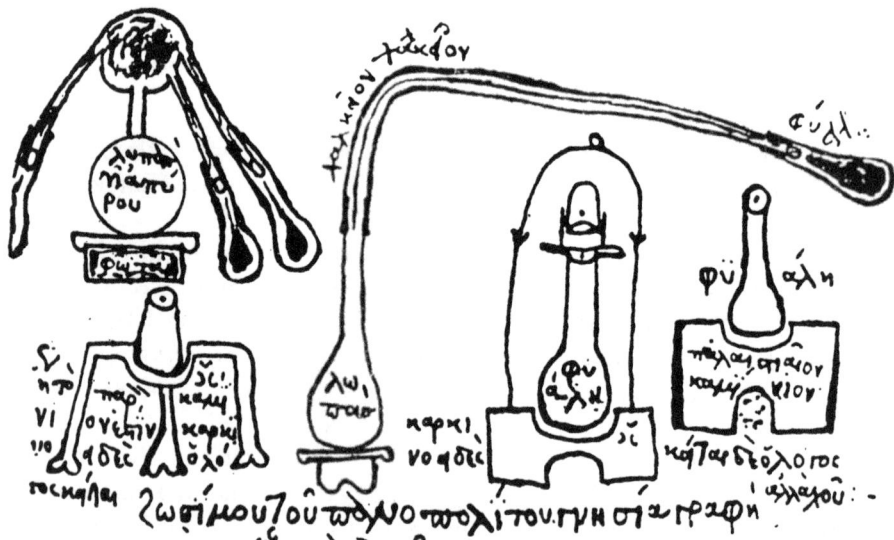

Fig. 8. Detailed reproductions of chemical apparatus used by the Alexandrian chemists (acc. to SHERWOOD TAYLOR)

apparatus. It has a special form because a rim or gutter is attached to the inner side of the neck. In this gutter the distillate is collected and carried to the receiver by the spout. Therefore the alembic is the true condensing element and the condensate is carried away from it in liquid form. There is no special condenser for the vapours apart from the alembic. This typical Hellenistic invention is called "bathos" (βάθος) ("cavity"), "phiale" (φιάλη) (originally an "urn", "kettle" or even "drinking cup") but usually we find the term "ambix" (ἄμβιξ) (ambikon) a Semitic(?) word which LUCIAN uses in the sense of "jar" or "pot", though DIOSCORIDES, PLINY and GALEN use it for "cover" like ATHENAEUS. It soon gets its definite meaning of alembic in Hellenistic chemistry. Only since the tenth century and later are both the term ambix and its derivative alembic (al-anbiq, the Arab for ambix) used for the combination of cucurbit and alembic as a whole.

The *outlet-tube* is called "solen", it is made of earthenware, glass or copper like the other parts. In the latter case one also finds the name "chalkeion" in the manuscripts. In one figure the neck of the cucurbit is called "likanos solen" and the outlet-tube of the alembic "anticheiros solen".

The *receiver* was generally a small vessel with a long thin neck, which we call "phial" like the ancients. But the receiver is also often called "bikos" or "aggos" (ἄγγος) (vessel) in the earliest manuscripts.

In several illustrations one finds distilling apparatus with two or more outlets and receivers, which according to the number are called "dibikos", "tribikos", etc. It is not clear whether these are meant to denote the distillation in two or more fractions or the simultaneous collection of distillate in more receivers. The latter is more plausible. It is strange to remark that there is no general word for distillation. Sometimes it is called "eos ekstrepho" (ἕως ἐχστρεφω) that is "turning inside out", in other cases the expulsion of volatile matter is called "exethatotis". The word "expaphlazo" also occurs, which originally means "boiling" or the "simmering of porridge" but not "distillation". Early Syrian manuscripts sometimes call mercury "anabibazon", that is the "rising".

As *lute* for the parts ZOSIMOS mentions fat, wax, clay, gypsum and an oil. Often the entire apparatus is covered with a layer of clay ("epidermis") to protect it against the fire. Recipes of the eighth century and later of Byzantine origin already give a recipe of lute on a base of lime, caseine, glue and white of egg (or yeast) which is quite similar to the famous "lute of the philosophers" (lutum sapientiae") of later ages. By luting or by covering the cucurbit with clay it was possible to overcome some of the bad properties of the glass of those centuries.

There were different *means of heating* at the disposal of these early chemists. First of all there was heating by direct application of fire, usually indicated in the illustrations by "phota", then of course the heating in an oven ("kaminion"). This oven, the later "altannur" of the Arabs and "athanor" of the Middle Ages, was a furnace that burnt for a long period like the type used by the glassmakers ("automatabion"). The oven is also called "phournos", which word is usually coupled with the baker's oven. The ash- or sand-bath ("thermospodion") was well-known, like the water-bath, for the Greek and Roman doctors used them to prepare their medicines and whoever opens a cookery-book of the period, for instance that of APICIUS COELIUS, will see that similar apparatus were quite common in the kitchens of those days. For the Hellenistic chemist adopted many types of apparatus from the kitchen and therefore The Work is often called "opus mulierum" in later days, not without some truth.

The *water-bath* is mentioned by DIOSCORIDES and PLINY uses it for the melting down of fat, resin and the like. PLINY, THEOPHRASTUS, CATO and GALEN advise its use in the extraction of perfumes from flowers with fats and oil. But DEMOCRITUS already used it for chemical work. We do not know why tradition mentions MARIA the Jewess as the inventor and why medieval chemists since ARNALD DE VILLANOVA call it Balneum Mariae. VON LIPPMANN tried to prove that the MARIA in this name was the Mother of God who took the place of an original "Isis Pelagia" and that therefore the name was originally "Bath of the Goddess Isis" (370). This seems very phantastical, as we have no certainty that the name MARIA is ever that of the Mother of God in these chemical texts. The evidence which VON LIPPMANN takes from HORAPOLLON and LEPSIUS has lost all its force in the light of modern Egyptian philology.

Apart from the water-bath we find the sun's heat or that of fermenting horse dung (CLEOPATRA) used for heating purposes.

Rushes, charcoal and wood in the form of billets are mentioned as fuel by ZOSIMOS, who wrote a long book entitled *peri organon kai kaminon* (*On the instruments and ovens*) that is adressed to one THEOSOBIA. Its sub-title is *True aphorisms on the letter Omega*. This book discusses many sides of the chemical work which we have mentioned, it draws extensively on older chemists. In general the apparatus is made of earthenware, glass or metal, sometimes these are combined. For a good distillate one should use (534) "a glass bikos, an earthenware tube of one ell length and a lopas or a phial with a long neck... This is the drawing (which however is missing from the manuscript!). One should also have a vessel standing on a stone destined to carry water to moisten the alembic. It is the same for dry vapours and mercury". Probably the receiver was placed in a basin or water and was cooled with sponges like the alembic. "It is also possible to fix the mercury and tinge it yellow in a phanos or similar vessel that has the form of a snake". The phanos seems to denote some kind of worm cooler as was already in use for sublimates. In that case the sublimate is condensed on the outer surface of the worm. Both ZOSIMOS and OLYMPIODORUS mention this phanos, of which no drawing has survived.

For the sake of completeness we must mention that fact that SYNESIOS in his letter to HYPATIA mentions a hydrometer, which according to some was already known in the fourth century A.D. to PRISCIANUS, that is a century before SYNESIOS and HYPATIA. So was the latter after all the inventor as is often said?

The above-mentioned apparatus could not be used to distil highly volatile liquids and distillates that easily solidified. Mercury-, sculphur- and arsenic-compounds and the elements themselves were distilled frequently. One should always take "sulphur" in a wider sense, often this "sulphur" means sulphur with a vegetable oil, forming sulphur substituted organic liquids, that distilled or tinted metals suspended in the apparatus to form thin surface layers of sulphides. Mercury, sulphur and arsenic-sulphides with lime form the base-material from which the Divine Water is prepared. The purifying action of distillation was well appreciated. HELIODORUS advises distilling twice and HIEROTHEUS and ARCHELAUS says that perfection is attained after three distillations. It is possible that JENSEN is right and that some allegories in the manuscripts cover the distillation of certain compounds (315).

Apart from distillation the *kerotakis-process* plays an important part in these texts. It is a kind of distillation combined with washing and reactions at higher temperatures. The orthodox kerotakis process is very intricate (308) and involves several "colourings" like Melanosis, Leukosis, Xanthosis and Iosis, which occur in series. "Kerotakis" originally meant the palette on which the wax-painters mixed their colours and it is here used for a vessel in which compounds are

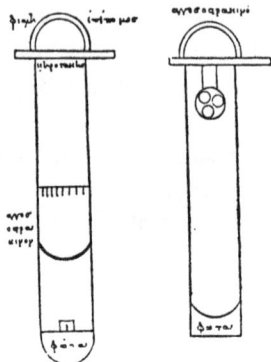

Fig. 9. The long form of the kerotakis apparatus (MS Marcianus 299)

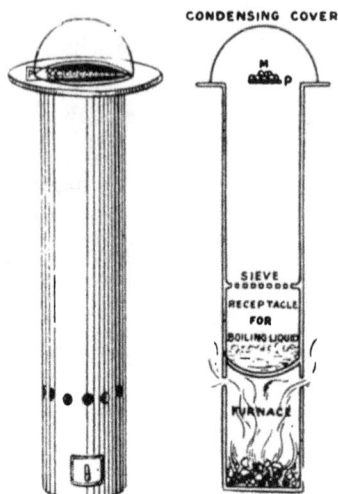

Fig. 10. Conjectural restorations of the long and the round form of kerotakis apparatus according to TAYLOR

boiled. The vapours rise, condense against the closed cover and the condensate falls on certain metals or alloys which are fixed to a palette hanging under the cover in the way a crucible is attached in our modern extraction apparatus. Here the reagents ("pharmakia")

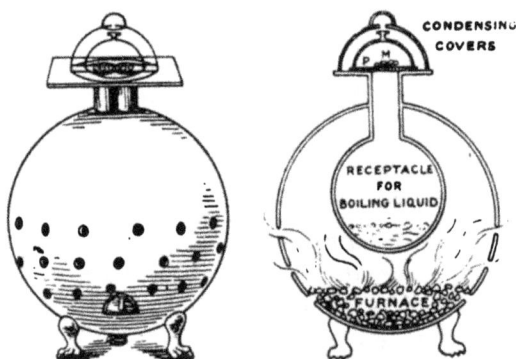

Fig. 11. See Fig. 10

can react with the metals. Fig. 9 give the drawing from the manuscripts and Figs. 10 and 11 depict the reconstruction given by SHERWOOD TAYLOR. There were a long and a round form.

Another intricate process of distillation, sublimation and heating combined is the *heating in a closed tube*. If the contents are only heated very carefully "to loosen the volatile components" the process is called "péphis" (digesting). If the condensate drips down from the cover, the apparatus is sometimes called "karkinos" that is "lobster". For the sublimation of certain compounds a "blind tube" is also used, which in this form was to reign in chemistry under the name of "aludel" (derived from the Arabic al-uthal).

In principle there is of course little difference between the aludel and the old *"pot-on-pot"* process which we met in DIOSCORIDES and which is later called "bût-bar-bût", "botus barbatus" or "destillatio per descensum", that is it is only called thus when the product desired is collected by "distillation" in the lower pot.

It is clear from the texts that the early chemists did not yet ascribe a certain specific action to each of these kinds of apparatus as is the case later in history. They are convinced that in each apart from "péphis" or digestion "analysis" (that is "solution" and not yet its later sense of "decomposing") and "teksis" ("melting") occur. But it is often difficult to find what the early texts mean because not only the terminology of apparatus but also that of operations is not yet by any means a fixed one. Though STEPHANIDES has thrown some light on the matter, one should always proceed very carefully in translating these texts. Thus "teksis" is originally used for both melting and dissolving. Later dissolving is called "diateksis" and Byzantine texts introduce the word "leiosis" for melting.

For that matter, in the Byzantine period, especially after the tenth century the terminology changes quickly, first under Arabian, then under Italian infuence. Thus the word ambix obtains its meaning of the combination of cucurbit and alembic, the word for alembic becomes "kapouzin" (Italian "capuccio"). The word "phournos" (comp. Lat. "furnus") begins to drive out "kaminion". Another word for alembic which we find in these writings is "kapitalon" (capitellum) and distilled liquids are called "hydor", especially acids and the like, the manufacture of which is learnt from Italy and exported to the East. The Greek "hydor kaustikon" is equivalent of the Western "aqua ardens" (alcohol).

The Byzantine chemists undertook very little original work notwithstanding the achievements of the Arabs which penetrated into their laboratories. Their books are little more than collections of recipes.

But returning to the Alexandrian chemists it is clear that they derived their apparatus from the older technicians, doctors and cooks. Probably the "balsam-cookers" (unguentarii or myrepsoi), the ancestors of our apothecaries have contributed their share too, though we do not know whether they used distilling apparatus as SCHELENZ claims. To these examples, however, they added many original inventions and cast the major operations of chemistry in a form which was to remain for many centuries.

We do not know whether such distillation processes were also used in chemical technology. They were much handicapped as it is certain that the Alexandrian chemists did not know either mineral acids nor alcohol. The insufficient cooling of their apparatus prevented their obtaining many compounds which gave a definite turn to the later chemistry and technology. The manufacturing of perfumes was continued in the old way. When the texts speak of "oil of roses", they mean decoctions but not distillates. Only in Byzantine texts of the tenth century and later a "rodostagma" (oil of roses) is mentioned, which is distilled from rose leaves in cucurbits. Such is the oil of roses of CONSTANTINUS PORPHYROGENETES and THEOPHANES NONNUS (tenth century) and of NICEPHORUS (thirteenth century). This distilled rose-oil was probably a product the manufacture of which was learnt from the Arabs, by which was meant Persia. Persia was always the country of roses and it possessed an industry of "oil of roses" as we shall see. The technical distilling apparatus was very simple. A Coptical papyrus of the ninth or tenth century found at Meshkaîkh in 1892 and edited by CHASSINAT (128) (470) describes the "heating of substances in a kind of oil-jar which was suspended in an earthenware pot" (a kind of air-bath). Probably this was the description of an apparatus to make "oil of roses".

CHAPTER THREE

THE ARABS

"A knowledge of natural science and an apprehension
of its facts form part of this subject." AVICENNA

If we speak of Arabs in this chapter we include all those that belong
to the civilisation of Islâm, which means Syrians, Persians, Copts,
Berbers and others too. As early as one century after the death of
Muhammed (632 A.D.) a large world empire has arisen from a
local Arabian movement, and its centre is transferred to Syria, and
later to Mesopotamia. The Islâm knocks at the doors of Byzantium
and menaces Italy and France. At the last moment CHARLES MARTEL
turns the torrent at Tours and pushes the Arabs back over the Pyrenees.

The Arabs take over the legacy of Hellenism. In this they are
aided by the fact that they harbour many who have fled from Byzantium because of the religious persecutions and many a group of
herectics settle in the Arabian Empire. Thus for instance the Byzantine
emperor had driven away the Nestorians who settled in the East,
especially in Iraq and Iran. The closing of their university at Edessa
was followed by the founding of a new one at Jundishapur. Again
the old Academy of Athens founded by PLATO was closed (529 A.D.)
and many Greek heathens moved to the hospitable cities of Iran.
Many Jewish scientific centres such as Nisibin were situated in the
Arabian Empire. The town of Harrân, the ancient Carrhae capitulated
in 639 A.D. In 830 the Chalif MAᵓMUN offered the inhabitants the
choice between Islâm, tolerated religion or extermination. They then
claimed to be Ṣâbians to save themselves and through them the
Ṣâbian lore influenced Arabian science. We can also distinguish a
direct influence from Egypt on Arabian science. We know that several
Coptic or Egyptian writings must have been translated immediately
into Arabic without passing the Greek stage.

To the East the Islâm made close contact with China and from
those parts too new influences penetrated science and therefore also
chemistry. We remind the reader of what we said about the "Philosopher's Stone and the Elixir of Life" which play such an overwhelming
part in chemical theory since Arabian times.

It may be true that Arabian science has not added much original thought to the ancient science but it had at least the great merit of collecting the ancient traditions and experience with filial piety, commentating and preserving them with care. A glance at the map (Fig. 12) will show what this meant to classical tradition and to the later flowering of science in Western Europe. We can say without exaggeration that the medieval scientists knew the classics only in their Arabian version with all the advantages and disadvantages connected with this transmission.

But apart from this preservative function Arabian science played

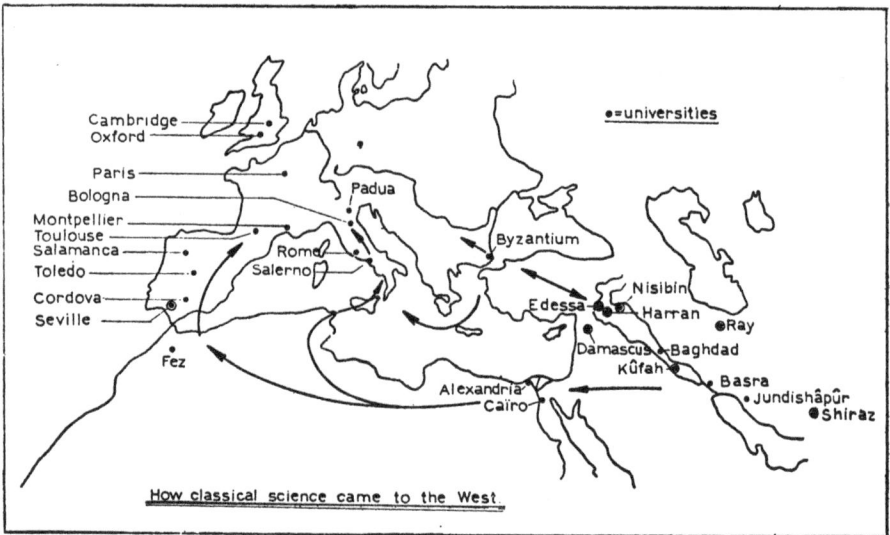

Fig. 12. Map of the spread of knowledge from the Arabian world to the West

perhaps a far more important part as an agent between East and West, a part which it fulfilled with more intensity than the older empires of the Near East that had gone before it. For not only is the Far East now drawn into the spiritual intercourse of the Islâm but science follows trade in both directions and helps to build up the international character of Arabian science at its apex.

The history of distillation is again closely interwoven with that of science and technology in this period, and again chemistry is that branch of science which stimulates the development of distillation more intensely. For though in Arabian chemistry sublimation was the major operation because of the reigning mercury-sulphur theory, distillation came into its own as the best and quickest way of obtaining pure chemical substances. Arabic alchemy was built upon Hellenistic

THE ARABS 31

and Iranian warps with strong Far Eastern woofs woven into the pattern. Though much fresh evidence is now available on its origin and development too few texts have been published. But it would seem that the major operations and principles were already well established before the twelfth century. Even if some of the earlier writings have been ascribed to apocryphal authors they can be traced back to the tenth century, the age in which the great chemist AL Râzî lived, whose books are not challenged as our knowledge of them grows. Muslims experimented and wrote, began to discuss and doubt the most advanced theories. They are the true continuers of Hellenistic traditions and the initiators of medieval alchemy. Even the more phantastical mystical type of Arabian alchemist may be of great use to us, for as SARTON said: "Even as war has been indirectly a source of progress in spite of its utter wrongness, even so treachery and even adulteration of science may lead incidentally to valuable discoveries. Historians of science must be very patient and tolerant and be ready to give men full credit for their intentions or disregard them altogether". Though the sceptical type of Arabian chemist has given more impetus to the development of chemistry and thereby has done more for the development of the distillation apparatus, it would not do to disregard entirely their colleagues of the "alchemist" type.

In chemical technology we owe much to the Arabs. If we disregard the discovery and early production of such substances as sal ammoniac, borax, soda, potash, nitre and cane-sugar, the technical improvements in certain branches of their industries and crafts have influenced the advance of distillation technique. Thus for instance the glass industry prospered under Arabian rule, used local traditions in Iraq where glass had been made even before the Assyrians and perfected their product. Arabian pottery specializes in all kinds of glazes. Both these branches of industry made it possible to make better vessels and containers for distillation technique and thus also made new operations possible to the chemist. Pharmacy and other branches of medicine could flourish and the Arabian knowledge of commodities was much larger than that of older generations. In their hands the distillation of water, vinegar, rose-oil and other perfumes and essential oils grew to become a true industry which we will have occasion to discuss.

Arabian authors use the word "distillation" in a far wider sense than we do (636). They include filtration, expressing of oils and extraction with water. It is also used for the ancient method of ex-

tracting essential oils from blossoms and herbs with fats and oils
(enfleurage). The Arabian technique, especially that of the preparation of pharmaceutical and medical compounds, was inherited by
the Salernitan school between the tenth and twelfth century, which
school formed an important gate for the theories and methods of the
East on their way to the West. It is not sufficient to study the general
works on Arabian medicine and pharmacy to get an idea of their
distillation technique. A book like that written by CAMPBELL (111)
would yield no evidence. It is necessary to read the sources themselves,
the most important of which are available either in Latin transcription
or in translations in one of the modern languages.

It will facilitate our discussion of these works if we state beforehand that no proof was ever found that the Arabs knew alcohol or any
mineral acid in a period before they were discovered in Italy, whatever the opinion of some modern authors may be on this point. So
here again chemistry means reactions in molten condition at high
temperatures which excludes a lot of preparations and operations
which now belong to the domain of chemistry.

Among the names of Arabian chemists which tradition has preserved for us we find that of JA'FAR AL SâDIQ, the sixth Imâm. He
lived in the beginning of the eighth century and therefore was a contemporary of the Hellenist HELIODORUS. Many books are ascribed to
him, though we have in no case proof that they were actually written
by him, even if they really belong to the eighth century.

In one of these, the Kitâb risâla Ja'far al-Sâdiq . . . . , the Book
of the circular of Ja'far al-Sâdiq on the science of the Art and the
Philosopher's Stone, the "calcination of gold" is discussed and the
following recipe is given: (533): "Then distill green vitriol in cucurbit
and alembic in a strong or average fire and take what you obtain as
distillate. You will find it pure (that is a white water will distill
over) with a greenish tinge". This description, which no doubt is a
preparation of sulphuric acid from vitriol, goes to show that this text
can not be earlier than the thirteenth century and that it is therefore
wrongly ascribed to JA'FAR. We should note that the word for distilling used in this passage is "qatara" which like the Latin "destillare"
literally means "dripping". There exists a Greek parallel "kata-stázo"
which is also used in the sense of "dripping, causing to drop, falling
in drops" but which is never used in the sense of "distilling", not
even by the Hellenistic chemists as we saw.

At the end of the same century we find another name famous in the history of chemistry, Abû Mûsâ Jabîr ibn Haiyyan al-Azdi, who is often identified with Geber, the author of the *Summa Perfectionis,* the main chemical textbook of medieval Christendom. This identification is quite incorrect, but as it seems very persistent in most books on the subject we feel justified in discussing this point in detail. Jabîr is often said to have been taught by Khâlid and Ja'far, he lived about 775 A.D. and five works are ascribed to him, but Holmyard recently found no less than eleven books which are authentic, but which have not yet been published (535). Yet the publication and study of this work would be most fruitful as Jabîr was the teacher of that flower of Arabian chemistry, al-Râzî. It seems that two series of legends were interwoven to form the traditional figure of Geber in older handbooks (427).

First of all there are several Latin manuscripts of the twelfth and thirteenth century which are ascribed to a certain Geber, mentioning aqua regia, sulphuric acid, nitric acid, etc. Some have thought that this Geber was Jâbir ibn Aflah Abû Muḥammed of Seville, whose son was a friend of Moses Maimonides, the Jewish philosopher and therefore Jâbir flourished about 1150 A.D. But this man was the astronomer Geber and the chemist Geber was probably a wholly fictitious figure, who was said to be "a great philosopher", "king of India", etc. His works were widely read and cited up to the eighteenth century. But his *"Summa Perfectionis magisterii"* gives the chemical knowledge of the twelfth and thirteenth century. This was already pointed out by Berthelot but it is often forgotten by later writers.

The second series of legends is of Arabic origin and arose in the sect of the Ismâ'îlîya, a Shî'ite sect which was formed after the death of Ismâ'îl, son of the imâm Ja'far al-Sâdiq (765 A.D.). After the death of his son Muḥammed further sects split off. One of them, the "Seven" (Sab'îya) took the name of Carmathians towards the end of the ninth century. It shows strong Neoplatonic, pseudo-hermetic and "Ṣabian" influences and universalistic tendencies, preparing Arabian science to absorb Hellenistic traditions. Their scientific and mystical writings are tinged with sufism and alchemy. Ja'far was said to be a scientist, whose possible companion and alleged pupil was the alchemist Jâbir. Ruska and Kraus have penetrated this legend (*Archeion* vol. XIV, 1925, pp. 425-435).

It is, however, certain that there was little Arabian science before the Abbasid era (e.g. the ninth century) when translations from the Syriac and Greek began to penetrate Arabian literature, and certainly nothing worth mentioning before the eighth. Abbasid science also shows marked Iranian and Indian contacts. RUSKA doubts any old Arabian or Latin manuscript not mentioned in the *Fihrist*. The commentary of the *Almagest* by JÂBIR as translated by GERARD of Cremona and other similar documents are of doubtful origin (532). But since STAPLETON found several authentic JÂBIR documents we must needs wait for their translation before pronouncing any judgement on pre-Abbasid science. One thing, however, is certain. There can be no doubt that the translation of these JÂBIR documents will not reveal any knowledge of the preparation of mineral acids, or of alcohol by distillation which is not yet found in any chemistry book, as JÂBIR is certainly not identical with the famous medieval author "GEBER".

This brings us to his alleged pupil AL-RÂZÎ after mentioning that AL-JÂHIZ know how to obtain ammonia by dry distillation of animal offals, which shows that experimental chemistry proceeded along the lines laid down by the Alexandrians in the early ninth century.

ABÛ BAKR MUHAMMED IBN ZAKARÎYÂ AL-RÂZÎ, also known as RHASES, born at Ray on August 28th 865, studied mathematics, philosophy, astronomy and philology. He became chief physician of the hospital of his native town and later in Baghdad. After devoting himself to the study of science during part of the period of thirty years travel from court to court, he returned to his native town where he died blinded by cataract on November 10th, 925.

His writings contain the most violent polemic against religion written in the course of the Middle Ages. He is above all inspired by criticism of the religion of Antiquity and believed earnestly in the progress of scientific and philosophical knowledge. Though tradition has it that he was a pupil of JÂBIR, others hold that he was not acquainted with JÂBIR's writings (535). Still without such inspiration what would be the explanation of the sudden appearance in the Arabian world of such a clear-headed thinker? Did AL-RÂZÎ invent a new science? We believe not, but the edition of the JÂBIR manuscripts should solve the question. AL-RÂZÎ's books like the *Madkhal* and the *Kitâb sirr al-Asrâr* present such a scientific spirit and excellent arrangement of facts, that they seem the culmination of a generation's efforts rather than the work of one man only. AL-RÂZÎ himself mentions JÂBIR as his master and we have no reason to doubt him. AL-RÂZÎ seems a most noteworthy follower of the Greeks even in other subjects than Kîmiyâ. He rejects everything but actual facts proved by test and experiment and translates the more mystical teachings of his masters into cold language. He dismisses the older Indian alchemy which was soon superseded by a "mercury" alchemy.

STAPLETON and his collaborators have given translations (601) of some important parts of Râzî's Madkhal at-Ta'limû (Instructive Introduction), in which we find the following accounts of the Instruments: "Some are instruments for melting the "Bodies" and the "Stones" ... Among them is a thing which they call Bût-bar-bût (botum barbatum) which is suitable for the process of Istinzâl (destillatio per descensum) ... As for the instruments of operation and treatment by means of which substances are so manipulated to become Elixir .... they are Qar and Ambîq and Qâbilah and the Blind Ambîq.

Qar' and Ambîq is that apparatus in which rose-water is made. The (Blind) Alembic is that which is fitted on the top of the cucurbit. As it has no spout through which the distilled water can flow into the receiver the water flows down, that is collects in the upper portion only.

The blind is a cup ("qadah") which is placed on the cucurbit in which collects whatever rises from the latter .... (Then a full description for the manufacture of an Uthâl or aludel is given) .... The cucurbit and alembic is for anything you wish to distill. The meaning of distillation is this. You put the substance into the cucurbit and fit the alembic over it and the joint having been luted with rags, the outside is then smeared with Khitmî (marsh mallow) and glue, the apparatus is then left to dry: Then you place it in a stove ("mustauqad") and adjust a receiver to it.

The meaning of the latter is this. The spout of the alembic is placed in a cup in such a way that what passes down, drop by drop, collects inside. After that you kindle a fire of the proper intensity under the pot. If it be a humid substance a gentle fire by means of charcoal or some other fuel should be used. If otherwise then a greater or lesser degree, which must be learned by experience, as our discourse would become too long. If properly heated for a long time the substance passes over in the form of water into the receiver. This is called the distillate ("muqattar") no matter what the distilled substance is. Sometimes the cucurbit if you desire it to be gently heated, is placed in a vessel of water. On the water being heated the substance distills by the heat of the water. Alternately a vessel containing ashes may be used or the cucurbit is simply suspended in a suitable vessel. These and similar processes are easier to understand by actually seeing them. Enquire therefore of these from adepts that you may view them".

In his *Kitâb sirr al-asrâr* (*Book of the Secret of Secrets*) (538) AL-Râzî remarks that substances of vegetable origin are treated like those of animal origin such as blood and hair. The latter substances were favourite materials for the production of sal-ammoniac by dry distillation. The tripartite apparatus for distillation is called alembic. The distilling flask is called qar'a (Greek "lopas", cucurbit), the alembic, ambix or ambikos returns as al-anbiq, the receiver or bikos is called qâbila. The helmet or alembic is compared with a leech in the *Mafâtîh* and therefore named "mihjâma", but a comparison with an old-fashioned glass fly-trap would be more to the point. Its channel collecting the distillate and delivering it to the spout of the alembic is called "handaq" which means "canal, ditch".

Filtering through cloth, which the medieval chemists call "destillatio per filtrum" is here mentioned as "washing on the râwûq". He then discusses washing, refining, dissolving, crystallising from a solution and distilling with cucurbit and anbiq. Distillation ("taqtîr") is defined in the *Mafâtîh* as "the process like that in making rosewater. It consists in putting the substance in a cucurbit, lighting a fire underneath, so that the water rises in the alembic and collecting this in the receiver". Tas'îd (Forcing up or elevation) is what happens in the U*th*âl, this word is used for sublimation and similar processes.

The second part of the book is devoted to instruments, its second chapter deals with the instruments used for non-metals. The following paragraphs are important for our discussion:

"§ 7. The instruments for treating non-metals are the qar', the anbîq, and the qâbilah, the anbiq (blind) and cucurbit, the u*th*âl (aludel), the oven, the beakers, bottles and phials, the rubbing stone, mortar and pestle, the furnace, the tâbistan, boxes and other things one wants.

§ 8. The qar' and the anbîq with a spout and the qâbilah are essential for distilling liquids. There should be no hole in the bottom and no pores anywhere. The alembic should be placed upon it. The earthen pot in which it is placed shall have the shape of a mirjal (copper cauldron) and the cucurbit shall be plunged in water to the height of the medicine (in it). There shall also be near the furnace a jug of boiling water from which to fill the pot when the water diminishes. Be careful that no water touches the cucurbit and fix it in such a way that it will not shake. The bottom of it shall not touch the bottom of the pot lest it be broken.

§ 9. Sometimes distillation is carried out with luted cucurbits hanging in a mustauqad constructed of clay and shaped like a dish cover. A gentle fire is lit underneath. When the mustauqad becomes hot and drops come over rapidly, stop the fire until it goes out and drops come over singly. Sometimes the cucurbits are placed in a pot containing ashes and a fire is kindled underneath. This is the best method for students. Sometimes the cucurbits are placed below the mustauqad while the bottom of the pot rests on big bricks. Ashes are heaped round them and the cucurbit is put on top and all its sides are surrounded with sifted ashes. Then it is heated. It is necessary that the sides of the lid of the cucurbit and the head of the receiving flask are well closed lest the smoke enter into it and the air spoil it (the distillate).

§ 10. Alembics are of four kinds:

1. One with a very broad diameter which is suitable for the distillation of the black constituents of substances that have to be calcined as well as for the volatilisation of sal-ammoniac.
2. The alembic which is not of very broad diameter and which is suitable for the distillation of Spirits and Tinctures.
3. The alembic of still less broad diameter suitable for the distillation of the Stone at the beginning of the Work.
4. The alembic of very small diameter is useful for the distillation of water and for its purification.

§ 11. The cucurbit and the blind anbîq are suitable for the solution of Spirits as well as of Bodies which have undergone the process of Ceration. It is a cucurbit which has an internal channel without a spout. Those things which you want to be dissolved are placed in the groove and in the cucurbit any strong water. The alembic is mounted on top and the joint is closed. The apparatus is then placed in a pot of water or ashes. It is not suitable for anything but the process of Solution. This is the al-Hammâm (The Bath) which is spoken of enigmatically by the philosophers.

§ 12. The blind cucurbit is a cup mounted on the cucurbit. You place in it the dissolved substances and the apparatus is hung in a symmetrical mustauqad. Below is placed either a burning lamp or lighted naphtha or charcoal or hot ashes. Take care that neither the fire becomes extinguished nor the ashes become cold, until dissolved substances are solidified."

He then goes on to describe the u*th*âl or aludel and the clay of the philosophers (lutum sapientiae) which is compounded of horse dung, pigeon dung and other substances.

The third book is especially devoted to the treatment of different substances with the instruments described above. Among the substances distilled for refining we find mercury, sulphur, sal-ammoniac, acetic acid, and crude petroleum. The latter when black is refined by mixing it with sal-ammoniac and distilling several times to obtain a colourless distillate. Even the higher fractions are distilled as the recipe mentions, that the distillation should be continued until the distillate does not burn immediately when coming into contact with fire. The vitriols are "dissolved by distilling", a procedure which involved melting, solidifying, distilling and pulverizing the residue, dissolving this in the distillate, etc. Refining urine means distilling it seven times (538).

STAPLETON (600) discovered a collection of alchemical manuscripts bound together in the Library of the Nawab of Rânpûr. Most of them were copied in 1283 by a copyist who used a manuscript of AT-TUGHRAI, the critic of IBN SÎNÂ. The collection also contained two unknown works of Râzî.

A chapter of the *Sirru al-Maknûn* (*Hidden Secret*) begins thus (600): „We therefore state that the oils which distill from all things are of various colours, red, yellow, green, blue and other colours akin to these. The water that is obtained from the oil is what is called "Spirit" and the "Tincture" that is in the water thus obtained, whether it be red, yellow, green, blue, is called "Fire", "Tincture", etc."

The second newly discovered book called *The treatise of the making of the Black Stone copied from the book of Abû 'Abdullah al-Bâkâwi* has similar definitions. It describes how the Black Stone is first broken up into water, oil and residue and says: "This is the water which the Sages have named "Spirit". The oil they called "Soul" and the residue "Body"". We should remark that this treatise ressembles several chapters of the Latin treatise *De Recta* ascribed to IBN SÎNÂ (*Theatrum Chemicum*, edit. ZETZNER, 1659, vol. IV, pp. 863-875).

In the *Kitâb al-asrâr* he also frequently refers to the "nafs" or "Soul" which is used to describe a very valuable product of distillation. It is the substance that rises when the distilled water begins to change colour. The phrase "to distill its water until its nafs rises" (538) occurs frequently. Mention is also made of animal hair out of which a white, red, yellow and black water is distilled as well as a nafs and an oil.

If we summarize the apparatus described in the writings of AL-Râzî (471) (531) (601) we find the qar' (cucurbit), anbîq (alembic), qâbilah (receiving flask), blind alembic, bût-bar-bût (descensory), qadah (beaker), kûz (glass cup, "the best kind of which is made in Irâq"), files, spatulas, shears, tongs and crucibles. The ranges of temperatures which AL-Râzî applies run from the "heat of a sitting hen" to that of a furnace. Heating is achieved by means of dung, sunlight, hot water, ashes, candles or lamps (qindîl), wax candles or torches, naphtha lamps (naffâta) and furnaces (atûn). Strictly speaking the atûn is a small model of a potter's kiln, but the tannûr (large baker's oven), the mustauqad (small cylindrical stove used more especially for the aludel), the kânûn (brazier or chafing dish), the nâfikhu nafsih (a stove with perforated sides on three legs, like the kerotakis apparatus), the tâbištân, mauqid and other types of furnaces are used too. AL-Râzî also used the word "furn", undoubtedly derived from the Latin "furnus".

The apparatus is luted with clay lute in the case of glass, this is composed of clay, rice, salt and chopped hair. A better type of lute is the "tîn alhikma", the "lutum sapientiae" or lute of the philosophers.

The main processes which he uses are distillation, decantation or filtration (all three indiscriminately described as "taqtîr), the descensory process (istinzâl), roasting (tashwiyah), digestion (tabkh), amalgamation, lavation, sublimation, calcination (taswîl) and lixivation.

In his medical writings AL-Râzî compares the stomach to the cucurbit, the head to the anbîq, the nose with the spout of the anbîq, when he discusses what happens to the digested food. This symbolic application of the elements of distillation to biological processes is far from rare in older books. ABÛ 'ALI AL-HUSAIN IBN 'ABDALLAH IBN SÎNÂ (980-1037), better known to the world as AVICENNA, paints a similar picture of the catarrh in his encyclopaedia Kitâb al-shifâ . . . . where he states that the body acts as a distilling apparatus, the stomach being the cucurbit, the head the anbîq which collects the "humors".

Returning to AL-Râzî it will be remarked, that nowhere does he mention alcohol nor any of the well known mineral acids, though one passage in the Madkhal might be interpreted as a very crude way of making hydrochloric acid, but he does not define the properties of this acid. There is absolutely no foundation for the statements of

PIQUE (483) and SCHELENZ (558) that AL-Râzî made alcohol by distilling and prepared araq and other alcoholic drinks from fermented sugar-solutions.

AL-Râzî's Black Stone is not the Philosopher's Stone but an aqua vitae, that is an Elixir of Life and certainly not an aqua ardens or some other form of alcohol.

The Arabian chemist in general is very fond of products obtained by dry distillation of materials of animal origin, those of vegetable and mineral origin are considered to give less potent medicines. He very often describes this dry distillation as "dissolving by means of cucurbit and alembic". Many recipes of this type are for instance found in the *Lapidary of Aristotle,* a book which enjoyed great popularity in the Middle Ages but which seems to have been composed by a Syrian chemist in the middle of the ninth century from Greek, Hellenistic and Iranian sources (529).

Apart from the *Fihrist* the most important contemporary source-book on Arabian science is the *Mafâtîh al-'Ulum* (*Key of the Sciences*) (4), written by ABû 'ABDALLAH MUHAMMED BIN AHMAD BIN YûSûF AL-KâTIB AL-KHWâRAZMI, the "scribe from Khwârizm" in the year 976 A.D. This Persian encyclopaedist published a classified vocabulary of technical terms and divided the sciences in two groups, the Indigenous and the Exotic. The latter including alchemy were derived from Greek, Syriac, Iranian or Hindu sources. In this encyclopaedia he mentions that the alembic "is shaped like the cupping glass (mihjâmah)". The Arabian form of alembic is obviously derived from the Greek form of the cupping glass, nipple-shaped with the lower half of narrower diameter and a hole at the top. The introduction of the cupping glass into alchemy was according to STAPLETON almost certainly due to magical associations. It removed the blood in which the soul of a person was thought to reside, while on the other hand the spirits of inorganic or organic substances were collected in the alembic. The definition which the *Mafâtîh* gives of "taqtîr" was discussed in connection with the views of AL-Râzî'. "Tas'îd" is the term more generally used for the volitalisation of dry substances. Distillation played a great part in medicine as early as the century we are dealing with. Though preparing distilled water with the classical classical "wool-condenser" the physician ALI IBN-ABBAS was one of the earliest of his caste to use distilled water regularly as did AVICENNA.

Distillation and sublimation are the most important processes

described by Abû al-Qâsim Ha'alaf ibn 'Abbas al-Zahrâwî, better
known as Abulcasis, in his medical encyclopaedia, the *Kitâb al-
tasrif...* (2). This famous Spanish-Arabian doctor was born in Zahrâ
near Cordoba and died there in 1013. As court-physician and surgeon
of al-Hakam II he compiled a large medical encyclopaedia in thirty
books. These works were translated into Latin by Gerard of Cremona
(1114-1187) and were widely read in the Middle Ages and after.
The pharmacological part was edited separately and printed fairly
often (3) as the *Liber servitoris*. Now Abulcasis is often said to
have mentioned the distillation of wine, acetic acid, etc. Whoever
peruses the text will see that this opinion is false. Abulcasis writes:
"Secundum hanc, disciplinam potest distillare vinum qui vult ipsum
distillatum (In this way wine can be distilled by anyone who wants
it distilled)", which means that wine might be distilled to obtain
a colourless product in the way acetic-acid was distilled, which product
was never transformed into a stronger acid by this operation, at least
at that date. Abulcasis therefore can never be accused of preparing
alcohol from wine.

There was some discussion on the "berchile" mentioned by Abulcasis (536) (539), which was said by Speter to be a self-filling
water-bath. But by close scrutiny of the text Ruska was able to point
out that the berchile was a vessel, probably lead-lined, of indefinite
form, in which the overflow of boiling water-bath was collected.

Another physician often prescribing distillates was Yahyâ ibn
Mâsawaih al-Mârdînî (Mesue the Younger), a Christian (Jaco-
bite), born in Iraq and working in Baghdad before he entered the
service of the Egyptian Chaliph al-Hâkim. He died in Cairo in 1015
ninety years old. Apart from a book on purgatives and emetics and
one on remedies relative to each disease he wrote a comprehensive
pharmacopoeia in twelve parts called *Antidotarium sive Grabadin
medicamentorum compositorum* in Latin, which enjoyed much popularity and was reprinted as late as 1602 (413) (414). Mesue is
often called the "pharmacopoeorum evangelista" in the West. The
editions of his works are mostly reprints of the standard edition of
1471 (Venice). He very often advises the use of empyreumatic oils
prepared by different methods of distillation. The second book of
his *Simplicia et composita* (405) gives details on the preparation of
rose-water without introducing any new views on the subject.

Mesue gives a clear description of the bût-bar-bût method when
describing the preparation of "oleum de gagatis", that is oil obtained

from some kind of bituminous shale. He says: "A hole is dug in the ground, the walls of which are plastered with clay and a glazed pot with a broad neck is put in and covered with a sieve-like iron plate with holes. On this plate the crushed shale (gagates) is put in a large pot glazed on the inside, having a narrow neck. This pot is placed on the iron plate mouth downwards and all the openings are closed with clay so that nothing can escape. Around the pot on top light a fire for two hours and then collect the oil that has dripped down".

Another book very popular in the West in later centuries was the *Turba Philosophorum*, printed since 1572 and known in three different versions. These all go back to an Arabian original written about the end of the tenth or beginning of the eleventh century by an author who was no professional chemist but who theorizes after having read many alchemical books of his period. The *Turba* (536) is a controversial pamphlet directed against the Greek alchemists, hoping to free chemistry from the pest of pseudonyms and mystical terms to build it up on the foundation of a general philosophy of nature. It reflects the religious and philosophical conflicts of the Moslim world of that age, though it also abounds with references to distillation and distillates prepared in various ways.

Among the important alchemical manuscripts discovered in Indian libraries and edited by STAPLETON we find the *Ainu-s-Sana'ah wa' Auna-s-Sana'ah (Essence of the Art and Aid to the Workers)* written at Baghdad by one ABÛ'L HÂKIM MUHAMMED IBN 'ABD-AL-MALIK AS-ṢÂLIHÎ AL-KHWâRAZMI AL-KâTHI in 1034. The incomplete manuscript of the Nawab of Rânpûr could be completed with the Persian version found in the Nîzâm's library (14) (599). A close study revealed that it seemed to be influenced by five sources: a) the *Rasâil* of the *Ikhwânu-s-Safâ* (970 A.D.), b) the *Fihrist* (988 A.D.), c) the second volume on Drugs of the *Qânûn* of IBN SÎNÂ (1021 A.D.), d) two Syro-Arabic treatises in the British Museum and e) section 10 of the *Mafâtik* of AL-KHWâRAZMI (976 A.D.). It was copied in 1283 and brought to India. In this treatise we find great importance attached to weight in chemical operations. The instructions, for instance, on the aludel and the furnaces are clear like those of AL-RâZî and GEBER. After the discussion of the names of substances, classification and division between Bodies and Spirits, the Qualities, the Proportions of Mineral Amounts we find chapters on the distillation of substances to obtain the "red" and the "white" and on instruments. The latter mentions the same apparatus which we have already discussed; files, glass cups, pestle, earthenware jars glazed inside with lids, glazed pots, the long uthâl, cloth for filtering purposes, glass funnels and bottles, crucibles, furnaces, bellows, pincers, ladles, cauldron and phials. AL-BALKHî, a friend of AL-Râzî, is often quoted as an authority, therefore much of this text may have been derived from the writings of AL-Râzî himself.

Similar apparatus and methods are described in the *Book of the Disclosure of the Secret Science of the Kâf* (the initial letter of Kîmijâ that is chemistry), the Arabic original of the popular *Tabula Smaragdina* (534) which KIRCHER ascribed to KHALID. Internal evidence, however, goes to show that it dates from the twelfth century. A special chapter is devoted to the distillation of the "Waters" from different substances.

It is interesting to read that the Arabian chemists were inclined to consider distillation an important process for agricultural industry. Thus ABû ZAKARîjâ YAHYâ BIN MUHAMMED BIN AHMAD IBN AL-'AWWAM AL-ISHBîLî who lived in Seville towards the end of the twelfth century was the author of a large work on agriculture. This work has 34 chapters, thirty of which deal with agriculture and four with cattle-raising. This careful work, giving more than 585 domesticated plants, cites Greek and Arab writers and embodies a lot of personal observations. The chapter XXX is a very heterogeneous one, giving not only the selection of the proper building site of farms and their construction, but also the description of the distillation of rose-water, preparation of vinegar, mustard, malt, etc., a summary of the work to be done in each month of the year and the description of the construction of levelling instruments. We cite the most important passages from the *Kitâb al-Falâha* XXX. 4 (7):

"According to az-Zahrawi (Abulcasis) there are different methods of distillation, e.g. with a wood or a coal fire and with or without a water-bath. A wood fire is more frequently used, for coals have too strong a smell for rose-water. The method of the people of Iraq is lengthy and expensive. A copper cauldron is placed on a furnace, similar to that used for a bath, against the wall. The smoke is drawn off to cause no changes in the rose-water. The cauldron is covered with boards in which holes have been made, and the cauldron is filled with water. Then insert preferably glass cucurbits and alembics. Fill the cucurbit with fresh leaves of roses and light the furnace until the water boils. If no others are available glazed cucurbits and alembics can be used. Then fill the cucurbits again after having detached the receiver. Always keep the cauldron full of boiling water and have water standing close-by to refill the water-bath. Adding cold water will cause the glass apparatus to break. If you use direct fire take an oven with a supporting ridge for the cucurbits, which should never be glass ones in this case. The spout of the anbîq should slope towards the receiver. The degree of heat is measured by touch. If it reaches the top of the cucurbit the right effect is sorted and rose-water with all its phlegm is obtained. Overheating means loss of phlegm. About one half or two-thirds of the contents

Fig. 13. A "mass-oven" for the distillation of rose-water, taken from an Arabian manuscript of unknown origine (Spanish?)

of the cucurbit should be distilled over. The furnace shall stand in a good room free of draught, not in an inner-court or the like.

(Next comes the description of a smaller furnace and a water-bath).

If the cucurbits are heated directly by the fire, "mass-ovens" containing 16 or 25 cucurbits are built (Fig. 13). These furnaces contain registers or openings to regulate the fire. The cucurbits should be placed at a distance of at least 12 cms from each other, they shall be at least 90 cms high including a good wide anbîq. The distilled water is collected in the channel of the anbîq and flows off through the spout. The receivers should have narrow necks to preserve the perfume well.

(Directions follow how to fill the cucurbits without damaging the rose leaves and diminishing the yield of rose-water).

The furnace is heated until one can no longer touch the anbîq and iwo-thirds of the contents are distilled over. Then shut the air-holes of the furnace with clay and let it cool down. Overheating means spoiling the smell of the rose-water. Then remove the residue in the cucurbit and wash it well lest you spoil the smell of the next charge. The residue can be diluted with water to a syrup and redistilled. Also dried rose-leaves can be moistened with water to be distilled and this gives a good water for medicinal purposes. With amber and with clay the smell can be adjusted and by decanting the colour too. By distilling rose-water with camphor, sandelwood, spices, musk etc. the smell can be changed. This is always done in glass apparatus on the water-bath.

According to al-Râzî the succes of the distillation depends on the use of an ample cucurbit with glazed inner walls. It should always be heated upto the height of its contents and inserted as far in the water-bath. Never let glass cucurbits touch the wall of the heating vessels lest they break. Let them remain in the water-bath until cooled down, the same is true for the sand-bath and the ash-bath which should not be heated higher than the water-bath. When distilling with direct fire earthenware cucurbits are most efficient. The lute is also very important. One can make "stone-oil" (using pieces of porous bricks to absorb some kind of oil in the cucurbit) by dry distillation (lit. "according to the dry way"). The qarah (cucurbit) is the part from which the vapour of the rose-water or whatever one distills escapes. The helmet put over the cucurbit is called anbîq and that part through which the distilled liquour escapes is the "tail" or coil. The receiver is the vessel collecting the rose-water at the end of the anbîq. The channel in the anbîq must be formed in the right way lest any distillate gets lost".

A similar description of the manufacture of rose-water is given by 'ABD AL-RAHMâN IBN 'OMAR ZAIN AL-DîN AL-DIMASHKî usually called by his nickname AL-JAWBARI, because he was born at Jawbar near Damascus. This wandering scholar travelled extensively throughout all the lands of the Islâm and the Eastern Caliphate as far as India (19) (20). He lived in Harrân (1216) and Kôniya (Iconium) ⟨1219) and then at the court of the sultan of Urtuk (1222). In his *Kitâb al Makhtâr... (Book of the disclosure of the Secrets...)*, a mine of information on quacks, alchemists etc., he speaks of "the people of al-kîmijâ (the chemists) who know three hundred ways of making dupes". But this "sceptical chemist" also discloses the preparation of

rose-water in this manner: "They take roses of Iraq, and macerate
them a day and a night in precious rose-water, put them in a distilling
flask and place a grain of musk in the spout of the anbîq. They have
also added to each ratl (over 400 grammes) of rose-buds 10 dirhem
(30 grammes) of cloves and 2 dihrem of cardamom. Then they
distill over a soft fire. The distillate is preserved in a glass phial
the neck of which is closed, wrapped in cotton and put in a box.
See that no air enters that the smell can not escape. Should one
wish to make common rose-water, then take pure fresh water, put it
in a cucurbit and distill off one third on a soft fire. Take it off and
protect it from dust. As it has cooled down anyone who wants to
make the elixir should take 3 dihrem of the first elixir (the pure
rose-essence) for every Baghdad ratl of boiled water. The mouth of
the cucurbit is closed and the whole placed in the sun for three days".
Does this boiling down the water mean that the process of sterilizing
was grasped as early as the thirteenth century?

Similar details are given by ABû MUHAMMED 'ABD ALLâH IBN
AHMAD DIYâ' AL-DÎN IBN AL-BAITAR AL-MâLAKI known as IBN AL-
BAITAR, the celebrated surgeon, botanist and herbalist, son of a veterinary surgeon, born at Malaga from an old family in the last quarter
of the twelfth century. He studied at Seville, travelled in North
Africa, Morocco, Tunis, Algiers, etc., entered the service of the Sultan of Egypt as chief botanist and transferred his services to the son
of this potentate, the sultan of Damascus, in which town he died in
1248. Apart from a book on materia medica he wrote the Kitâb al-
Jâmi . . . ., a collection of simple remedies (simplicia) from the
animal, vegetable and mineral worlds collected from Greek and Arab
authors. No less than 1400 medicaments are given in alphabetical
order.

All these authors describe the same apparatus, which were incapa-
ble of distilling low-boiling substances. As none of them ever mentions alcohol it is practically certain that this substance was unknown
to the Arab world upto the thirteenth century. Later on we find it
mentioned, not as "alcohol", but as "aithale, sudor" or "al-raqa"
(that is the "sweat") whence the name arak. This word is first used
for any distillate in general. Only much later such expressions as
"rûh-al-hamr" (spirit of wine) or "rûh-al-araq" (spirit of raki) found
are. This change occurs together with the introduction of the new
Western type of distilling apparatus, which enables the chemist to
recover low boiling distillates (362) (364).

About a century later the earliest references to mineral acids and their properties are found in Arabian manuscripts, this novelty comes from the West too, probably introduced from Italy, where both discoveries seem to have originated.

The term "alcohol", that is to say its Arabian original "al-koḥl" is up to that time still generally used for any substance attenuated by fine pulverisation, distillation or sublimation.

Fig. 14

Figs. 14 and 15. Arabian distilling apparatus compiled from different manuscripts

Upto 1300, therefore, we find the same types of apparatus all over the Arabian world. Illustrations in Arabian manuscripts are rare, a few of the most important apparatus are compiled in figs. 14 and 15. The distillation aparatus remained essentially the same as that of the Alexandrian chemist. Still the anbîq is cooled with sponges or wet rags, but even when cooled by the air only, the anbîq remained the cooling element where the distillate was condensed, hence the channel or gutter in the anbîq. No cooling of the spout nor any indication of the spout leading to some condensing apparatus are mentioned anywhere. These novelties were introduced with new types from the West into the Arabian world in later centuries. Only then could the cucurbit and the anbîq be combined to form the well-known retort, which is also of late date in Arabian chemistry. It is now called "mi'wagga".

The *manufacture of essential oils and perfumes* was one of the most important industries that applied distillation. Its centre was the Persian province of Sâbûr (426). Ten different types of essential oils were made there from violets, lotos-flowers, narcissus, lilies and other flowers. Another centre was the province of Babylonia, where Kûfah was famous for a still better quality of "water of violets" and "water of carnations". The neighbourhood of Shiraz was the centre of rose-water production, which was so important that it paid the Caliph AL-MâMUM a yearly tribute of no less than 30.000 phials. IBN HAUQAL tells us that rose-water was sent all over the world to the Maghrib, Spain, Jemen, India and China. It seems that the foundation of this industry was laid towards the end of the eighth or the beginning of the ninth century. A later but no less important centre was Damascus; details of its industry were discussed by AL-DIMASHKI whose story we will repeat in this chapter. It seems that the Arabians immediately tried the new and better method of distilling essential oils from the flowers instead of using the old classical way of extracting with fats and oils (enfleurage). As early as the ninth century AL-RÂZî could already write that the usual way of preparing rose-water was that using cucurbit and anbîq. Of course the essential oils as such were not prepared but the flowers were macerated with water and the oils distilled over with the water. There is no account of the separation of the oils from the distilled water. As early as 1266 the Syrian historian 'UMAR IBN AL-'ADîM wrote a guide on making perfumes. It seems therefore that the Muslim druggists were the initiators of this art and that the West imitated them. In the thirteenth century all the tricks of the already well-developed art of distillation were applied to the production of rose-water in the neighbourhood of Damascus.

We can study these in the *Kitâb Nukhbat al Dahr* of Abû 'ABDAL-LAH MUHAMMED IBN ABî TALIB AL-ANSÂRî AL-SûFî AL-DIMASHKî, the famous Arabic cosmographer born about 1256, who died as imâm of Rebwa, Syria in 1327 (16) (17). When describing Damascus in this book, AL-DIMASHKî sums up the seven branches of the river Barada. The last of these is the Nahr Mizzah, about which he says: "The Nahr Mizzah is called after a village al-Mizzah, the name of which was originally "al-Munazzah" (the Incomparable), because of the healthy air, the pure water, the beautiful pleasure-houses, the delicious fruit, the many flowers and roses and the production of rose-water; the residue of which is thrown on the roads, lanes and alleys of this place like dirt. Thus the smell is incomparable and finer than musk until the roses are overblown.

The production of rose-water is achieved with distilling apparatus. The method consists in digging a ditch in the ground two ells and a half square. Over this a barrel-vault is built of sun-dried bricks with a door on one side and a ventilating hole in the other. On top there is a ventilating hole too through which part of the smoke escapes. On the vault a pan is placed under which plentiful wood is burnt. Round the pan a circular rim is built up like the reservoir of a bath about half an ell high. On top of this a strong mat of firm and thick, fresh

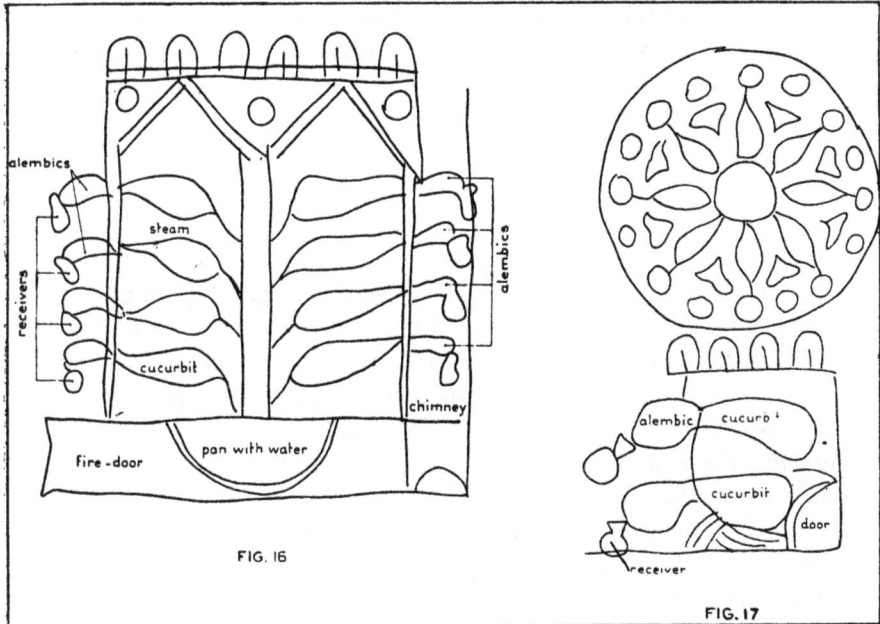

Fig. 16. AL-DIMASHKÎ's steam oven for the production of rose-water (MEHREN).
— Fig. 17. Cut and detail of the steam-oven (MEHREN)

stalks of Persian reed is attached. Then on the mat glass cucurbits are placed, the necks and mouths of which emerge from the furnace. When a complete circle of cucurbits is installed, they build up a further circular rim like the one below, to the height of four fingers. On top a second mat of Persian reed and another circle of cucurbits is built. Again the rim(wall) is continued until the construction is one and a half times as high as a man, every time a mat and a circle of cucurbits. Sometimes a wooden pillar is erected in the middle of the pan rising to the top of the construction and coped by a roof, the top of which is formed as shown in the figure (Figs. 16 & 17). Understand this well, if it is the will of God, for the aid comes from Him.

Then the receivers (qawâbil) also called "suckers" (ridâ'ât) are attached. This is done after the cucurbits have been filled with roses or the like from which the water is to be drawn. Everytime a "sucker" is full it is emptied in a large glass vessel called "qarâba" or in a large copper vessel called "qumqum".

There is another type of distilling apparatus which fired with wood enables one to obtain rose-water ("mâward") and other waters without using water (and a pan). This is done after filling the cucurbits with roses, water-lilies, flowers of the bân tree, the pomegranate, anemones and chicory or the leaves of the clove plant, that is grown in Damascus. It is shown in the figure (Fig. 18). Understand it well if it is the will of God for the aid comes from Him and He alone is sufficient for us and a splendid solicitor.

It consists of building a vaulted furnace for heating, the whole of which looks like an inverted well through which the flames and the smoke can rise like they do through a chimney (Fig. 18). Round this a wall is built, thus the whole looks like two concentric circles. Then glazed cucurbits are placed between the wall and the well, the bottom of which rest against the well while the necks project through the wall. Between the cucurbits holes are made in the well (chimney) through which the heat and smoke can enter and circulate under the cucurbits so that these can be thus heated according to need. The construction of well, wall and cucurbits is raised until it is higher than a man. Then the space between wall and well is covered, at the same time making the top of the well, that is the chimney bottle-necked and one heats with wood without anything else.

As to the so-called Baitûnî (rose-)water, that is produced, it is made in the "rose-furnace" and the "lead-brazier". This (the "rose-furnace") (Fig. 19) is built in the form of a small tower with two stories. In the first a fire of fine charcoal and other fuel and plentiful wood is burnt and the second is on top of it. It has many holes to allow the smoke and the heat to ascend towards the cucurbits, the number of which is four, three or less.

The "lead brazier" is made by casting in moulds of earth, its aspect is given in the figure (Fig. 20), when it is complete. The Greeks call it "athâl" (aludel?). It has a cover that is its anbîq. Sometimes this cover is made of glass and sometimes of lead. If one wants to work it, a layer of salt and hard loam is put beneath it, then the fire is lighted under it and the rose-water begins to drip regularly, beautiful in colour, limpidity and smell.

As to the H'akimi (glass of wisdom: water-bath) this belongs to the instruments of the Greeks and the philosophers. The cause of the dripping is nothing but the vapour of the water, that is boiled under it. Here is a picture of its appearance as you see (missing in the manuscript!).

The roses (probably the rose-water is meant!) that are produced in al-Mazzah are exported to all southern countries like al-Hidjaz and what lies beyond. At the same time the roses of al-Mazzah are ex-

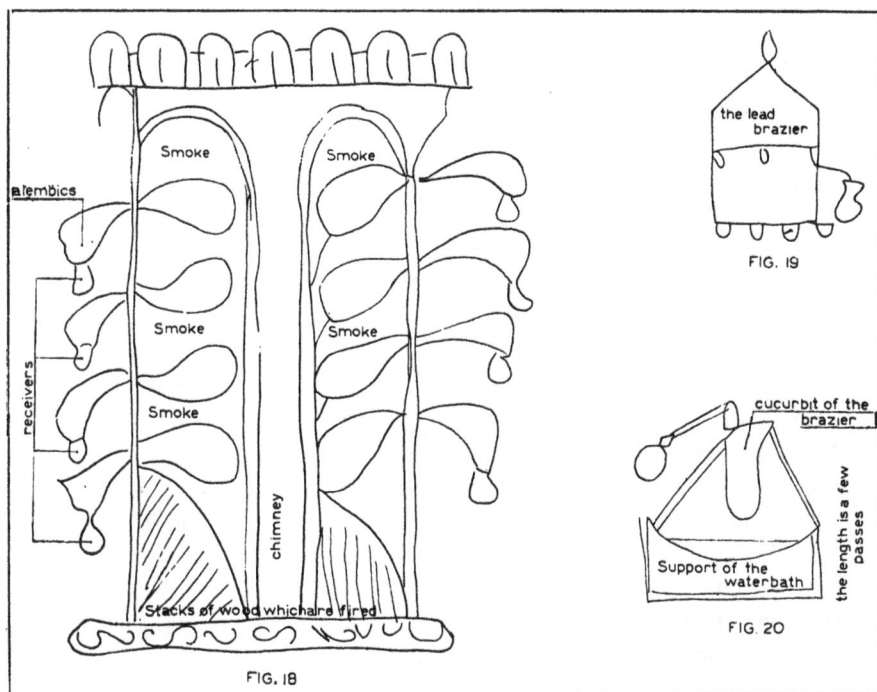

alembics

Smoke   Smoke

Smoke   Smoke

receivers

Smoke

chimney

Stacks of wood which are fired

FIG. 18

the lead
brazier

FIG. 19

cucurbit of the
brazier

Support of the
waterbath

the length is a few passes

FIG. 20

Figs. 18, 19 and 20. The "hot-air oven", the "rose-furnace" and the "lead-brazier" of AL-DISMASHKÎ (MEHREN)

ported to India, Sind and China and still further. There they simply call them "Flowers"."

MEHREN, who edited this text (St. Petersburg, 1866, pp. 194-198) and later published a translation (17), used two manuscripts, one from Paris and the other from St. Petersburg, both full of false readings and gaps. However, he does not indicate where these occur and it seems that he interpolated and interpreted quite freely as the translation given above deviates in several points from his French text though it was prepared directly from the Arabic text. It is not always quite clear what type of apparatus AL-DISMASHKÎ describes. His first apparatus is no doubt an oven with cucurbits heated by the steam of the pan or cauldron with water described. Figure 17 seems to include a top view of each mat on which the circular arrangement of cucurbits and ventilating holes are shown. The second furnace heats the cucurbits directly, it might be described as a magnified hot air bath. MEHREN took the "lead brazier" to be a description of the top storey of the "rose-furnace", but it is equally possible and indeed more plausible if we suppose that two apparatus are described of which the "rose-furnace" is a small hot-air bath for a few cucurbits, a smaller edition of the type shown in fig. 18 and more handy for the laboratory.

The "lead brazier" would then be a kind of ash-bath heater, though the legend of the drawing would imply that the cucurbit was heated by a water-bath, while in the case of the "rose-furnace" the upper storey is called "lead brazier" in the original drawing, the legends of which like those of the others are given in direct translation in the figures.

Lead does not seem a very suitable material for cucurbits or braziers when directly heated by a charcoal fire. If they did not actually melt they would become very weak and deform quite easily. Therefore the layer of salt and hard loam may have been meant as a protecting layer, the application of which was quite common in Arabic distilling apparatus as we have seen. However, the text seems corrupted and the only way of solving our difficulties would be the close inspection of the original manuscripts and their drawings.

From Coptic medical texts it seems that in those quarters rose-water was also prepared in similar primitive distilling apparatus and furnaces. CHASSINAT therefore translates the Coptic "weratostomon" (oϥpλ τocτoлoн) by "distilled rose-water" and "rotonon" (pωτoнoн) by "essence of roses". We shall see that the furnaces and ovens described by AL-DIMASHKÎ had a long life and return in more perfected form in the laboratories of the fifteenth century chemists, where we meet them under the name of "gallery-ovens" or "galley-ovens". The perfumes and "waters" which AL-DIMASHKÎ and his contemporaries described often have quite fantastic names, derived from the country of origin of the flowers or the road along which the final product was imported (659).

From the writings of AL-DIMASHKÎ we glean another interesting detail, the earliest reference to the industrial *distillation of crude oil or petroleum*. Crude oil or "naft" had been known for ages, in the Arabian period it obtained prominence by the invention of Greek Fire (603), that powerful weapon of the Byzantine and Arabian fleet and armies. The term is used to denote all kinds inflammable or burning compounds, some of which were claimed to burn in water. Sometimes it is self-igniting, mostly it is claimed that it can not be extinguished by water. Though the composition of these mixtures varied largely as far as our information goes, we are certain that crude oil or its distillates figured largely in them. We read about Greek Fire in warfare since the flight of the "inventor" KALLINIKOS from the Arabs to the Byzantine court where he betrayed the secret (about 650 A.D.). Now we have plenty of references on the different types of naft and their countries of origin, properties and applications. But AL-DIMASHKÎ is the first to tell us that it was distilled too. He tells us that the naft was used by the Coptic Christians and mixed with gunpowder and other substances to be burnt on New Year's Eve, and he continues: "Many types of naft are water white by nature and so volatile that they can not be stored in open vessels. Others are obtained from a kind of pitch (or bitumen) in a turbid and dark condition, but by further treatment they can be made clear and white by distilling them like rose-water".

It is possible that here again Damascus played a part as a distilling centre. May be it was a novelty in the days of AL-DIMASHKÎ for we find no further reference to distilled naft in earlier works and PLATEA-RIUS who translated many Arabic and Byzantine pharmaceutical works about 1150 does not mention distilled naft.

It is difficult to judge whether the *Hindu chemists* were original in their invention of distillation methods, as many of the native authors claim (501), or whether these methods were imported fairly early from the Arabian world. The problem is complicated by the constant, unnecessary ante-dating of Hindu chemical texts. Manuscripts of the thirteenth century mention the "patana-yantram" method, which is nothing but the sublimation apparatus of the Arabs, the aludel made of two crucibles fitted together. The "koshti" apparatus is a typical descensory, it is for instance used to prepare zinc from its ores. The "tiryakpatana" apparatus is a simple cucurbit with alembic, the receiver of which is cooled with water.

Such simple methods are also found among the primitive tribes of India such as the Bhils, who use the following apparatus for the preparation of their liturgical Mahuda liquor (478). This liquor is a country brew prepared of the flowers of the mahuda (Bassia lati-folia). The fermented mash of flowers, gathered in full bloom, is kept five to eight days in an iron pot or big earthenware vessel called hândî. It is two-thirds full and restes on three stones. Now a kind of calabash with the neck cut away is hung in the muzzle by means of a string. The "doî" being fixed, the muzzle of the pot is covered with a brazen dish called "vâtkâ". It is most essential that the bottom of this dish be perfectly spherical to leave no chink free for the alcoholic vapours to escape. The pot is then heated and water poured into the dish to cool the vapours which drip in the calabash. The result of this distillation is poor.

None of the Indian apparatus seems to equal the Arabic in effi-ciency and adaption to the specific purposes.

Only much later we find Arabic literature on the preparation of alcohol by distillation, for instance the *Ain-i-Akbari* of ABÛ'L FAZAL ALLANI (about 1600). Chapter 28 of this book describes the prepa-ration of araq from sugar solutions. The first method ressembles that of the Bhils quite closely. The second method collects the distillate on a kind of spoon with a hollow stem that allows discharge of the liquid into an outside receiver. A third method is similar to the second one except for the fact that the distillate is cooled by running the discharge tube through a water through. However, it seems neces-sary to redistil the liquid thus obtained at least twice (657).

The great achievement of Arabic chemistry is not the distilling apparatus but the knowledge expressed so well in AVICENNA's *De congelatione:* "As to the claims of the alchemists it must be clearly understood that it was not in their power to bring about any true change of species. The essential nature remains unchanged, they are merely so dominated by induced qualities that errors may be made concerning them".

CHAPTER FOUR

THE MIDDLE AGES

Lo! which advantage is to multiply
That slyding science hath me maade so bare
that I have no good wher that even I fare.
(CHAUCER, *The Chanouns Yemannes Tale*,
ll. 178-180)

It seems that very little of the Alexandrian chemistry penetrated to the West. Latin literature contains hardly any trace of these matters. Still on the other hand we should remember that Greek and Roman literature paid relatively little attention to technological or scientific problems. The Roman was a practical engineer not a scientist and anything of a speculative nature was foreign to his mind. Thence Roman technical tradition contained only those things that were in the limelight in classical times.

So it happened that after the fall of the Roman Empire the connections between the West and the scientific world of the Near East were severed for quite a long time, while the Eastern Roman Empire or the Byzantines profited by their geographical position and their trade connections with the East. In the West there remained nothing which one could call science if we except the few Latin texts on these matters. The main stock of knowledge which was handed over from generation to generation consisted of collections of recipes and directions of the useful arts, which in the early Middle Ages meant those arts which had any value for the church or the monastery as well as a few treatises on military strategy and siegecraft.

Among the latter we must count the treatise by MARCUS GRAECUS which will be dealt with in due order. To the former belonged the book *On the colours and arts of the Romans* by HERACLIUS, the *Mappae Clavicula* (to be discussed later on) and the book *On the different arts* by the monk THEOPHILUS, who gives detailed instructions on the casting of bells, on painting, gilding and all other arts useful to the inhabitants of those centres of civilisation in the Dark Ages, the cloisters. One art only, that of glassmaking, shows a gradual progress in this period. It was destined to play a part in the improvement of distillation apparatus in the later Middle Ages. Its early centre was in Murano (and Venice).

The earliest translations from the Arabic are two Latin treatises by one ARTEPHIUS. Shortly afterwarts ROBERT OF CHESTER translated several Arabic books (1144) followed by the Emerald Table translation by HUGH OF SANTALLA, translations of works of AL-Râzî by GERARD OF CREMONA, and others by VINCENT OF BEAUVAIS and ROGER BACON. IBN SÎNÂ was translated by ALFRED OF SARSHEL on the turn of the twelfth century.

Somewhat earlier are the works published by the Salernitan doc-tors. PLATEARIUS translated many Arabic pharmacologies round 1150 and we have the writings of BARTHELOMEW OF SALERNO and SALERNUS.

Though many books were thus translated or compiled from Arabic authors (up to the sixteenth century no less than 385 manuscripts are known) (486) it is important to note that Arabic science at its full development had already ceased to influence the West. More Arabic chemical treatises were translated in the course of the twelfth and thir-teenth centuries but their number is small when compared with the stream of astrological and astronomical books. Some alchemy was known in the West by direct tradition (286), but Western alchemy owes much to translations of the twelfth century when many Greek alchemical manuscripts came from Sicily too.

The second half of the twelfth century produced several original treatises by ROGER OF HEREFORD, ALFRED OF SARSHEL, MICHAEL SCOTUS and RICHARD OF WENDOVER, though the authenticity of MICHAEL SCOTUS as an alchemist is at the least doubtful (286).

But most Latin schoolmen were still more eager to discuss matters in the abstract. The experimental way of testing facts had hardly dawned upon them. As HASKINS put it very aptly: "In the medieval mind the science of magic lay close to the magic of science". Among the few who indulged in concrete observations and experiments were ROGER BACON and ALBERTUS MAGNUS. The latter felt the truth underlying alchemy but he was not prepared to state how much. Arabic alchemy had hardly developed when it conquered the West and the Latin alchemist guided by provisional hypotheses blundered along accumulating an increasing amount of experimental science, technique and knowledge and reducing these results to some sem-biance of order. The best treatises are those ascribed to GEBER, which will be discussed further on.

The thirteenth century represents the upper limit of alchemical knowledge in Latindom, but it is at the same time the awakening of chemistry in the West. Indeed we find in this thirteenth century the beginning of many threads leading up to important movements in later periods. Thus for instance HUGH BORGNONI and his son THEO-DERIC, both professor at Bologna, translated Arabic materia medica, but they toyed with chemistry in order to prove the medical value of many chemical compounds. Thus they were the forerunners of PARACELSUS and his iatrochemical school, in which distillation played such an important part.

But before we discuss the most important chemists of the thirteenth century and afterwards we must needs summarize the most important *technical achievements of the Middle Ages*. Apart from the discovery of the preparation of cast iron, which revolutionized metallurgy, and the discovery of many important chemicals such as nitre, alumn, different vitriols etc., which influenced the inorganic chemical industry and glass and ceramics, the Middle Ages bring the discovery of the mineral acids and alcohol. The former revolutionize chemistry because they make reactions in solution possible. The latter had a profound influence on the development of distillation apparatus and thence on laboratory technique and the art of the apothecary or pharmaceutical chemist. Gradually we see that the centre of chemical industry is shifted from the monastery and the home of the private artisan to a real industrial centre or to a chemist's shop. The rising capitalism of the later Middle Ages lead to a concentration of those trades which formerly formed part of the housework or belonged to the monk's work.

The earliest centres of the industries that concern us here were situated in Italy (Salerno, Venice and the Po valley), afterwards Southern France, Germany (Harz, Saxony and Bohemia) and the Hanse towns grew more important, while Flanders and Holland led after the fifteenth century, when Italy and Germany were devasted by wars and trade and industry fled to safer places.

We must now take stock of *the most important alchemists,* whose opinion is often quoted by later writers on distillation and find out what their opinion on this subject was, whether they made important discoveries in this field and what apparatus they used.

We have already hinted at the fact that alcoholic distillation was probably discovered in Salerno by members of the famous local school. In fact one of the earliest references to distilled alcohol is found in the writings of SALERNUS, though the *Mappae Clavicula,* about which more later on, antedates it. SALERNUS was a physician who lived at Salerno between 1130 and 1160 and who wrote a summary of pathology and therapeutics. It is strange to notice that no hint of alcohol is found in Byzantine literature which was profoundly influenced by Salernitan teachings. Thus NICHOLAS MYREPSOS (lit. "maker of ointments") (444), a Byzantine physician to the Emperor JOANNAS III DUCAS BATATZES (1222-1254) derived his information from the *Antidotarium* of NICHOLAS of SALERNO. But his large compilation of pharmaceutical recipes called *Dynameron,* composed towards the end of his life (1280), does not even mention distillation as a common pharmaceutical method. Still other Byzantine manuscripts, the so-called *Mount Athos manual* (469), mentions raki (that is raqi or alcohol) which is distilled four to five times, but then these writings contain interpolations which may date from anything between the eighth and the fourteenth centuries.

But alcohol must have been a relatively recent discovery in Italy for it is not mentioned by PLATEARIUS who compiled his recipes from Greek and Arabic pharmaceutical treatises in his *Circa Instans* round 1150.

We must mention BARTHELOMAEUS OF SALERNO, a physician who lived in the first half of the twelfth century, because he is often said to be the author of a booklet on distilled waters of which the book by MICHAEL PUFF is said to be a German translation or adaptation. However, this original booklet was never found and it is not included in BARTHOLOMAEUS' *Practica,* a treatise on pathology and therapeutics.

But the great philosopher and churchman ALBERT THE GREAT, ALBERTUS MAGNUS (1193-1280) did deal with distillation. Indeed, he considers distillation a most important method in alchemy and writes that "the alchemist requires two or three rooms exclusively devoted to sublimations, solutions and distillations". In his *De secretis mulierum* (15) we find two recipes for the distillation of alcohol (aqua ardens) under Sign. Dd 5 r and Dd 7 r. One runs thus: "Take thick, strong and old black wine, in one quart throw quicklime, powdered sulphur, good quality tartar and white common salt, all well pulverised, then put them together in a well-luted cucurbit with alembic; you will distill from it aqua ardens which should be kept in a glass vessel". In the other recipe he says: "When wine is sublimed like rose-water a light inflamable liquid is obtained". There is, however, some doubt whether this treatise was really written by ALBERTUS MAGNUS himself or whether it was compiled with the help of Arabic recipes by one of his pupils about 1300.

PIQUE's contention that the alembic was imported from the East by GERBERT OF AURILLAC (later Pope SYLVESTER III) is without foundation. Another authority often cited by later authors is PETRUS HISPANUS. PETER OF SPAIN (Pope JOHN XXI) was born at Lisbon about 1215, he travelled widely as he speaks of himself as familiar with all Italy, Burgundy, Gascony and parts of Spain. From 1246 to 1250 we find him a member of the faculty of arts at Siena. There he wrote his famous *Thesaurus pauperum* which is one of the few of his works which was printed and very influential it was. He became a cardinal in 1273 and was elected pope in 1276, but died only a year later. His correspondence with FREDERICK II, the Emperor, is famous. Among the authorities quoted by him we find PLATEARIUS, ALBERTUS MAGNUS, RHASES (AL-RâZî), AVICENNA and others. His *Waters* (481) is found in a number of manuscripts. Sometimes it appears to be the closing part of his first treatise on the diseases of the eyes *De morbis oculorum* and its first item is a "marvelous water to preserve and clarify the sight". It seems to include exactly twelve waters and so to conform to other medieval books *Of Twelve Waters*. Also its last two waters are an elixir of life (aqua vitae) and alcohol (aqua ardens). It is also found as a separate treatise in which directions are given for distilling various liquids which in at least one manuscript are accompanied by two figures of chemical apparatus (see the mss. mentioned under 481). If PETER OF SPAIN came from Compostella in Spain, he may have had something to do with a *Book of Compostella,* which treats of many waters, oils and salts of great virtue. However, Brother BONAVENTURA, a Franciscan, is said to have composed this book in the convent of the Brothers of St. Mary in Venice (Assisi ms. 292 of the fifteenth century).

Equally important is RAYMOND LULL, one of the most romantic characters of medieval alchemy. He was born at Palma (Majorca) about 1235 and led a wild life until his conversion in 1266. He was then impressed with the necessity of conducting the Moslim back to the bosom of the Mother Church. For nine years he studied Arabic and retired to Randa, then visited Montpellier and for another ten years taught Arabic and philosophy in the Franciscan convent of Majorca. He then visits Rome (1285-1286) and lives in Paris from 1287-1289. Returning to Montpellier he moves to Genova. In 1291 at last he sails for Tunis and preaches there but is taken prisoner and is expelled. In 1293 he is back at Naples and tries to get help from the Pope. In 1300 he visits Cyprus to stimulate the study of Near Eastern languages. Many of his works are written in Genova (1302-1305). After another unsuccesful expedition to Africa he teaches in Paris and achieves at least a nominal succes for his mission at the Council of Vienne (1311). Strengthened by this moral support he travels through France and Italy for help and then crosses to Bougie in 1314. There he is stoned to death by the Arabs under the town wall a year later.

The chief contribution of RAYMOND LULL to modern science or at least his chief step in the direction of scientific method was his use of letters as brief handy designations of various substances and concepts, thus e.g. I for Spirit, K for alembic, R for sublimation and L for digestion. His *Testament* (394) contains the following phrase: "Recipe nigrum nigrius nigro et distilla totam aquam ardentem in balneo; illam rectifibis quousque sine phlegmate sit". He thought that alcohol was known to the ancients but that the manufacture had been kept secret. The distillation should be conducted with a small fire or a water-bath. Usually three rectifications are sufficient, but seven are necessary, using quicklime to hold the water, if one wants to make a quintessence. This is probably the earliest reference to absolute alcohol. First the alcohol is "loosened" from the wine by cohobation or digestion, then rectification follows. The Testament belonging to the oldest members of the Lullian alchemical collection states definitely that mercury and sulphur are not changed by distillation. There is, however, no manuscript extant of this work and according to THORNDIKE it was not improbably fabricated after 1500.

In his *Potestas Divitiarum* or *Power of Riches* he describes a "retentorium", a kind of condensor which looks very much like a type devised by VON LIEBIG much later. But this work is in part at least identical with a work by ORTHOLANUS and many other parts of the Lullian texts are later compositions which are mistakingly or deliberately attributed to this famous alchemist (395).

A rather obscure figure in chemistry but a very important one in the history of distillation is THADDEAUS ALDEROTTI (of Florence) (620) (THADDEUS FLORENTIUS) of the University of Bologna (1223-1303). THORNDIKE gives 1295 as the year of his death. He was an author on anatomy and medicine who wrote on the medicinal value of alcohol in his *De virtutibus aquae vitae*. VON LIPPMANN published the most important part of this work (375) from the Vatican Latin codex No. 2418.156. This manuscript dates from the early thirteenth century. He distills wine in a tightly luted alembic and receiver taking off 30 % of the wine and redistilling this distillate, taking off 5/7 in every rectification. LIPPMANN maintains that it is easy to obtain 90 % alcohol in this way. Essential in his apparatus was the spout of the alembic ("which should be of the length of an arm") and the "canale serpentinum" which he advises to be used with a cooling trough and a regular supply of fresh cooling water. THADDEUS was therefore the pioneer of the method of cooling the distillate after it had left the still-head and he thus paved the way for the modern method of cooling the vapours outside instead of inside the still-head and collecting the condensate in the alembic itself. In fact his method was the only efficient way of producing low boiling distillates like alcohol.

His contemporary ARNALD DE VILLANOVA was a far more prominent figure, who is frequently quoted by later generations. He was a Catalan born about 1240. His early years were difficult. He studied at Naples and travelled about a good deal, observing and operating, writing many books on the way from towns in France, Italy and Spain and even Africa. He treated popes and kings and was often entrusted with diplomatic missions. In 1285 he treated PETER III of Aragon during his last illness and was granted a castle in Tarragona. He was professor at Montpellier at least upto 1309. In 1292 he writes a book on the tetragrammaton and gets into trouble with the French Inquisition (1299) but the pope protects him. Shortly afterwards we find him at Genova (1301). He treats BONIFACE VIII and is again in difficulties with the Inquisition (1304). Then he treats JAMES II at Barcelona. His testament drawn up in the year 1305 has been found. He meets CLEMENT V at Avignon but is then found in Naples where he sojourned with RAYMOND LULL. He died towards the end of 1311. He is said to have known Arabic too and to have helped LULL to bring its knowledge to Montpellier and other places.

In his *Rosarius* he takes up a new theory of transmuting mercury by a regimen of sublimating, dissolving and purifying. Here we find the first stress laid on relative weights of chemicals, the first stumbling steps of quantitative chemistry. The treatise is also called *Treasure of Treasures*. In another treatise called *Rosa novella* there are detailed chapters on dissolution, distillation and fixation as the four masteries of this regimen.

A *Distillatione sanguinis humani* is cited by SANTE ARDOINI DE PESARO in a big compilation *De Venenis* (*On Poisons*) composed between 1424 and 1426 and printed at Basel in 1562. Pure extracted blood is submitted to fractional distillation giving "water", then "yellow water", "air" and red water or "fire". The first fraction is redistilled thrice with the juices of fruit and flowers and thus acquires occult virtues against certain diseases. The same is done with other fractions.

Another book called *Modern Book of Inferior Astronomy* claims that the Stone can only be made from mercury, silver or gold by a method like distillation on the water-bath in a glass vessel returning the "water" and redistilling. Another frequent operation is the heating in putrefying dung or reboiling in a glass vessel with a long neck on a slowly burning fire of ashes.

In an *Artis Divisio* he distinguishes seven parts of stages of trans-mutation, viz. conjuction, dissolution, putrefaction, distillation, con-gelation, fixation and projection.

In the work *De aquis* ascribed to him we read primarily of medical and artificial waters, e.g. a marvellous water obtained by soaking all kinds of metals in diverse substances on successive days and then distilling these over a slow fire.

More important for our subject is his *Liber de Vinis* (641) (644) which is dedicated to FREDERICK II of Sicily. It contains no illustrations of distilling apparatus but only their description. Here again he extols the virtues of aqua ardens (alcohol) which he did not discover as some later writers claim. But he was one of the first to insist upon its virtues and perhaps the first to see in it a key to the preparation of the Stone. SCHELENZ claims that he already uses the term alcohol, but the perusal of his works has not confirmed this. In this *Liber de Vinis* and its sequel *Tractatus de aquis medi-cinalibus* (see above) he describes mainly pharmaceutical wines, though he gives nothing that is not already found in the Lullian manuscripts. Wine is often distilled with spices and sugar. Thus a distillate of wine and rosemary is mentioned that later became popular and was known as Aqua Hungarica. An Oleum Mirabile probably contained turpentine.

There is also ascribed to him an *Aqua vitae* (640) which bears strong resemblence to that of JOHN OF RUPESCISSA on the fifth essence (Basel, 1561, 1597) and also to EVONYMUS *De remediis secretis*. It is also attributed to GESNER in the English translation (London, 1576) of his works.

Gradually the stock of essential oils known to the Arabic chemists was increased. Thus SANCTUS AMANDUS (24) at the end of the thirteenth century wrote a commentary on the older *Antidotarium Nicolai* and there describes the preparation of oil of cinnamon, oil of burnt almonds and ruta-oil.

Nor were more primitive methods of distillation forgotten and we find a clear description of the Bût-bar-bût in the *Cirurgia* of LANFRANC. LANFRANCHI OF MILAN fled to Paris in 1295 where he was admitted by WILLIAM OF BRESCIA into the College of St. Côme. There he taught surgery and died about 1306. His *Cirurgia*, probably written about 1295-1296, mentions: "Make a fier about the pott that is aboue the erthe and there wole distille oile into the pott that is binethe".

The unknown person who composed the *Summa Perfectionis* (*Sum of Perfection* or *Perfect Magistry*) and whom we call GEBER must have lived about 1300. In this manuscript (Münich codex Lab. 353) metals are defined and their qualities are given. Methods such as distillation, sublimation, etc. are discussed and testing methods like cupellation, calcination, etc. demonstrated. We shall have occasion to discuss this work more fully, but we must mention that its influence was not felt earlier than about a century after its publication when it became the main chemical textbook of medieval science. It is now quite easily available in the edition prepared by RICHARD RUSSELL, "a Lover of Chymistry", who translated several other books on chemistry and alchemy, which edition was re-edited by E. J. HOLMYARD recently (London, 1928).

A less influential but still important book was VITALIS DE FURNO's *Pro conservando sanitate* (230). FURNO (DU FOUR) was a Franciscan who became bishop of Basel. He was created a cardinal on December 14th 1312 and became bishop of Albano in June 1321, but he died soon at Avignon (August 16th 1327). His authorities are PLINY, ARISTOTLE, RAZI, AVICENNA and PLATEARIUS. He tries out ARISTOTLE's method of making sea-water potable by filtering through an earthenware pot. Rose-water is according to him an "artificial water" that is made by boiling (decoctionem) by means of fire and more especially by distillation. Aqua ardens should be prepared from good red wine by sublimation or distillation in an alembic, like rose-water, after addition of some finely powdered sulphur to the wine. The quality of the aqua ardens is tested by burning it on a piece of linen. It is not only a good medicine but also a most useful solvent. By mixing saltpetre and copper sulphate with aqua ardens or wine and distilling the mixture in a glass alembic on an ash-bath one gets a very powerful solvent which tints and destroys a piece of cloth (formation of nitric acid).

WALTER OF ODINGTON (647), a monk from Evesham who is known to have lived at Oxford from 1330 to 1347 wrote an *Ycocedron* (*Twenty Books*) which treats of the separation of the four elements. Distillation provides the means of preparing "air" from any substance but he does not range it among the four main operations which are sublimation, congelation, solution and calcination. The "air", so he thought, "pours out its "humidity" rarified by heat into water", which is a rather roundabout way of describing condensation of vapours.

The various uses to which distillation was put were clear from the *Sertum Papale de Venenis* of WILLIAM OF MARRA (661). The writer who adresses this book to URBAN V (1362-1370) is possibly identical with the WILLIAM OF MIRICA adressing a commentary on the *Physiognomy* of ARISTOTLE to Pope CLEMENT VI (1342-1352). This Papal Garland was written in the year 1362 and warns against the dangers of poisoning to be averted. It also says: "Alchemists can distill water even from dry wood, horn, hair and the like substances".

It seems that distillation was also used to analyse water, thus for instance the *Liber de Venenis* of FRANCIS OF SIENA (224). The writer was born in 1343 and became rector at Perugia, then papal physician and professor of astrology at Bologna (1394), where he also taught practical medicine in later years (1396). In 1375 he dedicated this work to PHILIP D'ALENCON, patriarch of Jerusalem, archbishop of Auch. It is usually preceded by a "consilium balneo", a work on baths in which he says: "Alchemists have proved baths to consist partly of alumn, partly of sulphur and partly of iron by distillation of the waters".

An often-cited author was JOHN OF RUPESCISSA (ROQUETALLAIDE) (317) (318) about whose life we actually know very little. This chemist and prophet was a Franciscan living in the middle of the fourteenth century and often imprisoned for his prophecies. He was a Catalan but his books are as often in Latin as in his native language. He studied for five years at Toulouse, then became a monk and was imprisoned for the first time in 1346 in a monastery of his order. He wrote many further books with prophecies but was again imprisoned in 1349 and 1356. His principal work *De consideratione quintae essentiae* consists of two parts, canons and remedies. The supreme remedy against corruption is the fifth essence, that is pure alcohol. It is identified by its marvellous odour, very different from that of ordinary aqua ardens. The medicinal and preservative properties of pure alcohol seem to be the gist and back-bone of this book.

Thus this book really continues the medieval discussions of the aquae vitae or waters. Therefore not PARACELSUS but RUPECISSA is the originator of the doctrine of the fifth essence in each thing if VILLANOVA's *Aqua vitae* is not authentic. A discussion of the "aqua vitae rectificata" (or pure alcohol) is ascribed to him in a copy made in 1468 (MSS. Wolfenbüttel No. 3721). JOHN writes under the influence of RAYMOND LULL. His fifth inferior element or essence is rectified aqua ardens or alcohol but more strictly speaking a sort of cordial made by repeated distillations. This can be extracted from all kinds of substances, e.g. antimony. Over the latter JOHN waxes particularly enthusiastic.

In the later Middle Ages the alchemist was often looked upon as a heretic, no doubt with very good reason. Thus we find many discussions whether it is lawful for a true son of the Church to practise his art or to use his manipulations. One of them is BENVENUTO OF IMOLA (61), the commentator of DANTE's *Divine Comedy,* who died in 1391. According to him the alchemist aims at correcting the corrupt state of the metals except gold and silver by reducing the metal to its constituent first parts of sulphur and mercury by calcination and distillation. Hereby he certainly commits no sin (61).

A very influential author was MICHAEL SAVONAROLA, the grandfather of the famous reformer of Florence. He was born at Padua in 1384 and it seems certain that he was not yet a M.D. in 1405 though described as such by some papers, but he lives in the medical college at Padua. He became a scholar of arts in 1401 and in 1412 he was admitted as a member of the faculty of the university, receiving on July 13th a licence for examination in medicine. Here he stayed from 1415 to 1440 teaching and examining with a few interruptions of medical practice both private and in the service of the republic of Venice. In 1440 he became physician to NICCOLO D'ESTE and was connected with the university of Ferrara. He was created a knight hospitaller on December 5th 1452 and probably died in 1464.

At the suggestion of the jurisconsult ROSELLI he composed and dedicated to LIONELLO D'ESTE as marquis of Este a treatise on aqua ardens (554) in which he accepted the conception of alcohol as fifth essence and told of noted men who had lived long by using it. His principal work discusses the *Art of Making Waters* (556). After explaining what aqua ardens is, he discusses the type of vessels in which it should be prepared. Of old, tin or lead vessels were used, but wine can also be distilled in glass vessels like water. Luting the recipient well to the cucurbit is essential. The fire should not be too fierce, distilling in horse manure is best. Care should be taken that neither smoke nor fire shall touch the distillate. The cooling is either applied to the still-head or better still to the delivery-tube of the alembic which can be passed through a trough of water. This tube, which he calls "serpent" or "vitis" is not coil-shaped but has snake-like windings in a horizontal plane. Glass or glazed vessels are best, lead vessels give a poisonous aqua ardens, if one wants to use metal cucurbits these should be tinned on the inside. He goes on to say that one should first distill off one third of the wine and use this fraction for rectifications, each time taking off five-sevenths of the contents of the cucurbit. After ten distillations the aqua ardens is pure. He tells us that one of his friends wanted to avoid these multiple rectifications and constructed a cucurbit with a long neck so high that the flask rested on the ground floor and the still-head was on the first floor. He does not tell us whether this chemist succeeded in distilling pure alcohol (96 %) in one distillation as he proposed to do.

If one does not use horse dung for heating, wood should be used as coal gives off stench and smoke which might spoil the distillate. The distillate is tested by pouring a little bit on a piece of parchment and burning it. No traces of water should remain on the paper. He himself considers rectification necessary because the longer the wine is in contact with heat the better the quality of the alcohol. Like other medieval chemists he believes that the aqua ardens is actually formed by the heat of the fire from the wine.

The rest of the book is devoted to the preparation of simplicia and composita by mixing, macerating and distilling flowers and herbs with wine. The pharmaceutical properties of these "aquae ardens compositae" are discussed in much detail.

SAVONAROLA wrote a work on baths and mineral springs (555) adressed to BORSO D'ESTE and therefore written before 1450. For the analysis of these waters in order to isolate the minerals or other foreign substances contained in them in the form of a residue SAVO-NAROLA preferred the gentle application of heat known as alembication in which a lamp is lighted under a glass vessel and the water distilled slowly. He felt that experience had shown clearly that in strong boiling the finer particles of solids ascended more than in alembication. When evaporation was complete the solid deposit was removed from the alembic and SAVONAROLA advises drying it in the sun.

The recipe for the preparation of alcohol from wine as given by THADDEUS FLORENTIUS and SAVONAROLA is repeated by ORTOLAN in his *Practica* (459). ORTHOLANUS is recognised by contemporaries as one of the outstanding figures of the fourteenth century, but we know nothing of his life. Perhaps or possibly he is identical with GUILLAUME D'ORTOLAN, prevost of Apt (1389-1393), who became bishop of Bazas on January 27th 1395, bishop of Rodez in August 1397 and who died in 1417.

His *Practica* was composed in 1358. Usually we find prefixed to a commentary on Hermes by ORTHOLANUS a section on "the spirit of the fifth essence", one to congeal mercury and another to conserve human life. Here he explains that by "Stone" is meant anything from which the elements can be separated. If we take wine as an example we find that by distilling a spirit is first obtained which "takes on body in the upper nobler sphere of fire and which is the spirit of the fifth essence". By rectification we obtain from it successively aqua ardens, aqua ardens rectificata, aqua ignea rectificata and aqua vitae rectificata.

To make aqua vitae for medicinal purposes one employs aqua ardens rectificata but omits the rectified human blood which would destroy the virtues of the herbs used and also the aqua ignea which is too consuming. Vessels to be used are then described and the section closes with a prayer of ORTHOLANUS for those who go astray.

He also tells us, that "if a bit of vinegar distilled in an alembic is projected on a bit ferment of the stone, it will turn into purest wine from which aqua ardens and aqua vitae can be made."

The full recipe for the preparation of alcohol runs thus: "Put first quality red wine (or white) in a cucurbit with alembic which shall be heated on an ash-bath. The product of the distillation shall be divided in five portions. The liquid that distills first is stronger and nobler then the others because it contains much quintessence. The second distillate is much weaker, the third still less strong, the fourth is worth nothing at all and the fifth part remains as residue in the lees on the bottom of the cucurbit. The receiver is changed at regular intervals. Each of these waters is kept separately in a special vessel. The first three are aquae ardens, because a piece of cloth soaked in them burns without being consumed. If the cloth is not reduced to ashes, this is because of the phlegm which preserves it. To separate this phlegm from the aqua ardens each of these waters is submitted to a second distillation on a very slow fire and after twothirds have distilled one stops the work. What remains in the cucurbit is thrown away. This distillation is repeated three times until one obtains aqua vitae rectificata. This is recognised as perfect if a cloth impregnated with it burns up to ashes".

A *Libellus de distillatione philosophica* was written in the late fourteenth or early fifteenth century as it cites the writings of RAYMOND LULL and JOHN OF RUPECISSA (366). It states that the philosophical method of distillation was discovered in very recent times about the year 1371 (1351?) by a certain student of nature at the university of Toulouse. After discussing the virtues of herbs, it states that this method of distillation in alcohol will preserve them all the year round, so that apothecaries need not sell stale and outworn drugs. Herbs should not be distilled in lead vessels or allowed to loose their fragrance in the process of distillation. It describes in several chapters the furnace, the distillation of vernal and juicy herbs, vessels for distillation, the receptacle called the pelican, the erroneous popular distillation of aquae vitae and the philosophical distillation of the "root of life" (radicis vite), that is alcohol. Spaces were left blank in the manuscript for illustration of the text with figures of chemical apparatus.

The proper way of distilling alcohol from alcoholic beverages like beer and wine grew more and more popular. In a medical treatise of the Prague doctor JOHN WENOD (1420) we find the method of preparing an "aqua preservans ab omni dolori calculi" (609). The oven (fornax) contains several cucurbits with still-heads connected with receivers (receptaculum) by means of a cooling vessel through which the delivery tube of the alembic runs (vas cum aqua). From 12 pints of beer mixed with two hands full of salt half a pint of alcohol is obtained. The cooling tube was probably in the form of a coil or "serpent". Only a few instructions are given as to the method of distillation, as for the rest, "ut scis" (as is well-known to you). It seems that by this time the method of THADDEUS FLORENTINUS was common possession of all who had to use distillation apparatus. In this connection it is interesting to know that even LEONARDO DA VINCI had worked on a proper apparatus to cool the still-head and its discharging tube (see Codex Atlanticus leaf 79 v.) (Fig. 32) but he did not succeed in finding a practical solution.

The growing popularity of distillation can be seen in the works of ANTONIUS GUAYNERIUS (222). He taught medicine at Pavia in the years 1412-1413 and 1448 and spent the rest of his career in Savoia and Liguria. At some time he was professor at the university of Turin too. In his works we see the continuation of the experiments of the Borgnonis, the employment of chemical remedies. Therefore he too was a forerunner of PARACELSUS and the age of iatrochemistry.

In his pest tractate (especially *De Peste* II, ii, 3) we find a number of references to aqua ardens, aqua vitae, alembics, baths of Mary, distillation, coction and sublimation.

Selecting just one more example of the frequent use of distillation by the chemists of the later Middle Ages we refer to the *Compound* of *Alchemy* written by JOHN RIPLEY (515). He is said to be the uncle of ROGER BACON. His works were edited by ELIAS ASHMOLE in his *Theatrum chemicum britannicum*. He was born at Ripley (York) in 1415 and travelled as a Fransiscan monk through France, Germany and Italy. For a short time he was camerarius to Pope INNOCENT VIII (1477). Then he became a cannon of Bridlington (1478) and died at St. Botolph in 1490. Abroad he probably studied the latest distillation methods to which he refers in his book. Distillation figures among the twelve gates for the preparation of the Stone. Mercury and other substances are to be distilled frequently for purification.

This chronological survey will have given us hints on the gradual perfection and spread of distillation methods and the frequent use to which they were put by the alchemists. It is obvious too, that among the new low boiling substances which could now be recovered alcohol was very prominent. This substance therefore figures largely both in theory and practice. It is now time to turn to the *definition of distillation* as it was understood in the Middle Ages and to the apparatus devised for this new art.

It is not easy to define the word distillation as it was understood in those days, for one should never forget that the insight into natural phenomena was very vague, that science and magic touched closely. So even the nomenclature of quite common things remains vague and every manuscript should be read with great care.

There was no such word as *"distillate"*, a word which, in fact, did not come into use before 1860. Distilled liquids were called *"waters"* but even this term is not clearly outlined. We have already had occasion to point out that among the pseudo-literature current in the twelfth and thirteenth century there are treatises in medieval manuscripts devoted to marvellous "waters", medical and chemical. Such works were attributed to ARISTOTLE, PETER OF SPAIN or other famous men. At that time of course various liquid compounds and acids are known as "waters"; alcohol was called aqua ardens or aqua vitae and in one manuscript some of the "waters" are solid! In the case of the treatises ascribed to ARISTOTLE or PETER OF SPAIN twelve seems the favourite number, but the twelve are not always the same and sometimes a text will include more while the title says twelve. Such a collection of waters is sometimes ascribed to RHASES and once to VIRGIL, but often occurs anonymously. Other treatises on waters in general and the fountain of youth especially are ascribed to ALBERTUS MAGNUS, ARNALD OF VILLANOVA, TADDEO ALDEROTTI (FLOREN-TINUS). We also encounter *Nine Waters of the Philosophers, Physical Waters* and a *Book of St. Giles concerning the virtues of certain waters which he made while dwelling in the desert.* The saint would scarcely seem to have chosen the best place for the investigation of his subject, says THORNDIKE. In the later Middle Ages these series are multiplied and again the word "water" is used in a vague sense, though generally it denotes a liquid distillate (623).

Our verb *"to distil"* has different meanings, viz.:

a) To trickle down, to flow gently, to melt into or dissolve in;

b) To let fall (transitive);

c) To give forth, impart, infuse, instil (transitive);

d) The process of distillation, to extract the essence of a plant, transform or convert by distillation, to perform a distillation, to extract the quintessence off, to drive volatile constiuents off;

e) To obtain, extract, make, produce by distillation;

f) To undergo distillation, to distil over;

g) To melt or to dissolve (transitive).

The original meaning of the Latin "destillare" is of course "to drip down" and we find this meaning in many early texts, for instance:

"The sweat distilling with droppes abaundant" (BARCLAY, 1514);

„He hath caused holy oyle to distyll out of ye bones of his saints" (WYKYN DE WORDE, 1526);

"The malediction hathe distilled upon us... because we have sinned. (Daniel 9.11, Douay Bible, 1609);

"My speach shal distill as the deaw" (Deut. 32.2; Bible 1611).

But it is also easy to find early examples of the use of this word in the sense explained under d), e) and f). If we take the *Booke of Quintessence,* written between 1460 and 1470, we read:

4. Thanne must thee do make in the furneis of aischin a distilla-torie of glas....

10 ... Take the beste vynaygre distilled, than putte it in a lem-bicke and distill it at a good fier . . . . "

Or if we take that arch-liar, the Splendide Mendax of the Middle Ages, Sir JOHN MAUNDEVILLE (419), we find that he uses the word in several senses for instance in chapter XII of his *Voiage and Travayle*:

"And the Sarazines clepe the wood Enonch balse; and the fruit, the which is as cubebs, they clepe Abebissam; and the liquor that *distilles* oute of the branches they clepe Guybalse . . . . and some *distyllen* clowes of Gyloffre . . . . . and of other spices, that be well smellynge; and the licour that es *distilled* out of tham thai sell in steed of bawme; and thai ween that thay han bawme and thai have none".

We should note that the derivatives of this verb occur both in the full and the short form. Thus we have "Distillatory" and "Stilla-tory", "Distillery" and "Stillery" and even "to still" and "to distill".

In the older texts the word "distillatory" usually denotes a pitchers plant, that is a plant in which pitch was obtained by dry distillation of wood and coal. In the case of the twins "Still" and "Distill", we find that the latter expression is rather unusual, the common word for cucurbits and alembic in later ages is "still". The word "stillroom" which is not used before the sixteenth century denotes the distilling room or the room where the liquours, etc. are preserved. The crafts-men engaged in the art of distillation are called "Distillers", a word "Stillers" does not seem to occur. The word "Distillers" is, however, used in a special sense by the Royal Navy during last century to denote the apparatus used to prepare drinking water from salt water at sea. Most of these words are not used in the Middle Ages and do not occur before the sixteenth century.

Even our present word "distillation" (French distillation, Italian distillazione, Portuguese distillacio, Spanish distillacion, destilacion, German Destillation) has different meanings, viz. (WEBSTER):

a. gentle dropping, falling in drops and pouring or throwing down drops;

b. that which falls in drops;

c. volatilization of a liquid in a vessel by heat and its subsequent condensation in a separate vessel by cold as by means of an alembic or still and refrigeratory; the operation of extracting spirit from a substance by evaporation and condensation;

d. The substance extracted by distillation;

e. catarrh.

The latter meaning still occurs frequently in ancient medicine. A good example of this use of the word may be found in the *Castell of Helth* of Sir THOMAS ELYOT (208). ELYOT was a diplomat and a scholar, great friend of THOMAS MORE and ERASMUS. He wrote this popular medical book which was much decried by the physicians but eagerly bought by the public.

Here he defines distillation as a catarrh: "Destyllation is a droppynge down of lyquyde mater out of the head and fallynge eyther into the mouth or into the nosethrilles or into the eyes".

Examples of the meaning described under d. are first found in SHAKESPEARE's works, e.g.:
"Were not summers distillation left
a liquid prisoner pent up in walls of glass" (SONNET V) or
"And to be stopt in like a strong distillation with stinking clothes"
                                    (*Merry Wives of Windsor*)
"And in the Porches of mine ears did pour the leaperous distillment"
                                    (*Hamlet I, v. 64*).

The second meaning (b.) is very rare, as far as we know it is only used by JOHNSON in this sense.

By far the most common meaning is that formulated under c. It is certain that the medieval scientists understood the phenomenon of evaporation and condensation quite well and were thus aware of the principles underlying distillation. Thus we read in the *Buch der Natur* (*Book of Nature*) written by CONRAD VON MEGENBURG (422) between 1349 and 1350, one of the most popular books in the Middle Ages and reprinted as late as 1540 by EGENOLF OF FRANKFORT, the following explanation of rain in Chapter 16:

"The rain comes from the watery vapour which the heat of the sun has drawn into the middle region of the air, where because of the cold that reigns there, it pours itself out again in water, just like the vapour of the boiling pan on the fire: when the vapour touches the cold iron cover, it pours itself out in waterdrops. This occurs too to the vapour formed when roses are "burnt" or when wine is distilled; If that vapour touches the cold lead cover ("hat") it pours itself out in water and the same water tastes of the thing from which it originates".

Der regen kümpt von wäzzerigem dunst, den der sunnen hitz auf hât gezogen in das mitel reich des luftes, wann von der kelten, diu dâ ist, entsleuzt sich der dunst wider in wazzer, als wir sehen an dem dunst der von dem wallenden hafen geht ob dem feuer: wenn der dunst die kalten eisneinne hafendecken rüert sô entseluzt er sich in wazzers tropfen. Also geschiht auch dem dunst der dâ kümt von rosen

prennen oder von wein prennen: wenn der den kalten pleienne huot rürt sô
entsleuzt er sich auch in wazzer, und smeckt daz selbig wazzer von dem ding, dâ
der dunst kümt.

But the definitions given by various alchemists are clothed in more
cryptic terms. The *Buch der heiligen Dryvaltigkeit* says: "distillation
is catching water", and here we should remember that "water" means
"distillate". GEBER gives another definition, which is repeated fairly
often. We have already touched upon the GEBER problem and mentioned that he was not an Arabian chemist whose writings were
translated into Latin. The earliest date at which his works are mentioned is 1330, when PETRUS BONUS composed his *Pretiosa Marga-
rita Novella* (*Precious New Pearl*) at Pola, in which book he often
cites "Geber Hispanus". From RUSSELL's edition of his work we
know that GEBER was a practical chemist and a metallurgist and that
his work the *Perfect Magistry* was uncommonly sensible when compared with contemporary alchemical texts. Whether he really was
a Spaniard is unknown. Certainly much of the technique he describes
was the result of the development of Arabian science and its influence
on Western alchemy which we have described. It should be noted
that nowhere GEBER claim any originality and that this book, written
about or before 1300, was a most useful compilation of the ideas and
technique of the period. This is what he says on distillation:

„Distillation is an Elevation of aqueous Vapours in their Vessel.
And Distillation is diversified. For some Distillations are by Fire, and
some without Fire. Those made by Fire are of two kinds: one, which
is by Elevation into the Alembick; and the other by Chymical Descensory, by mediation of which the Oyl of Vegetables is extracted"
..... "But Distillation, which is made by the Filter, is performed
without Fire and the Cause of its Invention was Clearness of the
Water only".

From these lines it is perfectly clear that GEBER like most of his
contemporaries grouped distillation, descension and filtration together under the head distillation. The "Destillatio per ascens" means
both distillation and sublimation, the "Destillatio per descensum"
the descension already well know to us from the bût-bar-bût, or botus
barbatus apparatus of the Arabian and earlier chemists. Filtration
was considered to be a kind of distillation by the capillary action of
paper or felt hanging on one side in a solution, suspension or other
liquid and on the other side overhanging the rim of the vessel and
dripping into a receiver. In later handbooks on distillation such as

that of BRUNSCHWYGK and also the GEBER edition by RUSSELL we find a filtration apparatus appearing on the page illustrating distillation in general (Fig. 37). It is usually called "destillatio per filtrum".

That sublimation is classed with distillation is not illogical if we remember that the Latin "sublime" means "uplifted, lofty" and that distillation is often called "elevation" or "elevatio". The word "rectification" is not used until much later. It is derived from the Latin "rectus fascio" that is "pure, upright make".

There was a strong tendency among the medieval alchemists to group the chemical operations. In the *Summa* edition of 1541 we find eight major operations mentioned, but the number of four or seven occurs very often. Forced to group these operations in a preconceived number of categories the medieval alchemist often makes strange combinations, being also partly misled by the vague understanding of the principles underlying them. Thus for instance the *Spiegel der Alchemie (Mirror of Alchemy)* groups descension and distillation together under "ablutio" that is "ablution" (241).

As to the invention of distillation we already find traces of a story very common among later generations of distillers (492). They tell us that the principles of this art were discovered by a doctor who cooked cabbage in a closed pan and noticed the drops of water condensing against the cold cover.

If we now turn to the *apparatus* used in this distilling art we find that CHAUCER practically sums them all up in the following lines from his *Chanouns Yemannes Tale* (lines 238-242) (129):

> "and sondry vessels maad of erthe and glas,
> Our urinales and our descensories,
> violes, croslets and sublymatories,
> cucurbites and alembykes eek,
> and others swiche, dere ynough a leek".

The flask used by the alchemist was called *cucurbite* (concurbite, cocurbite, cucurbite, etc. from the Latin cucurbita, a gourd or cupping glass) or *matrass*. An example is shown in Fig. 21, a reproduction of a page taken from BRUNSCHWYGK's book on distillation. We must remark here that illustrations of apparatus in medieval manuscripts are neither very numerous nor clear. DAVIS published several of them (158) (159). For instance they are found in the *New Pearl of Great Price* written by PETRUS BONUS, to which work we have already referred. As soon as the art of printing was invented a stream of

Von Distillieren. xxiiij

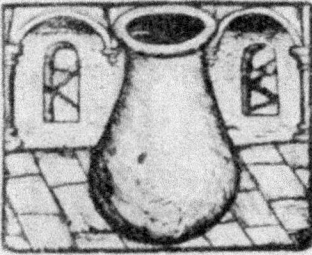

Darnach soltu haben fursatzgleser / so
man distillieren wil Aqua fort darein zü
entpfahen die spiritus vnd gersft bald nach
dem mandelo ⱨ sich weittern / auff das sich
die spiritus bald von einander thüt gnät
Acceptackel in Teutscher zungen fürseg‐
gleser darumb das darin entpfangē wirt
das gedistilliert des figur also ist

Sarnach solt du haben fursatzgleser /
die oben gantz vnd in der mitte ein rot ha‐
ben das durch das gedistilliert züentpfah‐
rff dz die spiritus
die sich auff subli
mieren nit verzo‐
chen werden / auch
genant Recepta‐
ckel / darumb das
sie fürgesetzt wer‐
den dere figur al‐
so ist als hier neben
stat.
Sarnach soltu haben krumme gleser /
genant Retort / darein zü distillieren was

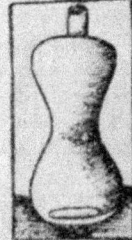

Darnach soltu haben
gleser / darinn zü digerie‐
ren vnd circulieren gnät
circulatoriū / deren ſ gut
mächerley ist / die erst als
hie neben stat.

Sarnach soltu habē
ander gleser / auch ge‐
nannt circulatoriim /
deren figur ist wie du
sie sichst.

Fig. 21. Receiver, cucurbit, retort, circulatoria and other vessels (from Brunsch‐
wygk's *Large Book of Distillation*)

books began to flow and the art of distillation is discussed but it is
always well to mistrust their illustrations as they usually refer to the
time when the printed edition was issued, which may be several centuries later than the original manuscript. It behoves us therefore to
admit, that we know only generalities of the medieval apparatus, and
depend mostly on the few descriptions given.

The cucurbit was made either of earthenware (glazed on the inside)
or of glass. Metal cucurbits (copper, lead) are mentioned but avoided
as it was known that poisonous distillates might be obtained from
them. Thus the apothecaries' ordinance given at Nürnberg June 7th
1555 says:

"As nobody can deny that the distilled waters, if they are made in
metallic stills and vessels, such as tin, copper or brass, are very
harmful to the bodies of mankind, the Town Council has ordered to
warn the apothecaries seriously, that it is their duty from now on no
longer to burn waters in such tin, copper or brass apparatus and they
shall remove such harmful apparatus and use only glass flasks to
burn their waters".

„Und nachdem niemandts widersprechen kann, dass die geprannten wasser, so
mans in Metallischen geschirren un gefeszen als in Zihn, Kupfer oder Messing
brennt, den Menschen in leid sehr schädlich sein kann, bey einem E. Rath bevohlen,
den Apothekern ernstlich anzuzaigen, dass sie nun hinfüro bey ihren Pflichten kein
Wasser in solchen Zihn, Kupfer oder Meszenen Prennzeugen prennen, sounder
tolche Prennzeuge als schädlich gar hinweg thun sollen und sich allein der gläser
zum prennen des Wassers gebrauchen sollen".

Such warnings are already found in VILLANOVA's *Antidotarium*
and other early texts.

The still-head was called *head, capital or alembic* (alembik, alem-
byk, alembike, alembyke, alimbeck, alembeke, alimbecke, alimbeck,
limbick) (French alambic, Portuguese elambic, Spanish alambique, Ital-
lian lambicco, limibicco). The latter word is of course derived from
the Arabic al-anbîq, which is in turn derived from the Greek ambix
as we have seen. The full form is generally used, but the form "lem-
bick", "limbeck", etc. had a certain vogue from the fifteenth to the
seventeenth century. The typical alembic of the Middle Ages is shown
in Fig. 22. A "blind" form without beak or spout occurs on the same
plate. It was used to shut off the cucurbit when digestions or sublimations were carried out. The same figure shows a lead ring which
was used to weighten the cucurbit if it had to be immersed in a water-
bath. The word alembic changed its meaning during the Middle Ages,
it gradually came to denote not only the still-head but the combination

of head and cucurbit. The latter meaning won on the long run, perhaps aided by the fact, that the technical evolution of the still led in this direction. Though the Arabic chemists preferably used glass apparatus the alchemist often combined earthenware cucurbits with strongly luted glass alembics. As the glass industry evolved, it be-

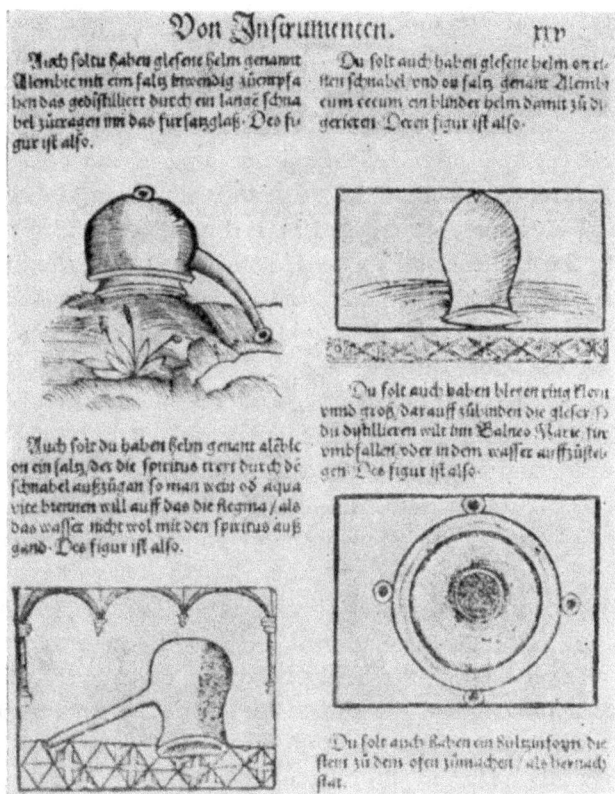

Fig. 22. Alembic "cum nasu", blind alembic and lead ring as counterweight for alembics in a water-bath (from BRUNSCHWYGK's Large Book of Distillation

came more and more common to use both glass cucurbits and alembics and gradually they were blown or cast in one piece.

The glass industry, an important factor in this art, received great impetus from the growing general use of glass for windows and chemical vessels. At the same time the existence of a flourishing industry at Venice and Murano must have influenced chemistry too (550).

This evolution of a combined flask and alembic had another notable effect, the evolution of a new type of distilling apparatus, the

retort (French retorte, Spanish retorta, from the Latin retortus, retor-
quere, to bend back). We find the word "Tortae", "Tortuosa" used
for bent and curved roots and plants since the early Middle Ages.
The word "retorta", "storta" appears in the *Experimenta* and the
Vademecum of RAYMOND LULL; VILLANOVA uses the word "distorta".
The retort is therefore a Western invention, it was used in Arabian
circles at a much later date only. A picture of a retort is found in
Fig. 21. The distillation in a retort is often referred to as "destillatio
ad latus" or "sidewards distillation".

A typical example of the change in meaning of "alembic" can be
found in a Greek codex of the fourteenth century, the Codicis graeci
Holkham Hall No. 290, folia 186-194. It contains many recipes which
go back to the Hellenistic Papyrus Holmiensis. When describing the
distillation of mercury (349) the head is called "alembikos" and the
entire apparatus "lambikos". It should be noted that the word "stala-
zon" is here used for "distillation".

The *lute* used by the alchemists was usually referred to as "the
lute of the philosophers" or "lutum sapientiae". It was compounded
of linen strips and flour or better still of mixtures of common clay,
potter's clay, horse manure and chopped straw or glass powder,
quicklime, powdered bricks, white lead and white of egg (349)
(303).

Very typical of the medieval alembic is the lack of an inner rim
for collecting the condensate formed in the still-head and delivering
it to the receiver through the beak. As cooling the delivery-tube grew
more and more common, the rim in the alembic disappears and we
find practically only the rimless form of alembic in medieval illustrations (Fig. 22). The art of blowing the old form was gradually
lost (611).

A host of other forms of vessels enabled the alchemist to execute
his digestions, circulations, cerations, etc. which he distinguished and
which he believed most necessary, often preliminary operations to a
good distillation. If we take the definitions given by GEBER for these
major operations, we find that:

"Sublimation is the Elevation of a dry Thing by Fire with adhe-
rency to its Vessel (Usually an aludel or cucurbit with a blind head
is used for this work)".

"Calcination is the Pulverisation of a Thing by Fire through Privation of the Humidity consolidating the Parts."

"Solution is the Reduction of a dry Thing into Water".

Das erſt bůch.

Darnach ſoltu haben andere gleſer zů digerieren vnd zu culieren genant cucula tonum/der figur alſo iſt.

lieren deren zwey in einans geen/ alſo was von ein auff/in das ander abgeen iſt.

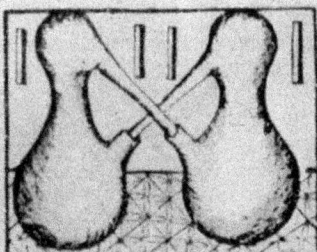

Auff ein andere maſß gleſer die man ha ben ſoll/darinn man digerieren mag/der figur iſt alſo.

Sarnach ſolt du haben gleſene trechter mit langen röten/erlich groß etlich klein/ Aqua fort damit in die gleſer zerthůn/ vnd öl vom waſſer zůſcheyden/deren figur iſt/ als hernach ſtat.

Sarnach ſolt du haben gleſer zů circu lieren vnd digerieren genant pellican/vnd ſeind die beſten vnder zu allen/deren form iſt /als hernach ſtat.

Darnach ſolt du haben viol gleſer der inn ölei von waſſer zůſcheyden/ſo ſie vmb gekert mit dem finger das loch verſtopffet aufgelaſſen das waſſer ſo der finger dar nen gethon wirt als lang bis das öl trumpe dan das glaß wider vmbgekert. Deren fi gur iſt alſo.

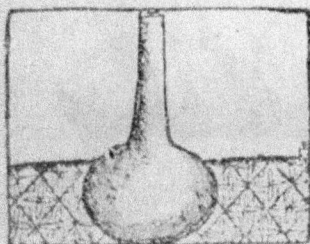

Sarnach ſolt du haben gleſer zů circu

Fig. 23. Circulatorium, vessel for digestion, pelican, "twins", glass funnel, phial (from BRUNSCHWYGK's *Large Book of Distillation*)

"*Coagulation* is the Reduction of a Thing liquid to a Solid Substance by Privation of Humidity."

"*Fixation* is the convenient disposing of a Thing Fugitive to abide and sustain the Fire."

"*Ceration* is the mollification of a hard Thing not fusible unto Liquefaction".

Fig. 24. Hydra   Fig. 25. Bear

Fig. 26. Turtle   Fig. 27. Pelican
(all four from PORTA's *De Destillatione*)

The correct vessels to be used for these and kindred operations can be found in Figs. 21 and 23. The *phial* often mentioned is a long-necked flask, the *urinal* has a wider neck. The latter was much in use in medieval medicine as it was believed that the colour, smell, taste and appearance of the urine of a patient was a good guide towards correct diagnosis.

Many of these vessels were compared with animals and in Figs. 24-27 we show some illustrations taken from the later work *De Destillatione* by PORTA (495). It explains the name "hydra" (a kind of fractional distillation tower), "testudo" (turtle), the alembic with a beak (here compared with a bear) and the "pelican" (a vessel very common for circulations and digestions). A type of "double-pelican", "twins" or "dyiotae" was used for circulation (Fig. 23). Not all the fanciful forms of PORTA's book occur in alchemical literature, but we see from the illustrations of BRUNSCHWYGK's book, that several of them are of medieval date, as they are shown there fully evolved.

The usual apparatus for the descensory, two pots placed upon each other and separated by a piece of cloth, gauze or a sieve, is the old form we have met quite often in these pages. We find it mentioned too in the early *Book of the Alumns and the Salts* (537) which was written by a practical Spanish chemist of the eleventh century as a kind of chemical compendium. Though no apparatus for distillation is mentioned there we find the "albot", "albae", that is the bût-bar-bût. We saw that LANFRANC too mentions it about 1295. In the later Middle Ages it was frequently used to prepare vegetable oils. Thus we read in the *Buch der Natur* by MEGENBURG (412) (Part IV, A. 19):

"About the juniper tree.

From the juniper tree an oil is made thus: Take two clean pots and put one on top of the other and the pot on top should have a hole in the bottom. This upper pot shall be filled with wood of the juniper tree, that is dry and it shall be closed well so that no smoke can escape and a great fire shall be lighted round it. If then the wood inside is heated the oil flows from the upper pot into the lower one, but there is very little of it".

„Auz dem kranwitpaum macht man Öl, alsô. Man nimt zwên häfen und setzet si über einander, und der ober hafen schol ein loch hân an dem podem. Den selben obern hafen schol man füllen mit kranwitholz, daz truckem sei, und schol wohle vermachen, daz ihts dar auz rauchs müg kommen, und schol ain groz feur umb die häfen machen. Und wenn dann daz holz inwendig erhitzt, sô fleuzt dasz öl aus dem obern hafen in den untern, aber des ist wenig."

The methods of heating were varied and in the course of the Middle Ages we notice the propagation of scales of *degrees of heat* to cope with the lack of proper means of measuring temperatures such as thermometers. ARNALD OF VILLANOVA distinguishes four degrees: the first in which the alembic can be easily touched with the hand is that of an ash-bath; the second, that of a coal-fire, heats the alembic so much that one can only touch it for a moment; the third and fourth degrees of heat are reached by direct heating of the still by the fire. For each degree of heat he prescribes the correct amount of logs of wood. Others distinguish as much as six degrees of heat: the first is attained by a lamp or horse dung; the second by "smokeless fire" at the moment of kindling the wood; the third by a coal-fire; the fourth, fifth and sixth are reached as the fire burns fiercer and fiercer.

The proper means of *heating* were lamps and candles, the heat of the sun or horse dung, which was thought to have a special effect. Heating in horse dung was thought to "digest and ferment" the con-

Fig. 28. An alchemist's laboratory from H. WAIBITS
PETRARKS *Trostspiegel*

tents of the cucurbit, thence the name "destillatio per ventrem equinem" side by side with the more common "destillatio per finum". Water-, sand- and ash-baths were often used. For higher temperatures or mass-heating furnaces of different types were used. One of the difficulties encountered was the breaking of the vessels, because the glass was still thick and irregular and of poor quality according to our standards. Hence the thick coat of clay or other protective coatings applied to the apparatus when the temperature was going to be high. Such coatings were made upto two fingers thick, but did not promote good heat transmission!

The *furnaces* were mostly built of clay or bricks and GEBER enumerates the following types in the *Book of Furnaces* usually appended to the editions of his *Perfect Magistry*. The calcinatory furnace was a simple square-walled furnace in which crucibles and pans were heated. The sublimatory furnace is made for the large vessels which are put on a perforated bottom-plate and heated by the flue-gases. The distillatory furnace is practically the same, nor does the descensory furnace differ much except for details caused by the apparatus to be heated. The melting furnace or fusory is built for the strong heating of crucibles. The fixatory furnace or athanor was made in the manner of the calcinatory furnace, but it contained a pan of ashes to harbour the vessels to be heated.

The change in *cooling methods* during the Middle Ages was most important and it must have been the prime factor in the preparation of low boiling compounds like alcohol. It is true that it is conceivable to distil alcohol in the ancient cucurbit and alembic without cooling the delivery tube and even when cooling the head, but only if the temperature could be regulated carefully. But usually the too fierce heating and the long digestion period before distillation drove off the low boiling fractions. As we have mentioned this digestion period was considered most important by the alchemist, because in this period the alcohol or similar compound was considered to be formed by the heat applied to the contents of the cucurbit.

The development of cooling methods tended towards more intensive and more continuous cooling than that with sponges and wet rags used upto this date. It was also important that the distillate was collected in portions and fractions and this must have influenced more careful distillation and incidentally the recovery of low boiling compounds. The old alembic developed in three directions. Two of these still employ the still-head as a condensing element, viz. the so-called "Moor's head" (Mohrenkopf) and the "Rosenhut" ("Rose-hat").

The *"Moor's head"* enclosed the still-head in a basin or container which was filled with the cooling water, either continuously or discontinuously. A well-developed form is shown in Fig. 30 taken from BRUNSCHWYGK's book. Seeing the very primitive forms still propagated in the early printed books on distillation and the absence of references to this method in medieval manuscripts it would seem that this was an invention of the later fifteenth century.

The *"Rosenhut"* (rozenhoed) (lit. "rose-hat") is certainly earlier as it is already shown in its fully developed form in the earliest printed book on distillation, that of MICHAEL PUFF (see Fig. 29). It is a high conical alembic cooled by air and very common in early apparatus used for making all kinds of liqueurs. It is fitting to the top of a wide-mouthed cucurbit and must have been built with an inner rim (never shown in illustrations) to collect the distillate rondensed in it. We read frequently of metal forms made of lead, copper, etc. and this is quite probably the cooler against which so many town ordinances, issued by the town councils, were aimed, as the cucurbit was more generally an earthenware vessel glazed inside.

The proper *cooling of the delivery-tube*, that is to say the development of a special condensing element in the distilling apparatus apart from the receiver, was already vaguely traced by us in the preceding pages. We met the first "wormcooler" or cooling-coil in the writing of THADDEUS FLORENTINUS who seems to have been the inventor

Fig. 29. Aquavitwoman with still equipped with Rosenhut, from the title page of MICHAEL PUFF VON SCHRICK's book

no longer relying on the cooling of a long delivery tube by air, but who led the tube through a tub of water. From the name "canale serpentinum" one is inclined to deduce that this tube wound wormlike through the cooling trough as is shown by many later pictures. Both SAVONAROLA and ORTHOLANUS mention "serpentes" and the earliest illustration of such a cooling coil is given by WENOD in 1420 (Fig. 31). LEONARDO DA VINCI's plan for a continuously cooled still-head would have meant a step back (Fig. 32), but his plans were never executed as his manuscripts were never published and in meantime better ways of cooling had been invented.

The *distillate* was called "spirit" or "water" according to its appearance, the word "phlegm" being applied to fractions containing much

Fig. 30

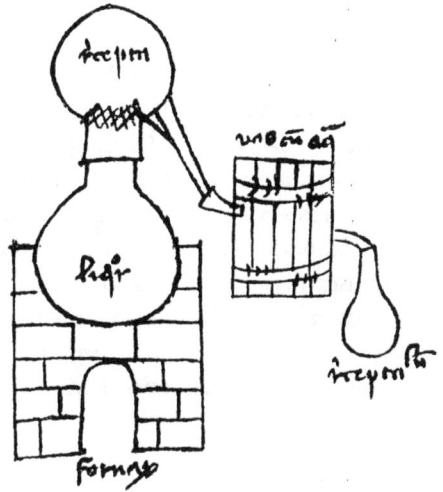

Fig. 31

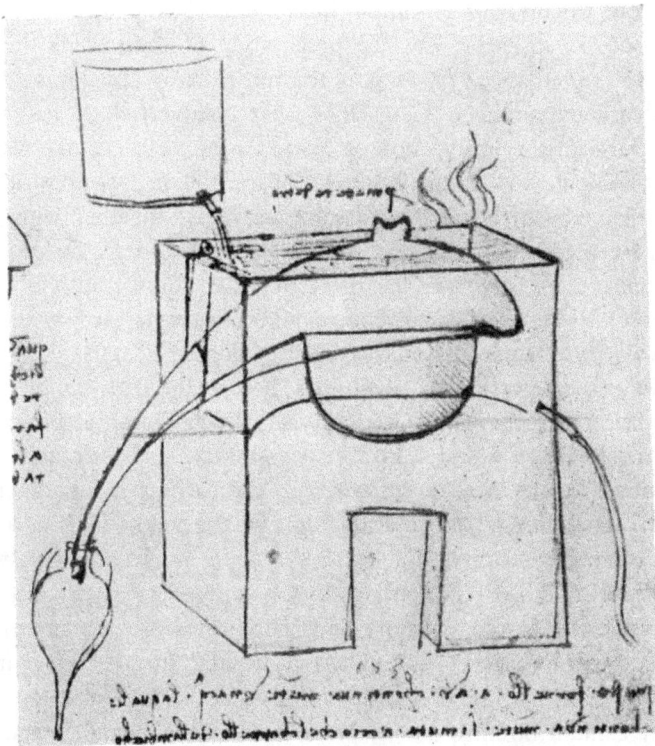

Fig. 32

Fig. 30. Still with Moor's head taken from BRUNSCHWYGK's *Large Book of Distillation* (notice tap of Moor's head placed too low and furnace with filling funnel called Faule Heintz or Slow Harry) — Fig. 31. WENOD's sketch of a distilling aparatus — Fig. 32. LEONARDO DA VINCI's still with cooled head (Codex Atlantico)

water. To separate the phlegm from the spirit a sponge is sometimes inserted into the neck of the alembic or delivery-tube, probably without much effect except when distilling small amounts of liquids in the laboratory. Cooling the delivery tube or "serpent" was quite common in the fifteenth century and this was the method that was developed by coming generations, while the Moor's head and the Rosenhut were destined to linger on for several centuries to be discarded afterwards. It should be noted that drawings of coolers often show the worm or coil outside the cooling trough whilst in reality it could not be seen as it was immersed. This is a small detail which has often caused much misunderstanding (609).

We must now discuss the application of distillation methods in industry and we first turn to the preparation of *mineral acids*. It is rather strange that chemistry had to wait so long for the acids as the Hellenistic chemists were already well acquainted with the vitriols and their calcination products, but the acid vapours were never condensed, so it seems. For the preparation of these acids the retort may have paved the way, as it is the most handy apparatus for such corrosive compounds. It seems that their discovery dates back to the early thirteenth century. For at the end of that century we find Byzantine manuscripts mentioning "hozos theion" (sulphuric acid) and "theion hydor" (nitric acid) (acqua regis) together with Italian names for parts of distilling apparatus. A medieval tract *De inventione sive perfectionis* gives under the title "De aquis solutivis et oleis incinerativis" recipes for nitric acid and aqua regis as does the *Liber fornacum* on "aqua dissolutiva nostra quae fit ex sale petrae et vitriolo". Both were written shortly before the *Summa* of GEBER, which mentions the same method and which is contemporary with ALBERTUS MAGNUS and VINCENT OF BEAUVAIS. Nitric acid is also mentioned in the *Rosarius Minor*. But the earliest useful description of the preparation of nitric acid occurs in the works of BIRINGUCCIO in the sixteenth century. The *Buch der heiligen Dryaldigkeit*, written between 1414 and 1418 mentions that it is prepared by dry distillation of mixtures of alunn (vitriol) and saltpetre or of sal-ammoniac and nitre. A century earlier both ORTHOLAN and FURNO know it well. The first says that the first drops of distillate should be thrown away as they contain nothing but water. It is tested by putting a few drops on a clean knife and watching the reaction. The distillation is achieved on an ash-bath, as is clearly shown in illustrations in the above-mentioned *Buch der Dryvaldigkeit* (479). As saltpetre or nitre had been known since about 1150 in Italy and was used for gunpowder and freezing mixtures the period of incubation necessary for the discovery of nitric acid seems fairly long. It seems, however, that this acid was very soon produced in larger quantities and as a separate industry or a side-line of the saltpetre-industry. Its manufacture on a larger scale was located (from the fifteenth century) in Italy (Venice) and at the end of that century or the beginning of the next France and possibly Germany began to produce it on a larger scale.

The discovery of *sulphuric acid* dates back to the same thirteenth century. Both the method of "roasting" or calcining vitriol or alumn and that of burning sulphur under a bell and dissolving the vapours in water (the latter "bell-method") seem to have been known. Contrary to what we have mentioned in the case of nitric acid, sulphuric acid seems to have remained a product of the laboratory for a long time. Even AGRICOLO and BIRINGUCCIO who describe its preparation in the sixteenth century know nothing of an industry. This sulphuric acid industry seems to have been a child of the seventeenth century when it developed as a branch of the vitriol-works.

We have no certain information about the preparation of *hydrochloric acid* in the Middle Ages, it seems to have been discovered at a much later date and though LIBAVIUS mentions it, we must wait until the days of GLAUBER to read the first proper description of its preparation from common salt and vitriol burnt together in an iron retort.

The discovery of the mineral acids and somewhat earlier (eleventh or twelfth century) that of *alcohol* may be said to mark the beginning of a new stage in history, the transition of the old and the new chemistry. But alcohol opened new ways too for the perfume industry and the preparation of alcoholic drinks and both developments had profound influence on the evolution of the art of distilling.

In the preceding pages we have shown that several alchemists knew alcohol and its properties and even waxed enthusiastic about them and praised alcohol or alcoholic extracts, cordials, etc. as elixirs of life. As far as we know alcoholic distillation was invented by Salernitan apothecaries about 1100. It might have been invented earlier, for example by Muslim chemists, but as we saw their writings contain no proof of it. In short, as long as no positive datable texts are produced the discovery of alcohol can not be ascribed to Oriental or Hellenistic chemists with certainty and the discovery in Italy (and probably Salerno) about 1100 remains the most probable solution of the problem.

One of the earliest direct recipes is contained in the writings of Magister SALERNUS (who died in 1167). Earlier Salernitan texts do discuss the manufacture of "beneficial waters" in the same way as rose-water using cucurbit and alembic but Magister SALERNUS is the first of his school to mention aqua ardens (379).

Two other and possibly earlier recipes in the form of cryptograms are found in the book of MARCUS GRAECUS and in the *Mappae Clavicula* both of which were the subject of fervent discussions.

The *Mappae Clavicula* or *Little Key to Painting* is available in two excellent versions, that of the convent of Wieszenau (162) which according to DEGERING belongs to the early thirteenth century on account of the type used, and the San Gimigniano Codex of twelfth century date published by PUCCINOTTI and RICHTER. On account of philological evidence we must date back the bulk of this book to the second half of the eighth century, but later editions contain many new paragraphs and additions. Only the twelfth century and later versions contain the recipe for the preparation of alcohol in the form of a cryptogram (403). According to BERTHELOT (626) this cryptogram, formula 212, reads: "On mixing a pure and very strong wine with three parts of salt and heating it in the vessels destined for that purpose, there is obtained an inflammable water which burns without consuming the material which it is placed upon". It was distilled "like rose-water".

Now both DEGERING (162) and DIELS (176) have tried to prove that the discovery of alcohol goes back to Antiquity. They have tried to show that there was no reason to suppose that the *Mappae Clavicula* was written in Italy and tried to make out that it was of Carolingian origin and written in France. They even considered it to go back to manuscripts of the Alexandrian school of the second to seventh century though they had to acknowledge that Arabic works manifest no evidence that alcohol was known then. DIELS mentions many instances of non-consuming fires used by priests and magicians of Antiquity to dupe the public and he holds that these men used alcohol. On the other hand VON LIPPMANN (377), RUSKA (530) and SUDHOFF (610) have proved conclusively that these tricks may have been executed with strong wine and that there is no evidence to cover the gap between the Hellenistic age and the twelfth century in the reasoning of their opponents. Arabic writers mention alcohol only after its discovery in Italy and the knowledge which the alchemists of the fourteenth and fifteenth century have of alcohol is proof of practical mastery of its preparation.

The second important source on this discovery is MARCUS GRAECUS' *Liber ad Comburendos Hostes*. The unknown author may just be a convenient label for this *Book of Fires for Burning the Armies* covering a compilation of recipes from all preceding periods. The Latin texts use words obviously derived from Arabic sources (alkitran, alembic) and the title suggest Byzantine transmission. The latest accretions are recipes for the distillation of terebithum, oleum laterinum ("brick-oil") and aqua ardens. BERTHELOT used manuscripts of the Bibliothèque Nationale No. 7156 (XIIIth-XIVth cy.), No. 7158 (XVth cy.) and CLM No. 267 (about 1300) and No. 197 (the very full version of 1438). In England there are several other manuscripts. The version given in HOEFER's *Histoire de la Chimie* is full of faults. Now the recipe for the distillation of aqua ardens dated back to the thirteenth (or possibly to the eleventh century) but not earlier. MARCUS GRAECUS mentions that aqua ardens burns on the finger or a piece of cloth without burning it, that is he did not yet know absolute alcohol, though his product was strong enough to burn up.

The above-mentioned three texts show clearly that alcohol was discovered about 1100 and the evidence points to Italy, where the school of Salerno was then the most important chemical centre.

The reason of the late discovery of alcohol was of course partly due to inefficient cooling and the unnecessarily long pre-heating period (636) but certainly also to the fact that even the strongest distillate which the early stills could separate in one distillation still contained so much water that it would not burn. The secret of the success after 1100 was not only the rectification of the distillate or the recovery of this distillate in several fractions, but mainly the addition of such substances as salt, tartar (potassium carbonate), etc. which absorbed part of the water and made the rest ready to distill. Now this enabled them to make alcoholic distillates which burn quite readily because they contain less than 35 % of water, and to obtain even absolute alcohol after several rectifications. RAYMOND LULL is probably the first to mention absolute alcohol as distinct from alcohol-water mixtures.

Now here again the nomenclature is vague, but it seems certain that no instance of a medieval manuscript is known in which the word "alcohol" is used in our sense. The Arabs had a word "al-kohl" which first meant the antimony suphide used for penciling the eyebrows and painting the eyelids and then it came to mean a very fine powder. Now SALADIN AB ASCOLA (547) says in his *Compedium aromateriorum*: "Quid est alcohol dicu atomi apparentes in sphera solis admodum subtilissimi pulveris qui vix tactu comprehenduntur", which is probably the earliest example of the use of "alcohol" extended from a ''fine powder" to something "subtle, something like a rarified vapour" which remained one of its meanings for many a century afterwards.

The common name for a fraction containing enough alcohol to burn was "aqua ardens", absolute or very strong alcohol being denoted by the term "aqua vitae" (though this is applied to all sorts of cordials and decoctions or alcoholic distillates too), "aquae vitae rectificata" and "quintessence". The latter term became more and more fashionable in the course of the fourteenth and fifteenth century because of the doctrine of alcohol as the fifth element next to air, water, fire and earth. Other terms for alcohol were "âme du vin, eau flagrante, eau permanente, eau éternelle, mercure végétale, air animal, lumière des mercures, prime essence, esprit subtil, esprit de vin, eau-de-vie, aqua vitis, aqua vini, menstruum vegetabilis, lucerna coelica, anima coelica, spiritus vivus, stelle Diana, sanguis menstrualis, urina sublimata, mercurius vegetabilis orus a vino rubeo vel albo", and many others.

We have mentioned frequently that alcohol was praised as the best medium for the preparation of pharmaceutical distillates if it was not taken as a pure distillate according to the prescriptions of THADDEUS FLORENTINUS, VITALIS DE FURNO, ARNALD OF VILLANOVA and RAYMOND LULL.

A typical example of the belief in the wonderful medical properties of alcohol is the fifteenth century manuscript of the Bibliothèque Nationale which is entitled *Cy après s'ensuyt les vertus et proprietez de l'eau-de-vie* (BN No. 7478) which gives only medical applications.

It seems that the *apothecaries* were the first to produce alcohol on a larger scale, though MAURIZIO maintains that vintners and innkeepers had been also making alcohol as "Hausbrand", that is as a home-brew (411) since the end of the twelfth century. Though this date seems rather early it is certain that this home industry satisfied the general demand for alcohol and alcoholic drinks. It is well not to forget the peculiar position of the apothecary in the Middle Ages. Upto the fourteenth century he was still closely connected with the town government or the local university, he possessed all the facilities for producing alcohol on a somewhat larger scale and he often took the place of the physician or doctor. From the thirteenth century onwards the apothecary becomes more and more a private owner of a shop and the more or less official connection with town council and university is severed. But the apothecaries still had their booth in the marketplace until the end of the thirteenth century where they sold not only their typical products, herbs and medicines but also confectionery. That they were the principal tradesmen in alcohol is clear from early police regulations such as those of the Town of Nürnberg of the thirteenth and fourteenth century in which "gebrannter wein, bernewin, brandwin, etc" that is brandy is specified as their special product. There is no doubt that brandy was not an expensive drink used by the higher classes only before 1500 as some authors have claimed. It was consumed by all classes and its spread can be read from the regulations cropping up from time to time, for instance in Frankfort, where we find regulations of 1361, 1391, 1433, 1456, 1487, etc. which intend to cope with the spread of drunkenness and unruly behaviour of intoxicated burghers.

It is possible that the Black Death (1348 and later years) favoured the spread of the use of strong alcoholic drinks which were then often prescribed as medicine for this terrible scourge. Gradually the preparation of alcohol passed from the hands of the apothecary to those of specialists like the vintner or the "water-burner" (Wasserbrenner), the distiller, who as DAHMS proved also practised medicine in the early days like the wise women, the "Wasserbrennerinnen" or "aquavit-women" for whom PUFF wrote his booklet. But the apothecary remained the specialist for all kinds of alcoholic medicines or alcoholates as the monasteries became less important centres of medicine and industry. The true apothecary still possessed his distilling apparatus and used it on a large scale as we will have ample occasion to prove. A very instructive proof can be found in the Vienna Codex No. 5400 which sums up the taxes paid by the chemist's shop of magister PÜRCKL VON SCHÖNGRABEN (1440) from which we read that he had always freshly distilled water ready for use.

The production of essential oils, perfumes and extracts of herbs and flowers for medical use was originally more or less concentrated in the monasteries, as far as the apothecaries did not take over this task. As early as the fourteenth century a country like Burgundy had large plantations growing lavender, borage, salvia, violets, roses, lilies, etc. which either were used dried (51) or distilled with water or alcohol. In Dijon no less than eleven stills were in use for this purpose. We have mentioned that several essential oils were discovered in the course of the Middle Ages. SALADIN AB ASCOLA (546) mentions in his *Lumen Apothecarium* written between 1442 and 1458, sandel-oil as a novelty added to the then known essential oils. Probably the editing of other manuscripts such as the *Elixir vitae sive confectione aquae vitae* which GOLDSCHMIDT signalled in the Library at Basel and other little known texts (268) will show us that more of these substances were known to the early alchemists than we suspected upto now. The monasteries in possession of botanical gardens were especially equipped to form centres of the production of perfumes and essential oils. The title page of BRUNSCHWYGK's book (Fig. 34) and another picture of a garden of herbs somewhat later in the sixteenth century (Fig. 33) show us a rather fanciful picture

Fig. 33. German botanical garden of the sixteenth century

of the different types of stills used in these botanical gardens. The Dominican monastery of Santa Maria Novella at Florence began a regular perfume industry in the sixteenth century after extensive trials. But the Benedictines and Carthusians were also producers of different aquae vitae, which came to stay as brands of liqueurs. Their competitors were the apothecaries and the distillers who also produced alcoholic extracts of herbs and bitters or decoctions which were sold as elixers of medical value under the name of Confortantia, Stomachia, Aromatica, Carminativa, etc. The iatrochemists with their use of chemicals for medical purposes stimulated this trade which grew to cope with the increasing demand of aquae vitae, Panacees, Athanasia or however they might be called. Special gardens of herbs, "viridaria", were laid out on the Mount Michael near Bamberg, near Würzburg (often called Herbipolis), in Languedoc, Burgundy and

Liber de arte diftillandi. de Simplicibus.

Das buch der rechten kunft
zu diftilieren die eintzige ding
von Hieronymo Brunfchwygk/Burtig vñ wund artzot der keiferliche frye ftatt ftraßburg.

Fig. 34. Botanical garden with distilling apparatus from title page of BRUNSCH-
WYGK's *Small Book of Distillation*

Southern France, but many of them belong to the sixteenth century
and later.

In surveying the role of alcohol in the story of distillation it is also

necessary to describe the change in *alcoholic drinks* which came about
as the manufacture of alcohol was discovered.

In Antiquity Roman and Greek wines were imported in the West
as civilised taste declared against those fermented drinks included in
the name of "mead" (146). Propagated and protected by kings,
noblemen and monks the vine reconquered part of the Imperium
which it lost and even new territories such as Ireland. From the
eighth century onwards the regions of Rhine, Moselle, Danube, Burgundy, Saintonge, Bordeaux, Narbonne and Spain regained much of
the old glory of wine-producing countries. In Merovingian and Carolingian times wine and corn were already products transported over
long distances between centres of production. Their trade is typical of
the ninth century (333) when for instance one of the great trade routes
led from England over Holland to the Rhine valley. Wines from the
Near East were brought to England and even to the north-east of
Europe. Monastic orders had cultivated vines and peasants and pro-
prietors followed. Wine along with corn and live-stock became one
of the great sources of wealth of the West (80). The vine appears
even in the least favourable countries such as the Low Countries, England, Germany and the North of France. It was and remained succesful all throughout the Middle Ages and later centuries along the
Rhine and Moselle, the slopes of France, Switzerland and Burgundy.
In France the vineyards covered Upper Burgundy, Soissonais, Limagne,
the Loire valley, Aunis, Languedoc, the Rhone hills and the sur-
roundings of Avignon. From Bordeaux the wine of Guyenne was
exported, in one year (1330) its total value was no less than
£ 50.000 and La Rochelle exported between 30.000 and 35.000 tuns
a year. Spain, Portugal and Italy rivalled and supplanted the brands
of Greece and Cyprus.

In the Middle Ages water mixed with wine supplanted pure water
as the common drink but light beer became a new competitor. Beer
brewing started in Flanders and England and the peasant of the north-
west soon began to drink beer which he brewed from barley or corn.
Beer became a common drink, rich men had theirs brewed, the poor
man made his own as a home-product. In French, German and Latin
countries, however, the poor drank a good deal of wine and its consumption spread far and wide in the village taverns of the West.

Home brewing remained quite common in the villages, but in the
town brewing became an industry side by side with the earliest industries in the West, soap and sugar. In Holland for instance brewing

was one of the most important industries in the fifteenth and sixteenth centuries. Its centres were Dordrecht, Delft, Haarlem, Utrecht, Gouda, Amersfoort and Amsterdam. In the fourteenth century Amersfoort had no less than 350, Haarlem 50 breweries (320).

A new and dangerous competitor of wine and beer arose when the distillation of alcohol and alcoholic beverages was discovered. We have already made many a remark to the point that medieval Europe was very rich in the lore of making cordials and essences and these soon became dangerous competitors of wine and beer. But there was not only a tendency towards stronger alcoholic beverages, these themselves tended towards lower viscosity and reduction of sweetness as the ages rolled on (146).

The production of alcohol as aqua ardens or even in its stronger form of quintessence since RAYMOND LULL had showed how to dehydrate alcohol stimulated its use as a medicine, for practically all the famous doctors praised it. At the same time distilling became more or less an industry, first in Italy, where we find a burgher of Modena producing larger quantities of alcohol for sale as early as 1320. TATE assures us (614) that home-made spirits were produced in England by apothecaries and vintners as early as the end of the twelfth century and that whisky was already made from corn, but there is no proof for these contentions.

For we can follow very exactly the path of the new strong alcoholic beverages over Europe. In the course of the fourteenth century they follow the Black Death on its path. It was known that strong alcoholic beverages give a feeling of warmth and so doctors prescribed it in the case of typhoid fevers, diarrhoea and similar diseases. Towards the end of the fourteenth century or the beginning of the fifteenth the manufacture of spirit from corn was discovered, which meant a cheaper product in those countries where wine had to be imported such as the Low Countries, England and Northern Germany. At the same time the use of sweetened alcoholic beverages spread again from Italy, where we find recipes as early as VILLANOVA. These "liqueurs" and the secret of their manufacture were brought to Paris by Italian distillers in 1332.

In general liqueurs consist of alcohol, sugar or syrup and some flavouring matter. At present they are made either by the old and best methods of distilling macerated leaves, seeds, etc. with strong spirit, which gives an "alcoholate", which is then mixed with the necessary amount of sugar and colouring matter; or by the essence method

(mixing spirit, essential oils and spirituous extracts); or by the in-
fusion process which mixes fruit, juices, alcohol and sugar. The second
method is quite modern, but the other two are as old as the art. The
most beloved liqueur which the Italians brought to France was the
"rosoglio", a liqueur with the smell of roses. From France the habit
of drinking liqueurs spread and though the amount consumed grew it
remained rather a luxury. An authority on the Middle Ages says quite
correctly that "the nobles at the end of the Middle Ages washed
down their meals with priceless wines, but the burghers and peasants
had their wine and beer for their feasts only, though very light beer
and wine mixed with water were drunk everyday" (80).

In the wake of the liqueurs brandy or aquavit came to France.
The art of the distillers of Modena travelled along the same road as
their product to Germany where spirit came into vogue in the
mining industry. At the end of the fourteenth century strong spirits
were drunk all over Europe and we see from compilations of town
ordinances like those of Frankfort (published by SENCKENBERG) that
as early as 1360 strong measures were necessary in rural districts and
in the towns to cope with the "brandy-devil" (Schnapsteufel). It is
often truly said that the vice of drunkenness became more pronounced
with the advent of strong alcoholic beverages. Heavy taxes in Frankfort, Hesse and other places were nothing but a provocation of the
consumption of the new strong drinks.

Soon a new method of manufacture was discovered, the fermentation of corn, which was especially important in those countries
where wine was imported, that is in the beer-consuming countries.
There is a strict relation between the waning of the brewing industry
and the growth of the guild of distillers in those regions.

Still, the grouping of the distillers into guilds did not take place
until much later. The *Livre des métiers* of ESTIENNE BOILEAU (1250)
for instance does not mention any such corporation. The name "ge-
prannter Wein, Brandewyn" ("burnt" or distilled wine) clung to that
spirit which was manufactured from corn too. There was great opposition against this new sin, which according to many medieval authors
meant the "waste of natural products which God had meant for food".
There are signs that the manufacture of brandy from sugar-beets was
known or at least suggested here and there at the end of the Middle
Ages.

Apart from the old centres Modena and Venice, which exported
large quantities of distillates not only to Germany but even to Turkey,

other local centres of distillation of wine or fermentation of corn, barley, etc. were formed. At the beginning of the fifteenth century it was discovered that spirit could be obtained from beer and yeast and though at first the yeast had to be obtained from the brewers, the distillers soon made themselves independent by discovering how to grow yeast. Again, though some countries like Saxony forbade the distillation of corn-spirit its manufacture spread far north and reached Scandinavia, where at first few understood the art of distilling, but the knowledge spread. In the reigns of ERIK XIV and JOHN II the production of spirit was quite common and GUSTAVUS I had to take measures to put a stop to its consumption. The many authors who propagated the use of distilled spirits as a medicine proved, however too strong for the law.

Thus the town ordinance of Nürnberg says in 1496:

"As many persons in this town have appreciably abused drinking aquavit the town council warns earnestly and with stress that from now on Sundays or other official holidays no spirit shall be kept in the houses, booths, shops or market and even the streets of this town for the purpose of sale or paid consumption."

"Nachdem von vil menschen in dieser statt mit nieszung gepranndts weyns ein mercklicher miszbrauch getrieben, eyne rat daran kommen, ernstlich und vestlich gepiettende, dasz nun fürbasz an einichen sonntag oder anndern gepanndten feyertagen gepranndter weyn hie in dieser statt von nymandt weder in den hewsern, krämen, läden oder an der markt, strassen oder suns yndert nyt veyl gehabt oder verkauft werden soll."

The codex Brandenburg I. 25.379, anno 1450, says: "Nobody shall serve aquavit in his house or give it to his guests". And a poet published a pamphlet (645) at Bamberg in 1493, warning people to use spirits moderately. The author is probably one HANS FOLZ, a surgeon of Nürnberg, who writes "As at present practically everyone becomes accustomed to drink aquavit ... one should remember how far one can stand it and learn to drink as far as one should" ("Nach dem und nun schir jedermann gemeincklichen sich nimet an zu trinken den geprannten weyn ......Darumb was er an idem schafft merck einer selber an im das und lern in trinken dester pas").

Though in the sixteenth century the number of apothecaries increased under the influence of PARACELSUS and his school of iatrochemists especially in Germany, Sweden, Russia, etc., their shops usually contained only simplicia and such composita as they could make without too great a knowledge of chemistry. Gradually the pre-

paration of brandy and other alcoholic beverages became the special
job of the distiller though the apothecary still made many alcoholates,
etc. for medical uses.

Map I. Towns that play a part in this story between 1000—1650 A.D.

CHAPTER FIVE

FROM BRUNSCHWYGK TO BOYLE

"The greater do nothing but limbicke their brains in
the Arts of Alchemy and Balancing".
(Sir EDWYN SANDYS, *Europae Speculum*)

When Sir EDWIN SANDYS, diplomat and founder of the colony
of Virginia, wrote the lines quoted above (549), he characterised
quite correctly the main ambitions of the European chemist, as he met
him when travelling on the Continent (1599). We must count ourselves lucky that a new edition of his work could be printed from the
few copies remaining from an earlier edition of his work, which in
its turn was printed from a copy of the manuscript stolen in 1605,
which edition SANDYS had managed to suppress almost entirely.

With the advent of cheap paper and the printing press in the
fourteenth and fifteenth century a new era was rung in, a period
with many commotions and disturbances, for new forces were at
work. The economic aspect of Europe was changed by two factors,
the barrier raised between Western Europe and the coveted products
of the Far East by the Turkish domination of the Near East; and the
incessant civil wars in Italy and Germany which gradually drove away
trade and industry to safer and more stable regions. In the course of
the sixteenth century the centre of industrial chemistry moved definitely to Western Europe, that is to France, the Low Countries and
England. There the great trade emporia and the new merchant navy
of the West brought riches and new materials to Europe undreamt of
by former generations. There alone the stability was found which
ensured a steady development of science and its technical applications.

Though there was a deepening cleft between religion on one side
and philosophy and science on the other, they were still linked by
many bonds. Only very gradually did a new generation of scientists
grow up who tried to build a system of natural philosophy based on
experiment and fact only. And though there was not yet any intimate
bond between theoretical science and industrial technology, the latter
was bound to profit by the large strides made by the former, at least
in the long run. Hower, we can deny that scientists became more
interested in practical problems. The new education, the new kind of

scientist strove towards an unfolding and esteem of a positive perso-
nality which was totally different from the type of self-effacing
medieval scientist, whose mainstay was authority. Humanism rent the
veils of credulity and replaced alchemy by chemistry.

PARACELSUS (THEOPHRASTUS BOMBASTUS VON HOHENHEIM, 1493-
1541), though a bitter opponent of the alchemists, adapted chemistry to medicine and tried to find the medicinal value of chemical
compounds, devising new methods of preparing and purifying them.
The iatrochemical school counts many scientists among its followers
who play a part in the story of distillation, such as OSWALD CROLL,
ANGELO SALA, ANDREAS LIBAVIUS, JEAN BAPTISTE VAN HELMONT,
OTTO TACHENIUS and BLAISE DE VIGENÈRE. They pave the way of
the new science of the balance, of analytical chemistry, which is
already found in the works of LIBAVIUS but blossoms to full bloom
in the writings of TACHENIUS. The practical results and methods of
the alchemists were neither lost nor rejected, but they were printed,
discussed, tried out and modified. In every direction we find a period
of amalgamation and transition, of trial and error.

New collections of recipes were printed side by side with the texts
of the older alchemists. New methods were developed but technical
traditions remained unbroken. Such operations as distillation, calcination, reduction and crystallisation were carried out with the same
skill in BRUNSCHWYGK's days as in our modern laboratory.

Apart from the true chemical textbooks of this era we find a second
series of books on practical technology, the so-called Berg-, Probier-
und Kunstbüchlein, mostly tracts on mining, treatment of ores, metallurgical and similar practical problems. The authors are mainly
Germans and Italians, but we find French names too. The most important of them were AGRICOLA, BIRINGUCCIO, ENCELIUS, ERCKER,
FACHS, LÖHNEYSS, MATTHESIUS and PALISAY, but there were many
anonymous pioneers, who wrote the earliest pamphlets of this kind
between 1500 and 1550. They too play a part in the history of
distillation but our main sources are a third class of texts called the
Arznei-, Kräuter- und Destillierbücher mostly written by physicians,
apothecaries and botanists, who discuss not only the distillation methods and apparatus but also the treatment of different herbs and
flowers and the medicinal value of their decoctions, "oils", and
"spirits". The most important German authors of these herbals are
MICHAEL PUFF, HIERONYMUS BRUNSCHWYGK, VALERIUS CORDUS,
WALTHER RYFF and CONRAD GESNER (EVONYMUS PHILIATER),

whilst among the French we find JACQUES BESSON and QUERCETANUS and among the Italians MATTHIOLUS and PORTA. All these men will find their place in our story.

It is strange to find very few books dealing with the distillers trade only, for the *manufacture of alcoholic drinks* developed rapidly. Some say that alcohol was introduced into *England* by RAYMOND LULL during his (apocryphal) visit to the English court in 1300 (636) but others like FAIRLEY contend that Irish imigrants founded distilleries in Pembroke and thus brought the art to England (216). But distilling was confined to the monasteries as centres of medical science, many of whom had a "stillatory". There was also a certain amount of home-brewing and distilling, but the prohibition of the alchemists or "multipliers" by HENRY IV in 1404 prevented the development in other centres. With the dissolution of the monasteries the monks were dispersed and thus we find them in the sixteenth century taking up the professions of doctors, brewers, apothecaries and distillers. Only then did the art of distilling become an industry. In Tudor and Stuart times a licence was exacted for the use of a still, only much later was duty paid for the amount of spirit produced.

By Tudor times Scotch whisky was already famous. READ (502) informs us that aqua vitae probably meant whisky at the time of JAMES IV of Scotland since the Exchequer rolls of 1494-1495 show that a certain pioneer of the Scottish distilling industry, known as FRIAR JOHN COR, was supplied with eight bools of malt for making aqua vitae. The preparation of strong alcohol by repeated distillation of wines and other fermented liquors was a well-known process in the England of the beginning of the sixteenth century. The spirit of wine which had reached "myche highnes of glorificacioun" by distilling upto a thousand times was identified, as we have seen, by some alchemists with the Elixir of Life, quinta-essentia and "mannys hevene". The Abbot of Tungland's interest in aqua vitae is easily understandable by students of alchemy if not by others. In the strange Scotland of JAMES IV it was rather still a rare drug, but he procured large quantities. But whether this raw material was wholly devoted to the Great Work or partly to "the refreshment of wearied Servants cf the Laboratories" is not clear from the Treasurer's account.

Brandy was imported into England by the Genoese from the fourteenth century. According to FAIRLEY (216) part of this product was not of Italian origin but acquired as arak from the East. At any rate in the days of Queen ELIZABETH this trade was already quite important being partly in the hands of English traders but mostly in those of French and Dutch merchants.

The older terms for alcohol were displaced by words like "brand-wine", "brandewine" and "brandy wine" which crop up in English literature between 1622 and 1657, to be replaced after the latter date by the present term "brandy", though one sometimes still meets the French "brandevin". Even now the common trade name for industrial alcohol is often "trade brandy".

The organisation of the trade in *France* was much farther developed, for not only was the spirit of wine known in this country at least from the thirteenth century, but the trade had been controlled since 1619 by the State and the distilling of corn spirit was forbidden. The art of distilling remained for a long time the official privilege of the apothecaries and spice-merchants but by statutes of September 7th 1624 and October 13th 1634 the distillers were grouped with the former in one guild; later a separate guild of distillers was formed (225). This guild was called "corporation des distillateurs et marchands d'eau-de-vie et eau forte", the term "eau forte" still meaning "redistilled strong alcohol" and not "mineral acids" as in later centuries. The manufacturers of vinegar belonged to the same guild. Its statutes were confirmed by LOUIS XII in January 1637 and their title changed to "distillateurs en l'art de chimie et vendeurs d'eau-de-vie", that is to say the producing and selling branches of the industry were still combined in one guild, which goes to show that the distilling trade was still mainly a home-industry. Two years later the statutes were renewed (April 5th 1639) and now masters of this guild were called "distillateurs d'eau-de-vie, d'eau fortes et autres eaux, huiles, essences et esprits", in short all the manufacturers of chemical products by distilling were grouped together.

In *Spain* state control of the industry dates from 1590.

The "waterdistilleries" (Wasserbrennereien) of *Germany* began as a form of home-industry. We found in France that the Benedictines of Fécamp first produced liqueurs for their own use, then gradually increased their production and sold their "water" on the market. In the same way the Carthusians of the "Grande Chartreuse" developed their own brand and the Benedictines of Epernay produced their "Champagne". But in Germany the new trade was not so closely bound to the monasteries but more to towns, which produced their own brand of "water". In the sixteenth century many people tried their hand at this new distilling trade and the town of Nürnberg

for instance had to forbid the distilling of alcoholic beverages by
the "Zuckermacherinnen und alten Weibern" (confectioners and wise
women), who combined the professions of distiller, confectioner,
fortune-teller and procuress. The aquavit-women retained their trade
for a long time, making their products in simple stills in their
kitchens or some special place. But as early as 1484 SCHRIEB (567)
had to warn the public against very serious adulterations of spirit
of wine, though he praises the pure product and gives details on the
proper pharmaceutical application. He is one of the first to mention
the use of bear-yeast in preparing alcohol. But the vice of excessive
drinking grew with the distilling trade and the consumption of wines
and strong drinks in apothecaries' shops soon became a public scandal.
In Cassel public protests were raised against the drinking bouts in these
shops, which in the year 1653 were said to start early in the morning.
Many towns had begun much earlier to restrict the sale of alcoholic
drinks by the apothecary, leaving him only the sale of certain specified
aquae vitae for medical purposes and restricting or forbidding him
the sale of wines altogether.

The sixteenth century was a period of many experiments. In
Germany LIBAVIUS and others tried all sorts of herbs and fruit in
order to find new base materials for the distilling trade, but apart
from cereals and wine or beer no new materials came to stay.
BRUNSCHWYGK and LONICER pointed out the possibility of making
alcohol from sugar-beets, but the times were not yet ripe for the
execution of their bold plans and we have to wait until the days
of ACHARD and NAPOLEON (421) for their execution. Wine had to
be imported from Italy and France and soon the fermentation and
distilling of corn became the mainstay of the German industry. The
new centres for the production of corn spirit were Wernigerode and
Magdeburg; they were founded about 1620. But there was still a
very strong public opinion against the abuse of corn for the manufacture of alcohol and several towns and districts forbade this manufacture. But they could not stem the tide on the long run and so we
can say that alcohol and brandy were made from wine and lees of
wine before 1500 and afterwards spirit of corn, prepared more
especially from rye, was the common form of alcohol in Germany.
Its production was often accorded to certain persons or groups of
persons in the form of a privilege or license. In several towns the
distilling trade was in the hands of the Jews. Two qualities of alcohol
were generally manufactured, an industrial spirit called "spiritus vini

rectificatus simplex" and similar name, or the stronger rectified "spiritus vini vulgo aqua ardens". In Italy both the spirit of wine and corn spirit were called "aqua vitae".

At first sight it is difficult to understand why the distilling trade in the *Low Countries* became so prosperous as these countries do not produce wine nor cereals in sufficient quantities to provide a cheap base material. Still there is a growing distilling trade in the sixteenth century which becomes a national asset in the seventeenth.

The emperor CHARLES V had favoured the rise of the towns in the Low Countries and the breweries had profited by his policy (32). Distilleries followed in their wake, often at the expense of the former. The breweries in these countries developed into somewhat larger industries because they required considerable outlay of capital in the form of apparatus and constantly required the rather expensive import of cereals and fuel. As early as the sixteenth and seventeenth centuries the Dutch distilling trade was working for export. In Amsterdam there existed an impost or excise duty on brandy in 1497 (91) which points to an industry of a certain importance. It is generally said that the use of strong alcoholic beverages came to Holland with the German soldiers serving in its armies in the Eighty Year's War against Spain (513). And indeed several distilling centres such as Schiedam often worked with German distillers who had emigrated from the distilling centres of their native country. For some time the distilling trade was dependent on the brewing trade of which it was first a side-line, then the brewers held the monopoly of the yeast production until the distillers learnt to grow yeast themselves. Gradually the brewing trade declined and the distilling industry took its place after the high prices of fuel had been lowered and the production of yeast was discovered in the seventeenth century. The early distilling trade was dependent on the import of French brandy and cheap German spirit and French wines long remained serious competitors of corn as a base material. Here again we find a strong dislike of the waste of cereals for the distilling trade, which here and there caused strong opposition (Delft, Haarlem) before 1600 as the distilleries tended to make corn more expensive for the poor.

But on the other hand the growing corn trade with the Baltic provided for an ample supply and this trade was practically in Dutch hands (455). For after the fall of the Roman Empire there was no longer a world trade in corn. The new states barred their frontiers for this trade and even towns and provinces forbade exports, or exacted exhorbitant duties from the merchants. Certain trade routes remained free as we have seen, but one of the most important obstacles to an international corn trade in the Middle Ages was the clumsy, unseaworthy type of ship still in an early stage of development. In the Mediterranean there was a growing corn trade among the countries of southern Europe which stood quite disconnected from the corn trade of the Hanse towns with England, Holland and Scandinavia. In the seventeenth century the merchant vessels were sufficiently developed to start a new international corn trade, which started with the corn-fleets of the north of Europe rushing to the rescue of Italy towards the end of the sixteenth century when famines devastated that peninsula. Amsterdam became the centre of intermediate trade linking up the corn-producing countries of northern Europe and the consuming lands of the Mediterranean. Danzig became the most important corn-shipping port of the wheat-fields of Poland; Amsterdam, after the opening of its corn-exchange (1617), became the financial centre of this trade, and remained so until the end of the eighteenth century.

Thus the corn-supply soon grew plentiful for the rising distilling industry. Books on distilling were published and though Meester PHILIPPUS HERMANNI's *Constelyc distileerboec* dates from 1552 only, that is nearly two generations after similar handbooks by BRUNSCH-WYGK and his school, this book was very often reprinted in a short period. But it describes the distillation of wine only and from these Dutch handbooks it seems that only at the end of the sixteenth century popular prejudice was sufficiently overcome to make the distillation of corn spirit the mainstay of the trade. But there were not only moral difficulties to be overcome, for COOLHAES who warns agains corn spirit also mentions that unsound corn was often used or the badly distilled spirit adulterated with sugar and honey and sold as "good water". There was a strong objection to the taste of corn spirit in wide circles and "French brandy" was long preferred to the home product from corn, but gradually this prejudice, partly founded on rectification defects of the older apparatus, was overcome.

Early in the sixteenth century we find ordinances regulating distilling, sales and retail trade in the towns of Amsterdam, Dordrecht, Leiden, Brielle, Haarlem, Delft and Alkmaar. By the end of this century the distilling industry was one of major importance. At the beginning of the seventeenth century we find only second quality wines, lees of wine, beer, prunes, raisins, etc. mentioned as base materials side by side with corn. But the trade reached the height of its prosperity only in the middle of the century when the first pro-hibition of French brandies (1651) was decreed culminating in the tariff of 1672. Corn spirit became the common form of spirit pro-duced in these countries or as COOLHAES has it: "In Dantzig, König-bergen and similar White Sea towns just like in our countries at Amsterdam and neighbourhood (Weesp?) and also at Rotterdam, Hoorn, Enkhuyzen and other towns corn-distilleries were founded".

„Tot Dantzwyck, Coninxbergen, ende diergelycke Oostersche Zeesteden, gelyk oock hier te lande tot Amsterdam ende daeromtrent, oock te Rotterdam, Hoorn, Enckhuyzen ende diergelycke plaatsen graanbrandewyn".

At Schiedam we find a real industrial development of the distil-lery. In that town the herring-fishery declined, but the distiling trade took its place and more than supplemented the loss of wealth from the sea. The town council acted extremely wisely by leaving the trade unhampered by any ordinance restricting either the manufacture or sale. The town itself was well situated for the import of cheap corn and the export of its product overseas (513), though the main export of the Schiedam gin was effected by the ports of Rotterdam and Amsterdam. These distilleries existed before 1630, in this town we find the first Dutch guild of distillers.

But not everywhere in Holland did the trade develop as early on an industrial basis, for if we turn to Amsterdam we find 400 small distillers in 1663, that is to say many small distilleries worked as home-industries against the local laws of retail trade in alcoholic bever-ages. Only many years later was a proper distinction made between the wine-merchant and the distiller and the trades were often combined. When the duties on French brandy became prohibitive several new small distilleries were erected in the district called Roeterseiland (91).

Rotterdam harboured but few "gebrandwynmakers" at the end of the sixteenth century (104). In the course of the next century we find more "brandewynbranders" later called "distillateurs". At first they are only few in number and the inventories of their distilleries show that we must still speak of home-industries. They usually possess two stills with the worm-coolers and other vessels and measures. For the rest they keep a stock of wheat, rye, buckwheat mixed with malt, spices, sugar, honey, juniper berries, aniseed, coriander seed, caraway seed, etc. A few specialize in the distilling of French wines, most have combined their trade of distiller with that of vintner. The wash is sold as food for the pigs. In 1674 Rotterdam counted 56 private distillers, though there were a few more attached to breweries. The distiller's guild was founded before 1631 and we know the names of several families of distillers who kept their "trade secrets" very jealously. But on the whole the distilling trade was very conservative and we shall have occasion to note that it profited only slowly and very slightly from the development of distilling technique by other scientists.

We also have mentioned that such *names* as "aqua vitae, aqua ardens etc." remained in use. The word "alcohol" is said to have been used in our sense by PARACELSUS and LIBAVIUS early in the sixteenth century. The meaning of the word changed considerably and to avoid misunderstanding we must point out that the word "alcohol" had two meanings in this period. It is either used in the sense of "a very fine powder" or in that of "the spirit obtained from fermented liquors by distillation" (164). FRANCIS BACON (1626) still uses the original meaning when he says: "The Turks have a black powder made of a mineral called alcohole (al kohl) which with a fine long pencil they lay under their eyelids". JOHNSON's *Lexicon Chymicum* (1657) gives both the meaning "powder of antimony and "spirit of wine" (29) (197) (334). In Dr. JAMES' *Dictionnaire universel de médécine* traduit par DIDEROT (1746) alcohol is said to mean "fine powder, volatile spirit". TRÉVOUX's *Dictionnaire* of 1771 gives "alcooliser" as "chemical term, to reduce to a powder that is very subtile and practically indivisible, or the purification of spirits and essences from impurities and phlegm which they might contain". BAUMÉ uses the double meaning of this word and even DAVY in his *Chemical Philosophy* refers to the "alcohol of sulphur" (1812). PHILLIPS defines (1706) alcohol as the "pure substance separated from the gross". In a former chapter we have sketched the general transition of the meaning "a fine powder" to "something subtile, a spirit". The word alcohol is gradually but not exclusively used more and more in the sense of "spirit of wine" and "spirit of fermented liquors" during our period, but the general use of the word in this sense only is due to the efforts of BOERHAAVE in the eighteenth century.

We have mentioned that two qualities of alcohol were generally produced. Whether the "strong rectified aqua ardens" was absolute alcohol is not certain, anyway we have no proof that the distillation of rectified spirit with dehydrating agents, such as we have mentioned before, always gave 100 % alcohol. The first indisputable preparation of absolute alcohol dates from 1796 when LOWITZ distilled strong alcohol with freshly burnt potassium carbonate. But at any rate the preparation of 90 % alcohol or even 96 % alcohol was quite feasible in the period we are discussing, at least in the laboratory.

Before, however, discussing the apparatus characteristic of this period we must review the different protagonists in the story of distillation and hear their views on the different aspects of our subject. The first author, whose work on distilled waters (498) was printed, was MICHAEL PUFF VON SCHRICK, often referred to as MICHAEL SCHRICK. He was born in Schrick (Austria) about 1400, got his M.D. in 1433 and died in 1473 as professor of the University of Vienna. It is known that his manuscript was already finished in 1455, but it was not revised and made ready for printing until 1466. The earliest edition was published by JOHANNES BÄMLER at Augsburg on the Friday after St. Vitus' day 1478; it comprised 15 folio pages. Many other editions followed, no less than 14 being printed at Augsburg, Ulm and Strassburg before 1500 and many more coming after that date from the presses of Strassburg, Nürnberg and Wittenberg. As early as 1484 a booklet was printed at Lübeck accusing Schrick of plagiarism and attributing the original to a Dr. BARTHOLOMAEUS OF BENEVENT (probably SALERNO is meant).

He mentions that he compiled these recipes from many books for the benefit of many persons. The recipes of these waters do not interest us except the last one, which deals with "geprannten weyn" or brandy. He gives details of the medicinal use of all the waters and states that "anyone who drinks half a spoon of brandy every morning will never be ill" and "when one is dying and a little brandy is poured into his mouth, he will speak before he dies".

„Auch wer alle morgen trincket den geprannten Weine ein halben Löffel vol der wirt nimmer kranck".

„Item wann einer sterben sol so geüszt man im ein wenig geprânts weins in dê mund, er wird redê vô seinê tod."

We saw that the title page of his book (Fig. 29) shows the conical cooler called "Rosenhut", which because of its size relative to the small still was probably sufficient to condense light fractions in its top. This form of cooler is also given by BRUNSCHWYGK and by the *Master of the Trostspiegel* in his drawings (1530) but it disappears from the scene pretty soon. Still HERMANN HAGER patented a similar "Dunstsammler" as late as the eighties of last century.

His opinion of the sworn "aquavit-women" appointed by state or town is not high. He says that one of the main reasons for writing his book was that "the abuse of medicaments is so great that no one would wonder if the stupid physicians, wise women, vagrants and apothecaries were to poison the whole world with them".

„Der Miszbrauch mit Artzneien ist so gross dasz kein Wunder wäre wen die Kelberärzte, alten Weibern, Landfahrer und Apotheker die ganze Welt mit Arztneien verderben".

The only way to combat this danger is to treat illness with distilled waters, though these should not be abused.

A far more important author is HIERONYMUS BRUNSCHWYGK (BRAUNSCHWEIG, BRUNSCHWIJG), who calls himself a native of Strassburg and surgeon of this Imperial Free City. He died at the end of 1512 or the beginning of 1513 at the age of sixty. STILLMANN gives 1534 as the year of his death without further proof. He was thus born about 1450 and descended from the famous local family, the SAULER. He studied medicine at Bologna, Padua and Paris and he is especially well-known to the historian of medicine for his *Chirurgia* of 1497. Here, however, we are concerned with his two works on the art of distillation, which had an enormous influence. Through many imitations, translations and works by other authors these two books gave rise to an extensive literature on distilled medicine. Distillates became much more prominent than the powders, syrups and decoctions of traditional medieval manuscripts, so that this new art meant a departure from scholastic medicine. STILLMANN points out that we do not find any claim to originality in BRUNSCHWYGK's books, they seem to represent the formulation of the practice of a considerable group of medical practitioners especially in Strassburg. Their goal was the preparation of a "spirit" or "essence" from a gross and perishable raw substance (606) (607). This was effected by distillation and here we see the influence of the "spagyric physician", PARACELSUS, who often discussed the value of distillation for separating the different principles in natural products. Thus he says in the *De Natura Rerum*: "The separation of things that grow from the earth and that are easily combustible takes place in many ways. Thus by distillation is separated from them first a phlegm, then mercury and oily parts, then its resin, fourth sulphur and fifth salt".

BRUNSCHWYGK's first book, the *Liber de arte distillandi de simplicibus* (99), or the *Small Book of Distillation* was published on May 8th 1500 and this edition is famous as one of the early printed works. As we shall see the first part of this work on simples was devoted to the construction of furnaces and various forms of stills, retorts, condensers, receivers, etc. and to various methods of distillation. The second part is a kind of herbal describing the plants, their useful parts and the correct time to distill waters from them, detailing also what purpose these waters served. The third part is devoted to all kind of diseases and their cures with the above-mentioned waters. The publisher JOHANN GRÜNINGER of Strassburg used part of the text for a new publication entitled *Liber de arte distillandi Simplicia et Composita*, in which book parts of other works are combined with chapters from BRUNSCHWYGK (Strassburg, 1509).

This first work on distillation was soon followed by a second by the same author, called *Liber de arte distillandi de Compositis* published at Strassburg on February 23rd 1512 (the so-called *Big Book of Distillation*) (100). This new book on the distillation of composita had a still greater succes, it was much larger than the former and contained no less then 79 illustrations of distilling apparatus, furnaces, etc. Not only was this *Big Book of Distillation* the starting point and the example for other writers such as ULSTADT, RYFF and many more, but it saw numerous editions (608), many of which were rearrangements supplemented by chapters from other writers as was the fashion of the time. We mention here *Das newe Destillierbuch* (Strassburg, 1519), *Das Buch zu Destillieren* (Strassburg, 1519, 1531), *Das Destillierbuch der Rechten Kunst...* (Frankfort 1553), *Hausartzneybüchlein* (Frankfort, 1594), *Thesaurus pauperum* (Frankfort, 1598), etc. The newest edition took the form of an appendix to UFFENBACH's *Comments on Dioscorides* (Frankfort 1610).

It was also translated into other languages. The earliest English translation is that of L. ANDREW of 1527 (101) called *The vertuose Boke of Distyllacyon of the Waters...*, in which distillation is defined thus: "Dystyllyng is none other thynge, but onely a puryfyeng of the grosse from the subtyll, and the subtyll from the grosse", a definition which as we shall see occurs in BRUNSCHWYGK's first book.

The earliest Dutch version is the *Die distellacien ende virtuyten der wateren* published at Brussels by THOMAS VAN DER NOOT in 1517. It is considerably shorter than the German version giving only 104 pages of the original 230, but these deal mainly with the distilling apparatus and methods, of which 23 figures appear, which originally belonged to the *Small Book of Distillation* (102).

The second edition of this Dutch booklet was considerably altered and it seems more or less a reprint of a *Die rechte conste om alderhande wateren te destilleeren*, published by WILLEM VORSTERMAN at

Antwerp in the same year. It had more the character of a pharmaceutical handbook, but one of the last chapters extols the virtues of the living water that is made from wine or lees (de cracht van den levenden water dwelck gemaeckt wort van den wine oft van den droesemme). Alcohol is said "to free the five human senses of melancholy and every impurity if drunk in moderation that is to say six or seven drops a day in a spoon full of wine on an empty stomach (Het suyvert den mensche syne vyf sinnen van melancolyen ende alle onsuyverheden als men t drinct bi maten. Te weten ses of seven druppelen deachs met eenen lepel vol wyns).

Returning to the *Small Book of Distillation* we find that its special purpose is the preparation of medical agents from non-essential matter by distilling "waters" from these substances macerated and mixed with water or alcohol.

After the title page comes a summary of the contents followed by a foreword. Here BRUNSCHWYGK states that he will tell from his own practical experience „how one should distill waters, use and preserve them". The ancients were masters in this Art and in thirty years he has compiled this book from the writings of HYPOCRAS (HIPPOCRATES), GALEN, AVICENNA, SERAPION and many other books from the libraries to the total of over three dozen books! The first book, dealing with the distilling apparatus and methods starts with an illustrated title page (Fig. 35) such as is found in many a medical book of the period.

The first chapter of this first book defines distillation as "the separation of the gross from the subtile and the subtile from the gross, the breakable and destructible from the indestructible, the material from the immaterial, to make the bodily more spiritual, the unlovely lovely, to make the spiritual lighter by its subtility to penetrate with its virtues and force which are concealed in it into the human body to do its healing duty". For distillation was discovered to serve mankind and to prepare precious medicaments, to be able to serve out pure medicines to control the mixture of the elements and "to serve mankind and to prepare precious medicaments, to be able to detail the different distillation methods which are discussed more fully in the following pages". We reproduced several pages (Figs. 21-23) on which this apparatus is discussed. The fifth chapter on the "bricks and lutes and glues" contains a curious poem on the building of a furnace (Fig. 36). The recipes for the lutes mention among other things a mixture of ox blood and quicklime. Compounds for the protection of glass apparatus are given.

Then the author proceeds to give some details on the different furnaces for direct heating or for heating water-baths. Some of these furnaces are built with a funnel through which new supplies of fuel

Fig. 35. Title page of the first book of BRUNSCHWYGK's *Small Book of Distillation*

are introduced, the so-called Faule Heintz or Slow Harry, a furnace especially designed for long heating periods. As to the material of which the cucurbits, etc. should be made, BRUNSCHWYGK is of the opinion that the best quality glass is bought in Bohemia or Venice.

Fig. 36. Poem on the construction of a furnace (Book I, chapter 5, BRUNSCHWYGK's
Small Book on Distillation)

For good chemical earthenware Hagen or Syburg (near Cologne) are
famous. Copper vessels can be used. Earthenware vessels should be
well luted but this is not necessary in the case of glass apparatus

Fig. 37. Destillatio per filtrum (Book I, chapter 8, BRUNSCHWYGK's *Small Book of
distillation*)

heated in a water-bath. Chapter 8, "de filtri destillationem" discusses
the filtration of liquids with a triangular piece of felt (Fig. 37).

FORBES, Art of distillation

8

"De soli destillationem" is the title of next chapter where distillation
by the sun's heat (still unaided by mirrors or lenses) is discussed and
chapter 10 gives details on distillation by the heat of rising bread
(de panis destillationem). Chapter eleven deals with distillation in
horse dung (fumi equini destillationem) for which a pelican or a
cucurbit with a blind alembic should be used. Chapter 12 "de destillationem tormice" suggests that the heat of an ant-hill could also
be used. Then (chapter 13) the virtues of the water-bath ("destillatione per balneum maris" or "in duplo vase"). The cucurbit should
be weighted with a ring of lead in order that it be immersed to the
correct depth. Heating in an ash-bath (per cinerum) means that the
cucurbit should be protected by a cupel upto three fingers thick. The
sand-bath allows us to attain high temperatures (distillationem per
arenam). Heating by direct fire (per ignem) is possible even with
a glass apparatus, in this case the cucurbit is often placed on an iron
tripod. Chapter 18 discusses distillation with a Rosenhut or with an
alembic using a glazed or copper cucurbit. One should use coal fires,
wood fires are not clean enough. Glass cucurbits always give better
distillates. If one is forced to use a copper alembic, one should take
care that it is tinned. Chapters 19-23 discuss the proper methods for
each of the herbs, etc. mentioned in the further books. One should
seek carefully for the method proper to each individual herb, often
digestion and circulation are necessary before distilling. This is true
both of the vegetable and animal substances. This is necessary for the
preservation of the properties of the waters prepared and applies to
decoctions with water and alcohol. Most of the waters should not be
kept near a stove or in a moist cellar. Then they can be kept upto
three years if distilled with water, but the alcoholates have a far
longer life. They can be "strenghtened" by circulation in the cucurbit with blind alembic (lembicus cecus) in horse dung or a water-bath.

    Books II and III deal with the different herbs and their medical
application, they too are illustrated (Fig. 38).

    We do not know whence the illustrations of the distilling apparatus
came, whether BRUNSCHWYGK or his publisher had them expressly
cut for this edition or whether, as was the custom in this period, they
were woodcuts from other publications of the same editor. For the
latter is certainly the case with the illustrations of the "herbal" part
of the book. The first printed herbal, the *Hortus Sanitatis Deutsch*
of PETER SCHÖFFER (Mainz, 1485) had been a great succes. SCHÖF-
FER was a friend of GUTENBERG (555). His *Herbarius* was soon

translated into Dutch, Italian, etc. and then the larger *Hortus Sanitate* of JOHAN VON CUBE was published, again by SCHÖFFER (Mainz, 1485). This CUBE was physician to the town of Frankfort (1484-1503). The woodcuts of his book were sold in Basel but a reprint of his book with similar (the same?) blocks was published by SCHÖNSPERGER at Augsburg and also by JOHANN GRÜNINGER, BRUNSCHWYGK's publisher at Strassburg. In meantime a still larger, the so-called "Large" *Hortus Sanitatis* was edited which was an elaboration of the *Buch der Natur* by CONRAD VON MEGENSBERG, which we have already discussed. This was published by JACOBUS MEYDENBACH at Mainz (1491).

GRÜNINGER, however, used the woodcuts of his *Hortus* edition to

Fig. 38. Figures from the Second Book of BRUNSCHWYGK's *Small Book of Distillation*

illustrate BRUNSCHWYGK's *Small Book of Distillation* in 1500 and again the *Newe Distillierbuch* of 1531.

This use of the same woodcuts for editions of books by different authors is quite common in this period and we will have occasion to discuss other examples of this practice.

We have already remarked that the *Large Book of Distillation* is really a fuller edition of the earlier work also includes the preparation of composita. In this work BRUNSCHWYGK says that distillation was invented by peasants who studied the fermentation of grapes and collected the vapours in a vessel. This was PORTA's opinion too. In this book we find similar illustrations, of which we reproduce a water-bath for several cucurbits, heated by a furnace and where the flue gases heat the bath through a copper dome which is seen emerging from the middle of the water-bath (Fig. 39). Fig. 40 shows a water-bath for four cucurbits equipped with a Rosenhut. The furnace has a filling funnel emerging from the centre of the brickwork and four small chimneys are shown at the corners of the furnace. Exactly the same picture is shown in the earlier book.

When studying the Moor's head which BRUNSCHWYGK describes it is interesting to note that he does not yet quite understand the proper circulation of cooling water and proposes to tap the water mantle from the bottom (Fig. 30). Alcohol, he says, should be prepared by distillation per ascens, because "the spirits driven overhead are much more subtile and pure. For in rising thus everything that is earthy, heavy and phlegmatic remains behind. Therefore the spirit of wine should be distilled

Fig. 39. Water-bath distillation according to BRUNSCHWYGK (1512)

Fig. 40. Water-bath with Rosenhut according to
BRUNSCHWYGK

overhead but other material that contains phlegmatical humidity should be distilled per descensum".

„Dieweil die Geister, so über sich getrieben werden, viel reyner und subtiler seynd denn in solchem aufsteigen alles so schwer, irdisch und flegmatisch ist, nit hinaufkommen mag. Darumb die geister des weins am flüchtigsten über sich, aber andere materi so mehr mit flegmatischer feucht behafft under sich getrieben werden."

One of the earliest disciples of BRUNSCHWYGK in England was THOMAS GALE. He was born in 1507 and learnt his trade from RICHARD FERRIS, court physician to Queen ELIZABETH. Then we find him as a surgeon at the battle of Montreuil (1544) under HENRY VIII and as surgeon to King PHILIP of France in the battle of St. Quentin (1557). He returned to London where he died in 1586, after reaching such fame, that he was called the "English Paré" because of his miraculous treatments of wounds. In his *Antidotarium* (235) we find many traces of BRUNSCHWYGK's books and he was the first Englishman to use the word "still" for a cucurbit and alembic combined.

Before discussing other disciples of BRUNSCHWYGK we must turn to GIOVANBATTISTA DELLA PORTA, the Italian BRUNSCHWYGK, though he lived a full generation later. He was born at Naples in 1545 and travelled extensively through Spain, France and Italy, collecting specimens and making notes, after which he settled at his country-house near Naples. In 1603 he was appointed professor at the Academy of Naples, where he died on February 4th 1615. He was a great admirer of PARACELSUS' medical theory of the "signatures" but very sound as a student of physics. There is considerable confusion about his *Natural Magic*. In fact there appeared two works by PORTA called *Natural Magic*, the first in four books (Naples 1558) (493) and a second in twenty books (Naples 1569) (494). Both were very often reprinted up to the eighteenth century. The later *Natural Magic* was often published with his *De Destillatione* as an appendix, which latter work, in nine books, was also printed separately (Rome, 1608) (495) and even translated into German by PETRUS UFFENBACH in 1611. In the English translation the *De Destillatione* appears as an appendix too (496).

It is this *De Destillatione* which interests us most, for here PORTA deals with our subject as fully as BRUNSCHWYGK and LIÉBAUT. According to to him "distillation is running drop by drop and just because the alembic delivers the distillate little by little the operation is called thus". On the other hand he recalls the fact that other authors often mean catarrh when they speak of distillation: "Vidimus ex Avicenna in catarrho quomodo vapores ex corpore a naturali calore ad cerebrum diffisu sua frigiditate in aquam coguntu et diffugia quaerentes per narium canales corriventur et defluxes facientes quasi per alembicus rostrum experimentur" thereby comparing the human body with a distilling apparatus.

PORTA tried to explain the form of different distilling vessels by comparing them with animals (Figs. 24-27) and it is worth while to quote part of his discussion. "Both the vessel and the receiver must be considered according to the nature of the things to be distilled. For if they be of a flatulent nature and vaporous they will require large and low vessels and a more capacious receiver; or when the heat shall have to be raised up to the flatulent matter and that find itself straightened in the narrow cavities, it will seek some other vent and so tear the vessels to pieces (which will flie about with a great bounce and crack, not without endamaging the standers-by) and being at liberty will save itself from further harm. But if the things be hot and thin you must have vessels with a long and small neck. Things with a middle temper require vessels of a middle size. All which the industrious artificer may easily learn by the imitation of nature, who hath given angry and furious creatures as the lion and bear thick bodies and short necks; to show that flatulent humours would pass out of the vessels of a large bulk and the thicker part settle to the bottom; but then the stag, the estrich (ostrich), the camil-panther, gentle creatures and of thin spirits have slender bodies and long necks; to show that thin and subtile spirits must be drawn through a much longer and narrower passage and be elevated to purify them" (496).

Apart from the vessels illustrated he also mentions the "twins" or "double pelican" which he calls "Dyiotae" or "Amplexantes". After having discussed the definition and general meaning of distillation he treats of the distillation of herbs macerated with water in parti-cular. The third chapter is devoted to the distillation of aqua vitae from wine, for which operation he advises the use of a water-bath and a cucurbit with a long neck (like the hydra) or a long tube, such as he illustrates elsewhere in the work (Fig. 41). Here we see a long, air-cooled glass tube running from the cucurbit to a kind of condenser, where the heavy distillate will condense leaving the lighter fractions to run through a second air-cooled tube into a receiver or flask. The condenser shown in this picture is not a "reboiler" as some believe.

To purify the aqua vitae repeated distillations are necessary. The next chapter discusses the distillation by the sun's heat (Fig. 42).

Fig. 42. Destillatio per descensum by the sun's heat, PORTA

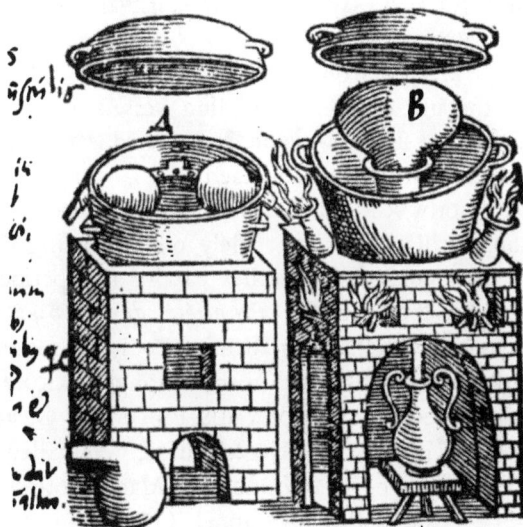

Fig. 41. Still with condenser and receiver, PORTA

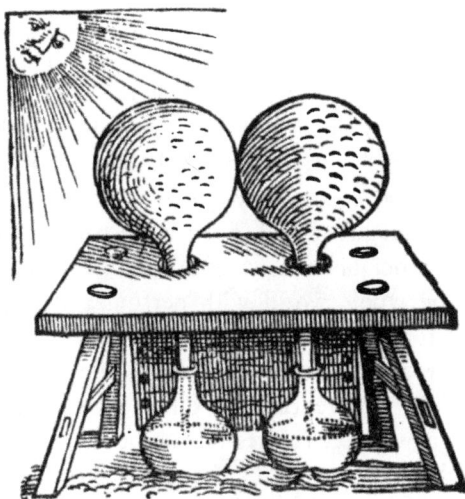

Fig. 43. Descensory in water-bath, PORTA

This is a type of "destillatio per descensum". The following chapters treat of the manufacture of vegetable oils, especially essential oils. This is done by pressing, by decoction with water and with a kind of descensorium in a water-bath which is shown in Fig. 43.

The whole remainder of the book is devoted to the preparation of essential oils by various methods though chapter XX treats of "aquis fortibus", of mineral acids and their preparation.

PORTA is a very good authority on the manufacture of essential oils in his period, because he had occasion to observe the industry in the neighbourhood of Naples, which as he says "could easily supply the whole world with perfumes and spices for the kitchen". To get strong perfumes he applies the "destillatio super recentiores flores irroratur" that is he redistills the essential oils over a fresh batch of flowers to enrich the oil and get better yields. PORTA is the first to give correct figures on the output of essential oils and the yield from different materials, though in later editions of his work these figures have often been changed by some amateur and give entirely faulty impressions of PORTA's exact figures.

In discussing the different types of distilling apparatus he tries to design the best apparatus to obtain an alcohol of arbitrary strength and thinks that a coiled glass raised over the stillhead is the best solution. With a "hydra" he hopes to get fractions of different strength at once, for as he says the "more spiritual parts ascend higher than the phlegmatic parts".

PORTA is well informed on the abuse of "distilled waters" and he warns against the use of metal distilling vessels as their distillates will cause diarrhoea. If aqua vitae is rectified seven times a pure alcohol is obtained, which he calls "aether".

Neither BRUNSCHWYGK nor PORTA gives any attention to water-cooling of the distillates, relying solely on air-cooling. This is not uncommon in the books of the earlier part of this period. Thus the pharmaceutical handbook called *Propriété des Choses* written by BARTHOLOMAEUS ANGLICUS (25) mentions only air-cooling too.

But the matter is fully discussed in the writings of CONRAD GESNER who is often referred to as EVONYMUS PHILIATER. GESNER, one of the most famous scientists of his generation, was called the "Pliny of Germany". He was born at Zürich on March 26th 1516, but as his father fell in the battle of Cappel (1531) he found himself destitute at an early age and had to take the position of famulus (assistent) to the theologian CAPITO at Strassburg. There he learnt

Latin and Greek, and gained a scholarship which permitted him to visit France (1533). In 1535 he took a teacher's post at Zürich and studied medicine, then became a professor of Greek philology at Geneva and gathered enough money to study for some time at Montpellier. This interrupted medical study was crowned by a graduation at Basel in 1541. He went to Zürich as a private physician and professor of philosophy to be highly honoured in his native town. In 1554 he became "Oberstadtarzt", in 1559 canon. He died on December 13th 1565 leaving the unfinished manuscript of a *Historia plantarum,* for which work he had already prepared many blocks and plates. 1500 of them were taken over by his publisher CASPAR WOLFF who later sold them to CAMERARIUS. We will follow their fate further on.

GESNER was a widely-travelled man and a profilic reader. His works contain an incredible number of references to personal experiments and to authors on his subjects. We are here especially interested in his *De remediis secretis,* a book in two parts, the last of which was published after GESNER's death by his publisher WOLFF from the finished manuscript (246). Both were often reprinted and translated into several languages, German (247), French (248) and English. In the latter language we first find the translation by PETER MORWYNG (MORWINGE) (1530?-1573?) who obtained his B.A. degree at Magdalen College, Oxford in 1550, of which college he became a fellow in 1552. As a Protestant he vainly tried to get his M.A. degree and was even expelled by a visiting bishop (1553). He then went to the continent and travelled in Germany until he was allowed to return in the reign of Queen ELIZABETH to get his M.A. degree at Oxford in 1561. Afterwards he was rector at different places and died as prebend of Pipa Minor in the cathedral of Lichfield. He translated several important works, among them the works of GESNER which he had collected when travelling in Germany. His translation reached two editions under different titles (249) (250). Another important translation of GESNER's works was prepared by GEORGE BAKER, court physician to Queen ELIZABETH and president of the College of Physicians in 1597. BAKER was the first to use the word "distillatour" (distiller) (251). The 1576 edition contains an "aqua vitae" which is usually attributed to ARNALD OF VILLANOVA but which is there given under GESNER's name. The fourth edition of the *New Jewell of Health* was published under an entirely different name *The practise of the new and old physicke* (252).

It is clear from these few lines that GESNER's works enjoyed a considerable popularity in many countries. It is therefore necessary to give some details on the contents of this *De Remediis secretis*.

The work begins with a short historical introduction which says that the Greeks and Romans could not yet distil and claims that the art was invented by the Barbarians, Carthaginians and Arabs shortly after the famous Hellenistic physicians. He gives a few short notes on Arabian scientists like MESUE, AVICENNA and BULCASIS, describing their methods of making rose-oil.

In the bibliography that follows he mentions AETIUS, ALBERTUS MAGNUS, ARNALD DE VILLANOVA, ABULCASIS, AVICENNA, EPIPHA-NIUS, GEBER, WALTHER RYFF, HIERONYMUS BRUNSCHWYGK, CAR-DANUS, JOHN OF RUPECISSA, NICHOLAS MYREPSUS, MATTHIOLUS,

Fig. 44. Destillatio ad latus, GESNER

ULSTADT, RAYMOND LULL, TRITHEMUS, RASIS, REMACLUS LYMBUR-GENSIS "who wrote of the distilled waters in common use", ROGER BACON "whose book on the twelve waters is often falsily attributed to Arnald de Villanova" and ADAM LONICER "who summarizes Brunschwygk and Ryff".

According to GESNER "destyllatyon not distillatyon (as lerned doe write) is the drawing forth of a thinner and purer humor out of a juice by virue of heat". Destillatio per ascens is the "distilling of a water when the vapours have been elevated and congealed". He also gives a description of the "destillatio ad latus" (Fig. 44).

After a short summary of the different distillation methods GESNER discusses the use of distillates in medicine and the use of the water-bath in preparing distillates of vegetable and animal matter. He gives a good description of the lead rings used for weighting the cucurbits (Fig. 45) and describes different types of water-baths (Fig. 46). After enumerating several recipes for simplicia he discusses the pro-

 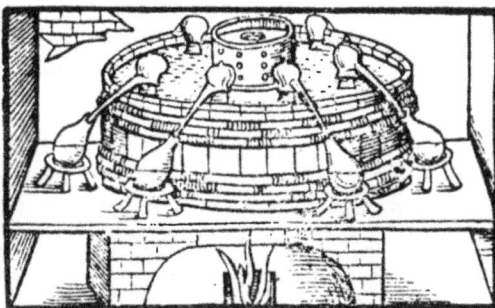

ig. 45. Lead rings for cucurbits in water-
bath, GESNER

Fig. 46. Water-bath on furnace, GESNER

Fig. 47b

Fig. 47a

Fig. 47c

Fig. 47d

Fig. 47a, b, c en d. Destillatio per fimum, GESNER

per material for cucurbits and other vessels. He does not favour metal and warns more particularly against the use of lead. Glass and earthenware are better, especially those made at Schaffhausen near Basel or better still those of Venice.

He uses a kind of fire-screen if he wants to look at the fire of his furnace to judge the temperature; this he calls "dioptra, escrime or eclipse". The construction of the furnaces and the preparations necessary before starting a distillation form the contents of a special chapter. Distillation by the sun's heat and with the help of horse dung are discussed in detail (Fig. 47). These are said to be the best form of heating when one prepares "subtile spirits" like perfumes. In this

Fig. 48. Bulcasis "woolcondenser", GESNER

chapter he mentions the "wool condenser" of ABULCASIS, which should be used to "purge and refine troubled waters" (Fig. 48). Special attention is given to the distillation of aqua vitae, by which word GESNER means all kinds of composita. Raw alcohol bears the name of aqua ardens in his book, he discusses it after having treated the principles of rectification and distillation "per filtrum". It is clear from Fig. 49 that he uses a cooling tube which runs with several bends through a tub of cold water. Elsewhere in his work he gives a simpler form consisting of a straight tube running either horizontally (Fig. 50) or diagonally (Fig. 51) through the cold water. There is no sign of any continuous cooling apparatus. He also applies the cooled still-head (Fig. 52) of which he gives a simplified form by depicting an alembic cooled by a water-filled bladder bound to it (Fig. 53).

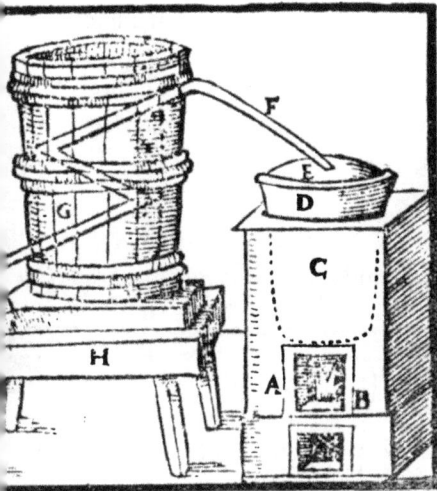

Fig. 49. Still with water-cooler, GESNER

Fig. 51. Water-cooler, GESNER

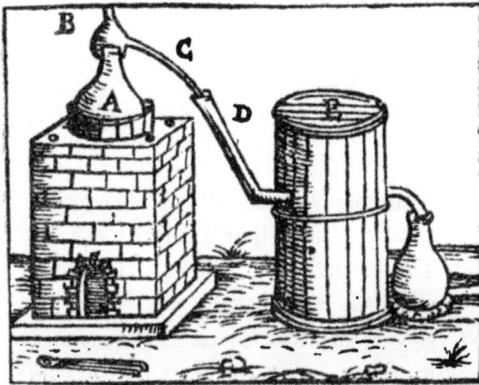

Fig. 50. Straight cooler, GESNER

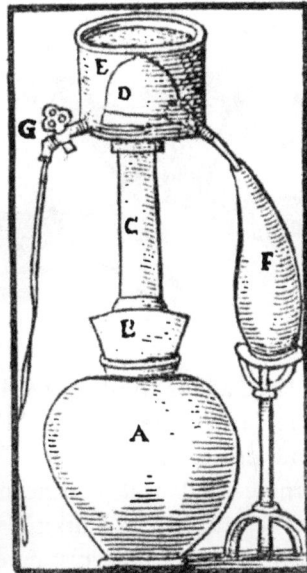

Fig. 52. Moor's head, GESNER

The virtues of aqua ardens and its simplicia are discussed in great detail using many arguments of VILLANOVA. Rectified alcohol is called "quinta essentia", different recipes for preparing it according to LULL, ULSTADT and RYFF are given and then he treats of "de aquae vitae compositis" and "de aquis destillatis compositis", giving a host of recipes from many quarters.

The second part of the *De Remediis secretis* is devoted to detailed study of the methods of distillation, furnaces, cucurbits, alembics, receivers and other instruments. Here he mentions a new type of distillation (Fig. 54) which introduces the principle of 'reboiling". These are his own words: "In the little furnace on the right, from which three flames are seen to emerge, pure sand is put and a fire

FIGVRA RARA QVAEDAM
Alchymiftarum, defumpta ex libro veteri
a lchymico manufcripto

Fig. 54. Circulation ("reboiling"), GESNER

Fig. 53. Simple Form of cooled still-head, GESNER

as large as the third degree lighted therein. On the other furnace that is in the middle one should put sand too and light a fire of the second degree as the flame from the furnace-door shows. The third furnace shall heat a balneum Mariae (or water-bath) and contain a very small fire." Working carefully with this apparatus one could obtain a fairly pure distillate could, in so far as the fires of the different furnaces could be regulated with care. Further recipes deal with the distillation of plants, herbs, animals, minerals,etc. in the course of which he gives a laboratory form of the "destillatio per descensum" (Fig. 55).

A contemporary of GESNER was VALERIUS CORDUS, born at Erfurt on February 18th 1515. He studied medicine and botany at Wittemberg and taught "Dioscorides" there. His uncle, the apothecary RALLA, advised him to publish his *Dispensatorium* (141), a pharmacopoeia, which became very popular in Saxony. It was printed at Nürnberg by orders of the Senate as the first pharmacopoeia. In 1542 CORDUS travelled to Italy and visited Padua, Bologna, Ferrara and Venice. Then he went on a botanical walking tour and reached Rome, after many detours, so exhausted that he died there on September 25th 1544. His commentary on the *Herbal of Dioscorides* (142) was re-edited by GESNER after its first successes. In the editions of Nürnberg 1540 and Frankfort 1549 the appendix on the distillation of vegetable oils is missing, but is however given in the editions prepared by GESNER. Here he gives details on the distillation of macerated herbs in glass cucurbits on a sand-bath. He also discusses the preparation of sculphuric acid from vitriol in a glass retort (143) which should be resting on a glazed tile to protect it against the effects of direct fire. The acid shall be re-distilled changing the receiver after the first distillate, which contains mostly water, has been recovered.

Somewhat less famous was the *Historia omnium aquarum* by REMACLE FUCHS (FUSCH, FUCHSIUS, REMACLE DE LIMBOURG) (217). FUCHS was born at Limburg in 1510, he studies at Liège, travels in Germany and studies medicine and natural science. In 1533 he becomes a canon in the place of his brother and dies at Liège on December 21st 1587. Apart from his book *On all waters* he wrote a *Methodus curandi morbi Gallici,* a nomenclature of plants and a *Lives of the Illustrous Physicians of the last century* (Paris 1541). His book on the waters gives nothing new as regards the apparatus, it merely contains many recipes of different distillates beginning with "de aqua absinthii", a type of book that is quite common in the sixteenth and seventeenth century and even several generations later.

The *Coelum Philosophorum* of PHILLIP (VON) ULSTADT is of more importance. ULSTADT was a physician and professor at Nürnberg who seems to have been intimately connected with BRUNSCHWYGK, as both profit freely from each other's works and use each other's blocks as illustrations. His *Coelum* was often reprinted (632) and translated into many languages (633) (634). It was probably written

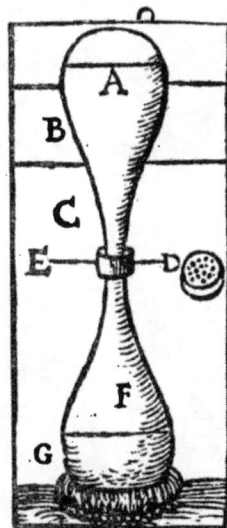

Fig. 55. Destillatio per descensum, GESNER

in 1525, and though the year of its first publication is usually given as 1528, the British Museum possesses a copy dated 1526. The work is mainly based on these of JOHN OF RUPECISSA, RAYMOND LULL,

Fig. 56 ULSTADT'S cooler

ARNALD OF VILLANOVA and ALBERTUS MAGNUS, apart from the strong influence of BRUNSCHWYGK'S works. He describes the different distillation methods which we have already met. His illustration of the still with cooling-coil (Fig. 56) was inserted in later editions of BRUNSCHWYGK's works and it seems that this was also the case with the picture of the Moor's head given in Fig. 30, which does not occur in BRUNSCHWYGK's earlier editions.

He tests alcohol by burning it on a piece of cloth where it should not leave any water after having been consumed nor burn up the linen. This shows that he did not make very strong alcohol. He gives many details of decoctions and distillates of herbs and plants with wine. He gives recipes for claret, hypocras, aqua vitae Frederici II and other spiced wines, which were very much in vogue in those days but which would not suit our taste nor our stomachs.

ULSTADT had some vague idea of dephlegmation in the stillhead, if we read the following passage: "Take a pure thin sponge and cut this up in pieces so large that they touch the neck of the cucurbit on all places (They shall be attached to a string so that they will not fall into the cucurbit). And the same sponges shall be immersed in cottonseed-oil beforehand and then pressed out a bit lest some of the oil fall into the cucurbit. And through this sponge the raised spirits of the aqua vitae simplicis are distilled very carefully and finely. And everything gross, unclean, earthy matter and substance will not be able to penetrate and pass the sponge because of the oil".

„Nimm einen reinen dünnen Schwamm und zerhaue denselbigen in soo grosze Stück, welche in der Grösse sygend, dasz sie für an allen Orten inwendig des Kolbens mogind anrüren (sie sollen mit Schnüren so befestigt werden das sie nicht in den Kolben fallen können) Und dieselbigen schwämm sollend vorher in baumöl gesetzt werden und danach wider ein wenig ausgedrückt, damit nicht etwan das baumöl in den Kolben herabtrieffe ...... Und durch den schwamm werden die aufgetriebenen geister der aquae vitae simplicis seer wäsentlich und fein destilliert. Als was grober, unreiner, yrdischer und ungedöuvter materie un substanz ist, mag von wägen des oels nicht durch den Schwamm gen und durchtringen."

In ULSTADT we find a good description how to cut glass after the operation is finished and the residue or "caput mortuum" is to be extracted from the cucurbit. "If you want to open it, wind round the neck of the glass six or seven times a thread covered with sulphur, then light it carefully with a wax candle and when the thread is burnt up, the glass breaks easily on the spot. Or have three or four iron instruments made, two elbows long, of which each has a ring on each end. Make one of these red hot and force it round the neck of the glass, then it will break and even snap off."

„Wenn du aber es öffnen wilt, so umbwickel den Halsz des Glasz mitt einem Schwefelfaden sechs oder siebenfältig herumgewunden, den zünde dann an hübschlich mitt einem Wachskerzlein und so der Faden gar auszgebrannt ist, so bricht das Glasz daselbst ab. Oder lasz dir drei oder vier eisne Instrument machen zweyer

elenbogen lang, deren jedes an jedem Ort zween ring hat. Aus welchem mach einen
dir glüend heisz und zwingen umb den Halsz des Glas so bricht es oder knallt
gar ab."

One of the most profilic German authors of this generation was
WALTHER HERMANN RYFF (RUIF, RUFF, RIFFUS, REIFF, GUALTHE-
RIUS H. RIVIUS). He was born at Strassburg but in 1539 we find
him as a physician at Mainz, then from 1549 to 1551 as town-physician at Nürnberg. He died before 1562. He was the editor of the
works of DIOSCORIDES, PLINY, ALBERTUS MAGNUS, RAYMOND LULL
and others and at the same time a most impudent plagiarist who
published German translations of works of LEONARD FUCHS, GESNER
and others under his own name. Of his own works the *Teutsche
Apotheck* (541) and the *Kleine Teutsche Apotheck oder Confect-
büchlein* (Strassburg, 1552, 1559, 1562) are perhaps the most famous,
but we are especially interested in his *New Gross Destillierbuch*
(542) in which he borrowed and copied freely from others without
ever blushing. Again the main substance of this book is the description of medicinal herbs and decoctions or distillates but in a chapter
"On the manifold force, virtue, properties and efficacy of aqua vitae
simplicis, that is brand-wine" he gives the recipes of sixteen "golden
waters" which are said to be "costly, wonderful, strong, very good,
etc.". In this chapter in which we find many examples of stills fitted
with worm-coolers (which illustrations are taken from LONICER!) we
read an interesting discussion on the relative merits of different cooling
systems.

A still fitted with an alembic with two spouts leading to two straight
cooling tubes (Fig. 61) is said to have little condensing power. For
volatile compounds one should cool well to avoid overheating and
burning of the vapours for which purpose one should use a coil
called serpentina (Schlangenrohr). These cooling coils receive the
heated vapours and lead them a long way through the water that
cools them (."durch sonderlich instrument recht digeriert oder gekühlt
und von den unmessigen hitz unnd verbrennung solcher geyster abge-
zogen werden, als nemlich mit den rören vilen krümmen wegen
Serpentina gennat, das ist Schlangenrohr. Solche rören empfahen die
erhitzigsten geyster des weins, so von der werme auffgetrieben werden und füren sie durch die vilen krumb lini, und wider durch das
Wasser, damit sie genügsamlicher gekült werden").

"To make a good still buy a copper or earthenware cucurbit and let
the potter make a good alembic to this flask glazed inside and outside.

The alembic should fit well on to the flask and no smell should be detected for the whole is built into the furnace. On top of the alembic there is a hole for the serpentina which shall fit closely and the coil shall be led through water that must be kept cold to cool the vapours quickly lest they should be overheated. These cooling coils (Fig. 65) have a vertical coil of a horizontal, but the most efficient coils have a vertical coil in a deep cooler. The French use "sideways" (horizontal) coils made of tinned iron which are led through wooden tubs as used for bathing children (Fig. 57)."

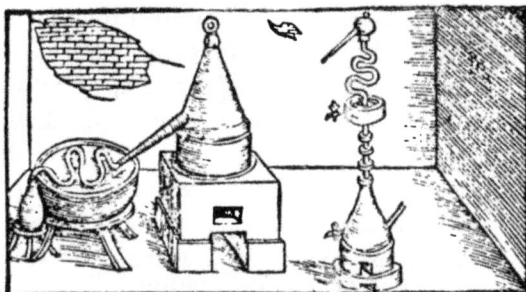

Fig. 57. Serpena cooler, : ᵣRINGUCCIO

„Schaff dir ein kupfferin oder irdin Kessel .. Auf diesen Kessel lasz dir bei einem Hafner einen Helm bereyten von guten 'ᵉn, innerhalb und ausserhalb wohl glasiert. Dieser Helm sol sich auff den obₑ ᵢnelte Kessel wohlschliessen in den absatz, also dasz es nit möge ausriechen den sₑᵢt du in aller masz einmauern, wie von den andern gemeynen öfen gesagt, darein nun ein Kessel gestellt wirt, dieser Helm sol oben ein Loch haben, darein du die rören oder Serpentina stecken, unnd auch auf das best vermachen mögest, welche Serpentin durch ein wasser gericht werden soll, das alle Zeit kalt sei, damit die geyster, so fast rein und subtil, und ganz leichtlich und verhitzigt und verbrannt werden, und underlasz külung und erquickung empfahen. Solche Serpentin magstu nach mancherley art und manier bereyten, also dasz die geyster undersich oder obersich getreiben werden. Aber diese hier nach gesetzte form und proportion bedunckt mich die bequemste in aller obgemelter Operation, die magstu also zurichten wie sie steht. Die Welschen brauchen ihre Serpentin nach der seit, bereytten einen gemeynen irdin oder kupfferin Distillierkessel, den stellen sie an allen ofen auff einem gemeynen dreufusz, under eine camin, stellen einen uberlengten hültzin zuber, wie mann hie zu land die kinder darinn zu baden pflegt darzu, in welchem die Serpentina eingefaszt allein von blechen rörlin gemacht, wie du solches verzeichnet siehst".

If one wants to separate "spirits" from "heavier vapours" he advises using a double-spouted Rosenhut: "They make this conical alembic higher and make a separate spout (and inner rim) so that the more subtile spirits which rise higher will find a separate exit to the receiver, which "water" is much stronger and more subtile than that collected from the under spout".

„Sie füren den spitzigen Helm in obgemelter proportion höher hinauf, solche
höhe verordenen sie einen sonderlichen absatz, der die subtilen geist, so etwas höher
hinauf steigen um sich daselbst zu resolvieren, empfahn, und durch einen sonderlichen Ausgang hinweg führn zu der versammlung, welches wasser vil subtiler und
krefftiger wann das underst, so vom undern schnable gesammelt".

Here again we see an early appreciation of proper rectification.

If we now turn to the Italian adepts of the art we must first
mention HIERONYMUS CARDANUS (GIROLAMO CARDANO), born at
Pavia in 1501. He studied philosophy and pharmacy at Venice and
medicine at Padua (1525), where he became rector of the university
(1526). CARDANO was a genius, who travelled through France, England, Scotland and Germany and then settled in Milan, Bologna and
finally in Rome where he died in 1596 leaving no less than 222 books
from his own hand. His *De subtilitate* discusses distillation methods
in the second book *On the elements* and in the sixth book *On
the plants* without introducing much new matter. GESNER, however, took over his ideas on distillation in horse dung and RUBEUS
borrowed one of his illustrations of an alembic.

Of more importance are the details on distillation given by VANOC-
CIO BIRINGUCCIO, the technologist. He was born at Siena on October 20th 1480. By patronage of the local dictator PETRUCCI he became
administrator of the silver-mines in the Avanzo-mountains and in
1513 the arsenal was entrusted to his care. But he had to fly with
PETRUCCI and his followers because the town-jeweller and he had
faked coins. In 1516 he was outlawed and in the next year we find
him in Rome, Napels and Sicily. In 1523 he returned with the PE-
TRUCCIS and was entrusted with the manufacture of nitre, but had to
fly again within the same year. Then he joined PETRUCCI in the siege
of Siena, commanding the artillery, but was beaten off and had to
wander for another three years. In 1530 peace was restored and
BIRINGUCCIO returned to Siena being entrusted among others, with
the building of the cathedral. In 1538 he was called to Rome by the
pope as manager of the papal foundries, but he died before April 30th
1539. BIRINGUCCIO became famous as the author of the *Pirotechnia,*
a book often edited and translated (72) (73). Most of his material
is taken from German authors like AGRICOLA, but as regards distillation he has freely used PORTA and BRUNSCHWYGK.

This chapter on distillation is rather lost in the Ninth Book
(chapter 2), and is called "the art of distillation in general the
methods of making waters, oils and sublimates". Some of his information comes directly from GEBER, especially that on the manufacture

of acids, which we will discuss later on. BIRINGUCCIO supposes that by distillation one obtains the "airy and watery parts", the composition of which depends on the base material. Some give no distillate at all nor any sublimate. The quality of the distillate also depends on the apparatus and the heat of the fire. After driving off the "water" the rementum remains from which an "oil" can be distilled (containing more of the "fiery parts and more of the spirit that is the essence of the matter distilled"). The residue is called "caput mortuum". Some say that they separate something different from the four elements, the "quintessence", but BIRINGUCCIO doubts this as much as the Philosopher's Stone.

He then describes the alembic called "campana" (bell) which is nothing but the Rosenhut. For heating purposes ash-, sand- and dung-baths are used. The cucurbits are made of earthenware or tinned copper. They shall only be exposed to direct fire. They are strong enough. There are two types of heating; the first, "hot and dry" permits three degrees of heating. The second, "hot and moist" is the water-bath.

The distillation with dung can be speeded up by introducing steam into the dung, or mixing the latter with quicklime, sawdust, lees of wine or aiding putrification by reflecting the sun's heat onto it with a mirror. The distillation of alcohol from wine is achieved with a Rosenhut and a flat horizontal cooling coil (Fig. 57) or with a Moor's head type of cooled still-head. To refine the "spirit" it is circulated in such vessels as pelicans and here BIRINGUCCIO borrows a woodcut from BRUNSCHWYGK's book. Oils are usually distilled in retorts in a furnace, the distillation is continued until a black or coloured smoke is noticed. Many oils are better prepared by distilling "per descensum" or by "subliming" in a Rosenhut.

He then describes different types of furnaces and distillation towers combining many cucurbits (also called "galley" or "gallery" ovens). Then he suddenly breaks off and goes on in the next chapter with the discussion of a totally different subject, coinage.

Similar distillation towers are described by PETRUS ANDREAS MAT-THIOLUS (PIERO ANDREA MATTIOLI), born at Siena on March 12th 1500. He studied and graduated at Padua (1523) and practised medicine at Siena, Perugia, Rome (1527) and finally at Triente. In 1540 we find him at Görz, then he became court physician to the Emperors FERDINAND and MAXIMILIAN II. He was knighted in 1562 and left the Imperial service, travelling through Tirol and dying of the plague at Triente (1572). His commentary of DIOSCORIDES'

Fig. 58. Galley furnace, MATTHIOLUS

Herbal was often printed (416) and translated (418). The German editions often contain illustrations made from blocks originally in possession of GESNER and sold after his death. This Commentary contains an appendix on Distillation, which we also find in editions of the complete works of MATTHIOLUS (417). Here he defines distillation as "an Art and means by which one extracts the liquid or humidity of things by virtue of fire or similar source of heat. It is

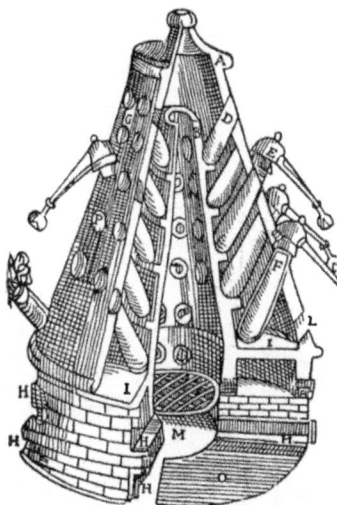

Fig. 59. Cut of sixteenth century
mass-oven for distillation

true, that the word distilling is often used in a wider sense including the things that distill without heat (meaning filtration)." The appendix is illustrated with six woodcuts the first and sixth of which are pratically identical. The first depicts a "galley-oven" (Fig. 58), a cut of such an oven is given in Fig. 59. The second gives a curious picture of a water-bath on a furnace with a cooled still-head. Both still-head and water-bath are fed discontinuously from reservoirs in the wall, through which the chimney runs! (Fig. 60). Of the further figures only the fourth has some interest for us, as it depicts a still with a water-cooled straight delivery tube running into the receiver (Fig. 61).

A famous iatrochemist like ANDREAS CAESALPINUS (1519-1603) uses distillation very often in his numerous experiments, but has very little to tell as regards improvements of new methods (109).

There is more to be gleaned from the works of HIERONYMUS RUBEUS who wrote a special book on distillation (524). GERONIMO

Fig. 60. Cooling a still-head, MATTHIOLUS

DE ROSSI was born at Ravenna in 1539. By his studiousness he attracted the attention of a high official in the order of the Carmelites, who was a member of his family and who offered to guide and to pay for his studies. He graduated at the university of Padua in philosophy and medicine and became court-physician to pope CLEMENT VIII. Later

Fig. 61. Still with cooler, MATTHIOLUS

he retired to Ravenna where he wrote a commentary on CELSUS and his book on distillation, dying in 1607.

He states that the Greeks did not know distillation but that in his time noble amateurs such as COSIMO DE MEDICI, the dukes of Ferrara and many Austrian princes were not ashamed to distil vegetable oils, aquae vitae, essential oils, etc. He too includes filtration under the head "distillation". One cools the distilling vapours in a coil or "by raising the copper cucurbit with a glass alembic", but he prefers the first method which is able to give a pure aqua vitae in cne distillation. But aqua ardens is only obtained by repeated distillations or rectifications.

Among the most important pupils of GESNER we must mention ADAM LONICER (LONITZER). He was the son of a theologian and philologist born at Marburg on October 10th 1528. He obtained his Magister degree, after studying at Marburg and Mainz, at the early age of sixteen and pushed his medical studies further while teaching mathematics at the university of Marburg (1553). He obtained his M.D. in 1554 and became town physician at Frankfort where he died on May 29th 1586. He wrote a *Natural History* (384) which gives some information on distillation but discussed this subject in greater detail in his *Herbarium* (385) which was soon translated into German (386) and often re-edited with additions from other writers on the subject (387) (388). In this book we find all the illustrations which GESNER had prepared for his unfinished *Historia plantarum*. For LONICER married MAGDALENA EGENOLFF, the daughter of the famous publisher of Frankfort. EGENOLFF had often used other books to edit new editions or books. Thus he was involved in a difficult lawsuit because he had copied BRUNSCHWYGK's book and illustrations as an appendix to the *Kreutterbuch* of JOHANN CUBA and edited by EUCHARIUS RÖSZLIN, a friend of LONICER, BRUNSCHWYGK and GESNER. CHRISTIAN EGENOLFF lost his law-suit against his competitor, the editor SCHOTT, who seems to have obtained something like a copyright, for the latter published a second edition of the book in 1535. At Marburg EGENOLFF had printed a new edition of DIOSCORIDES' *Herbal* (1543) which was severely criticised by WALTHER RYFF and defended by EGENOLFF's friend FUCHS. EGENOLFF then took a shrewd step and engaged RYFF as his adviser for medical editions, publishing RYFF's *New Gross Destillierbuch* with the cuts which he had obtained from GESNER's estate. He was LONICER's editor before he became his father-in-law. His firm published no less than 4 editions of LONICER's *Kräuterbuch* between the years 1557 and 1577, for LONICER became its director after the death of EGENOLFF (1555). Though we can thus trace the origin of the woodcuts belonging to the *Herbal*, it is very difficult to find out whence the illustrations of the distilling apparatus came, as writers borrowed freely from each other's works and there was no official copyright in those days (565).

LONICER's *Herbal* states that apart from the classical authors JEAN RUELLE, VALERIUS CORDUS, PETRUS MATTHIOLUS, HIERONYMUS BRUNSCHWYGK and CONRAD GESNER have supplied the author with much of the information presented. His motive for writing this book was the fact that the use of distilled waters from herbs and plants is so common now, that everybody studies and applies distillation. That is why he gives a special appendix on distillation in this book, which shows the usual picture of a botanical garden with distillation apparatus on its title page.

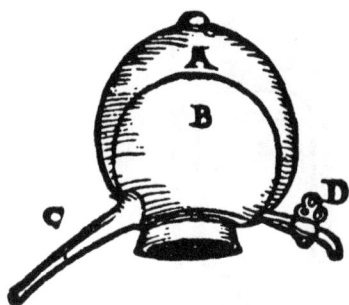

Fig. 62. Discontinuous still-head   Fig. 63. Rosenhut with two spouts,
          cooling, Lonicer   Lonicer

This invention is not old, as the Greek and Latin doctors know nothing about it, but it was made by the Arabians, Geber being the first to give details. Even nature distils when it rains and he too compares the human body attacked by a catarrh to a still:

"By the heat of the liver the inner vapours rise by natural or external heat from the cucurbit of the stomach into the head which functions as a still-head where driven close together by the cold these vapours fall from the nose, acting as a spout, in the form of slime, driven out by sneezing and coughing".

He starts his description by giving details on the correct construction of still and furnace. Good earth such as that from which the jeweller prepares cupels should be used, mixed with salt and potsherds of old cupels. The distilling apparatus should be made of copper and well tinned. Different furnaces are described for use with an ash-, sand- or water-bath. He gives two simple forms of Moor's heads, both cooled

Fig. 64. Double straight-tube condenser, LONICER

with water (discontinuously) (Fig. 62). He also mentions the double Rosenhut (Fig. 63) as a means of obtaining two fractions with one fire and one still. Several types of stills fitted with cooling coils are given; the less efficient still ("It is too weak to use with so much liquid as the contents of the flask") with two spouts and two condensing

Fig. 65. Cooling coil, LONICER

tubes (Fig. 64) is given together with several "vertical coil" condensers (Figs. 65 & 66) and in a larger picture he delineates a cylindrical cooling column filled with water through which a cooling coil ascends! (Fig. 67). For distilling in a water-bath he advises resting the cucurbits on a metal sheet with holes through which the steam escapes. One can also heat the water by immersing a ball composed of one part of quicklime, half a part of sulphur, ¼ part of nitre and ⅛ of alumn, a recipe which he drew from RYFF.

After describing the distillation of wine with alembics he discusses how one should fit lead or earthenware alembics on cucurbits, for the former give the best aquae vitae. The description of the galley-ovens and water-baths of different types fills many pages. Finally he describes the curious apparatus which PORTA gives too (Fig. 41). LONICER's

Fig. 66. Worm-condenser, LONICER

picture is practically the same, except for the fact, that the final distillate runs through a water-cooler before entering the receiver (Fig. 68). He gives a lengthy explanation in the course of which he states that the cucurbit is half filled with water and the powdered herbs are added, then the fire is lighted and the distillation is finished after three to four hours. The oil that swims on top of the water in the receiver can be separated by means of a siphon (which he describes too). This means that he collects the essential oils in concentrated form.

„Man bereitet einen gemeinen Destillierofen, wie zu einem einfachen balneo Mariae pflegt gemacht zu werden, darin setze man ein kupffern Blase so gros ist dasz sie ein gemeine Masz oder sechs halt. Solcher Blase Hals und Mund sol oben handbreit weit seyn und ber den Ofen heŕausgehen. Solcher Hut sol oben ein Rörlin haben so eines Fingers dick weit ist und eines halben Fingers lang über sich geht. Darin stecket man die blachen rören so uff die art wie folgende Figur ausweiset, bereitet sein, dasz sie gehet in ein andern kupffern Kolben, so auch

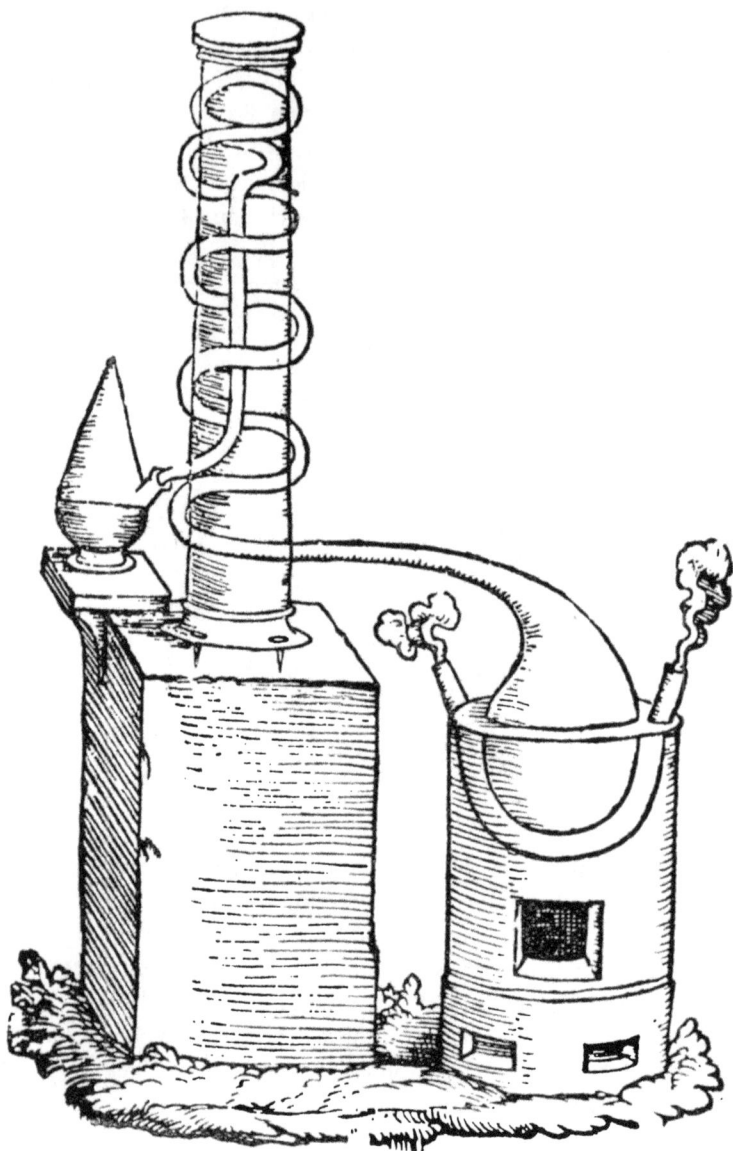

Fig. 67. Cooling column, LONICER

einen Hut mit einem Rörlin hat oben. Darauf setzet man eine andere, auch dergleichen rören oder serpena welche durch ein Vasz, in ein fürlegerglasz, darin die destillierte Materie fliesset ausgehet. So man nun von Gewürzen oder von Samen die Olea distillieren will, sol man die kupffern Blase so in den Brennofen steckt voll

Fig. 68. Still with condenser and receiver, LONICER

Brunnenwasser füllen und darnach die Gewürze oder Samen davon man das Oel abziehen will so zerstosen derselben ein Pfund oder zwei darin tun. Die Instrument oder Rören an allen Orten da sie zusammen gesteckt werden, wol gehabt mit Ochsenblasen und Meel verwaren und das Feuer undermachen. Erstlich sanfft und danach je lenger je hefftiger regieren. Solche destillation gehet geschwind naher in drei oder vier Stunden. Wann nun die beste Spiritus herausgeflossen und abgelauffen sein sol man das Oleum so oben in das Glas schwimmt sauber davon in ein besonder Gläszlein absondern."

Next we read of different chemical vessels used for distillation, circulation, putrification, digestion, etc. and the means of cutting and fusing glass. The following chapters discuss the correct way of collecting herbs and the season in which they should be collected and distilled, together with details on the conservation of the essential oils.

The last part of the book gives the herbs, the waters distilled from them and their properties.

As to distillation in general he says that the first fraction contains "humidity" (Wässerigkeit), then "phlegmatic humidity" (Ungedäwte Phlegmatische Feuchtigkeit), "wet, delicate, subtile matter" (nassgedäwte, zarter und subtiler Materi), "finest oiliness" (Feynste Olität) and finally the fifth water "quintessentia" distill over.

Many descriptions of distillations of herbs and plants are found in the *Hortus medicus philosophicus* by JOACHIM CAMERARIUS (110). KAMMERMEISTER was born at Nürnberg on November 6th 1534 and studied at Wittenberg, where he lived with MELANCHTON the reformer, Leipzig and Breslau. He obtained his M.D. at Bologna (1562) and settled at Nürnberg (1564) where he founded the Collegium Medicum (1592). He died there on October 11th 1598. CAMERARIUS was not only a good physician but also a famous botanist. For his *Hortus* he bought GESNER's blocks and manuscripts from the editor CASPAR WOLFF at Zürich and he also used them for this edition of MATTHIOLUS' Commentary on Dioscorides. His *Hortus*, which is practically a catalogue of his own botanical garden unfortunately contains no illustrations of distilling apparatus. His book was also published in German (*Kräuterbuch*, Frankfurt, 1600).

We must now turn to a group of French authors who tried to teach distillation methods to the farmers of France, who, until then had not distilled their own liqueurs or brandy.

The oldest French book for the small farmer was the *Theatre d'agriculture et Message des Champs* written by OLIVER DE SERRES, seigneur de Pradel. The author of this agricultural handbook was born at Pradel in 1539. As he became a Protestant like his father he had to fly to Geneva. He lived for some time in Switzerland and Germany where he married in 1559 and where he probably studied the new distillation methods. He was recalled to France by HENRY IV to start the silk-culture in this country. After the publication of his *Cueillette de soie* (1599) he was appointed director of the new industry. In 1600 his book on agriculture was published (571) which was often reprinted in the sixteenth and seventeenth centuries. He died at Pradel on July 2nd 1619. In this book on agriculture he advises the use of earthenware or glass alembics as metal apparatus will give a bad smell to the distillates. As the correct degree of heating is difficult to realise he advises the use of a water-bath. He then gives recipes for the distillation of plants and herbs by "destillation per ascens", "destillatio per descensum" or by the sun's heat. Though he talks of the grape and viticulture he does not mention aqua ardens, which in his time was still the product of the apothecary or distiller. When he speaks of "quinta essentia" he means perfume or essential oils.

Another typical specimen of this group of handbooks is *L'agriculture et maison rustique* written by CHARLES ESTIENNE and JEAN LIÉBAUT. ESTIENNE was born at Paris in the beginning of the sixteenth century. He belonged to a Lutheran printer's family and obtained his M.D. in 1542, but as his brothers were taken prisoner as heretics he tried to continue their business (1551), but was himself imprisoned, dying soon afterwards (1564). JEAN LIÉBAUT was his son-in-law who had studied at Paris and who was a most successful doctor. He died at Paris on June 21st 1596.

This book on agriculture contains a chapter on distillation on "the art of distilling waters and the oils and quintessences from all kinds of rural base materials" giving figures of the common types of stills. We can read from this book that even in 1647 but little brandy and liqueurs were made on the farm itself.

But LIÉBAUT, the author of a *Thesaurus sanitatis* (367) gives much more information on distillation in his *Four Books of Secrets of Medicine and Philosophical Chymistry* (368). In this book he borrows freely from such German writers as GESNER, LONICER, etc. and also from PORTA and BIRINGUCCIO. His book and the French translation of GESNER's work are the oldest full handbooks on distillation in the French language. LIÉBAUT describes himself to be inspired by the French edition of the second part of GESNER's book "which was translated by Gaspard Wolphe, a German physician" (he probably means CASPAR WOLFF, GESNER's editor) to "give the apothecaries a taste of distilling and stimulate them to be more and more careful in preparing their medicines".

He defines distillation as "a conversion and recasting of the subtilised humour, which first is a vapour by virtue of the heat, then is condensed and thickened by the cold... In an earthenware, glass or tinned copper vessel, placed on a furnace the ardour of the fire raised from the heated substances a vapour, which is then condensed and thickened by the cold in the vessel which we have named capital, by the air which surrounds it, and is converted into a liquid which drop by drop runs into a flask suspended under the beak of the capital; this we call distillation." Other forms of heat are the sun or putrifying matter. "But there are also other ways of distilling which are often used such as "by the felt" (filtration), "by the sponge" and "by pressing", but of these we will talk only if necessary."

Fig. 69. Solis destillatio, LIÉBAUT

Fig. 70. Continuously cooled Moor's head, LIÉBAUT

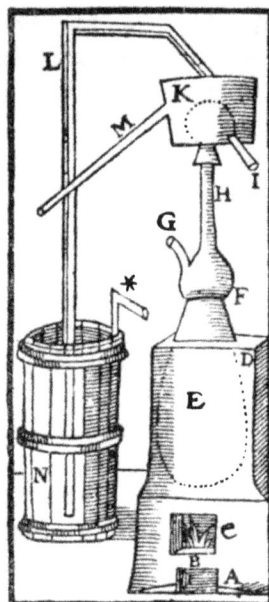

Fig. 71. Water-bath, LIÉBAUT

He then discusses the different vessels and apparatus, saying that the cucurbits should be long to obtain good distillates. Different forms of the Moor's head are given of which we reproduce one "seen at Florence" (Fig. 70) in which the cooling water is pumped from a tun of fresh water and which is said to use only little cooling water. His "solis destillatio" is taken from LONICER (Fig. 69). He also describes different forms of water-bath distillation, one of which shows

(Fig. 71) a water-bath which is kept full of boiling water by supplies from a pot of water placed on a furnace, a kind of indirect heating. Some forms are especially designed for the production of alcohol from wine. He gives an interesting picture of the dung-bath with steam heating (fig. 72) "taken from the author of the Pyrotechnia" that is from BIRINGUCCIO. His "ice-distillation" is nothing but cooling a solution or mixture for the sake of congealing the larger amount of diluent. His "galley-ovens of the Saracens" is taken from RYFF and he gives a strange combination of air- and water-cooling combined in a picture (fig. 73) taken from GESNER. Finally we must mention a "most rare figure of the alchemists taken from an alchemist's

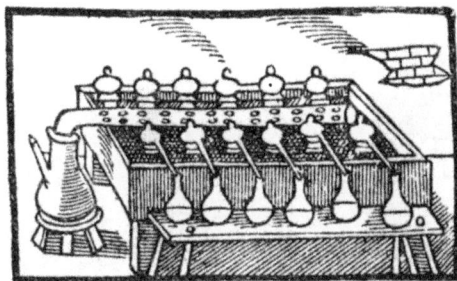

Fig. 72. Steam-heated dung-bath, LIÉBAUT

Fig. 73. Serpena cooling with reflux, LIÉBAUT

book written by hand" which is nothing but the distilling apparatus for essential oils fitted with a condenser between still and receiver such as we met in the works of PORTA and LONICER. Finally the first book describes the "Lute of the Philosophers", "the "destillatio per descensum", filtration and rectification. The second book is devoted to the "waters", the third to the oils distilled from vegetable matter, animal matter and minerals, giving also recipes for the preparation of vinegar, "oil of vitriol" (sulphuric acid), etc. The fourth book discusses the aquae vitae composita and simplicia and the virtues and defects of aqua ardens.

Another interpreter of German distillation methods was BLAISE DE VIGENÈRE, born at St. Pourcain in the Bourbonnais (1522) and secretary to the famous BAYARD at the age of eighteen. After BAYARD's death in the battle of Pavia he travelled through Germany and attended to the Diet of Worms (1545) and in 1547 became a secretary to the Duke of Nevers, accompanying HENRY III to Poland. He died at Paris in 1596. VIGENÈRE was a profilic reader and a remarkable classical philologist. The different distillation methods which we have met are all discussed in his book (642).

JACQUES BESSON is the author of a book *On the extraction of oils, waters and medicines* (*simplicia*) (70), about whose life nothing is known. Perhaps he is identical with the French mathematician, born in the sixteenth century at Grenoble, who taught mathematics at Orlé-

Fig. 74. Water-cooled
Moor's head, BESSON

Fig. 75. Copper tube-condenser, BESSON

ans and who was very interested in mineral waters and their natural sources. His book, which was also later printed as an appendix to LIBAVIUS' *Praxis Alchymiae* (Frankfort, 1604) discusses the proper form and size of the furnaces, the different types of stills and still-heads (Fig. 74), etc. He mentions a cooler of the tube-type made of copper and soldered to the spout of the alembic (Fig. 75). The other vessels are of the familiar types.

Similar methods were propagated by CLAUDE DARIOT, a pupil of PARACELSUS. DARIOT was born at Pommard (1533) and he studied medicine under RONDELET and SAPORTA at Montpellier. He became town-physician of Beaulne and died at Dijon in 1594. Inspired by PARACELSUS he uses distillation to prepare the "quintessentia" or the purest substances from plant, animal or mineral or as he says in his *Grande Chirurgie* (154) and his earlier *Preparation of medicines*

Fig. 76. Still with Moor's head and condenser, Dariot

(153) that one should give one's patients "only the pure substances separated from all impurities". This is why he describes distilling apparatus in the second of the *Trois Traités* appended to his *Grande Chirurgie* in which he says that his cousins ETIENNE PERUCHOT and CLAUDE PÉRARD both apothecaries at Dijon, have used them with succes. Wine and its spirit are most proper to extract the "faculty of the medicines" because of their subtility. In the German translation

Fig. 77. Still for gums and resins with steam injection, DARIOT

of his work (155) and its later edition with many additions (156) he remarks that the distilling apparatus should be made of copper, but that there is no need to discuss them in detail as everybody in Germany uses them to make aqua vitae. The normal type of still recommended by him is shown in Fig. 76. Here again we notice that the manner of replenishing the cooling water is entirely wrong, though the principle of continuous cooling seems to be understood. The still is used to distill herbs with water, aqua vitae or bad wine. For the distillation of wood and similar materials a descensory should be used, but gums and resins are distilled in a retort, in an apparatus shown in Fig. 77 in which one still contains water and the other the gum, or in the still already discussed. A third type of still is recommended for compounds containing both light and heavy fractions (Fig. 78).

One of the worst types of Paracelsist was LEONHART THURNEISSER ZUM THURN who was born at Basel on August 6th 1531. He was an apprentice in his father's jeweller's shop but embezzled money

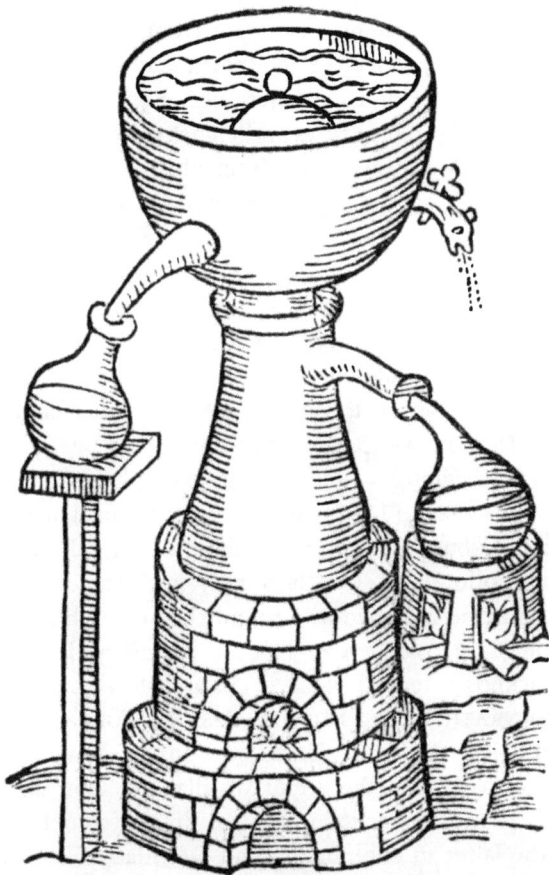

Fig. 78. Still for two fractions, DARIOT

and fled in 1548 spending some years in France, England and Germany where he became a soldier under the Markgrave of Brandenburg. From 1553 to 1555 he works in the mines of Tirol, then becomes a jeweller first at Strassburg then at Konstanz. In 1558 he buys sulphur mines in the Tirol and attracts the attention of the Archiduke FERDINAND who sends him as an expert to study the mining conditions in the Orkneys (1560) and later (1561) to Spain, Portugal, North Africa and the Orient whence he returns to the Tirol in 1565 to travel in Bohemia and Hungary in 1568. When looking for artists to cut the wood-blocks for his works he meets the Elector JOHANN GEORG of Brandenburg at Frankfort and cures his wife. He then gets the position of court physician at Berlin (1578) and grows very rich but his third wife is so extravagant that he has to fly from his creditors (1584). He travels to Rome, is converted to Roman Catholicism and wanders as a "goldmaker" to die in a monastery near Rome on July 9th 1596. His works are full of small pictures and vignettes of distilling apparatus (624) (625) (626) mostly copied from older MATTHIOLUS editions but some new blocks were designed for him by the goldsmith FRANZ FRIEDERICH of Frankfort and cut by the artist PETER HILLE from that same town.

A Paracelsist of the better type was ANDREAS OF LIBAU better known as ANDREAS LIBAVIUS. He was born about 1546 at Halle and studied medicine and philosophy at Jena where he also became "Poetas laureatas". Then he was teacher at Ilmenau (1581) and Coburg (1586). From 1588 to 1591 he was professor of history and poetry at Jena, but then went to Rothenburg where he was town-physician (1591) and inspector of the schools (1592) until he was nominated rector of the Latin school at Coburg (1607) where he died on July 25th 1616.

He did not follow PARACELSUS blindly but often voiced his own opinion. He must be regarded as a sound chemist with much common sense and one of the pioneers of modern chemistry. In his *Alchymia* (362) he acknowledges LONICER, PEDEMONTANUS, VILLANOVA, GESNER, KHUNRATH, RYFF, RUBEUS, BESSON, PORTA, ANDERNACH, BRUNSCHWYGK, RUPECISSA, ULSTADT, LULL, SAVONAROLA and CORDUS as his masters in the art of distillation. This art belongs to the Encheria, group Extractio, which he divides into Expressio and Prolactatio, the latter in Sublimation and Destillatio. The first part of the *Alchymia* discusses the operations, the second part the furnaces and instruments, but here we find nothing new or startling. In the *Commentarium Alchymiae* the technological methods are given which are mostly derived from AGRICOLA and other technological authors.

Pars I. Liber I. caput XXXIV gives details on the still-heads, phials, etc. and Pars I. Liber IV. caput VII the distilling apparatus for all kinds of purposes.

His *Praxis Alchymia* (364) too begins with an extensive treatment of "de destillationibus et extractionibus". Though he theorizes on cooling he has nothing definite to go on but the idea that, in the case of volatile spirits, the course through the air traced by the cooling tube ascending from the alembic should be as long as possible. For slow distillation he recommends heating with candles or oil-lamps.

The eighth and last book of the *Syntagma* (365) is concerned with the extraction of gums, resins, balsams, herbs and minerals. Their "quintessence", "elixir" or "tincture" is distilled by the well-known methods. For the laboratory he designed all kinds of retorts and glass apparatus (Fig. 79), several of which are able to give different fractions at a time. A fractionating tower with separate cucurbits built into the furnace is shown in Fig. 80. He also depicts the still

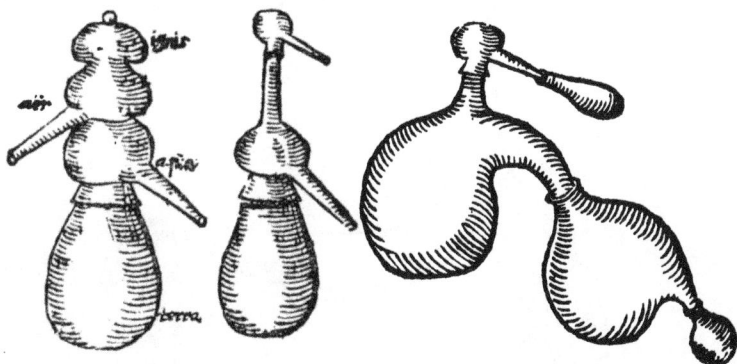

Fig. 79. Glass cucurbits, LIBAVIUS

and the dephlegmator of PORTA (Fig. 81) and gives two drawings of stills with double water-coolers which in one case cool the two fractions obtained (Figs. 82 & 83). Finally he also mentions the "reboiling" system, which we have already discussed when reviewing the work of GESNER. It is here given in a more logical and efficient form (Fig. 84).

The KHUNRATH mentioned by LIBAVIUS is CONRADUS KHUNRATH, a chemist from Leipzig and brother of the famous alchemist HEINRICH KHUNRATH. He lived for some time in Holstein and Schleswig. His *Medulla* (325) was translated into German in 1680 (Leipzig). The first book treats of the distillation of wine, lees of wine and of all kinds of snakes, honey pearls, amber and other ingredients distilled with wine. He generally used a simple cucurbit with an alembic and receiver cooled with wet rags! The second book discusses the method of obtaining brandy from corn, rye, etc. with malt and the use of this alcohol to make aquae vitae. His "waters" include syrups and decoctions. But he invented no new apparatus and the pictures taken

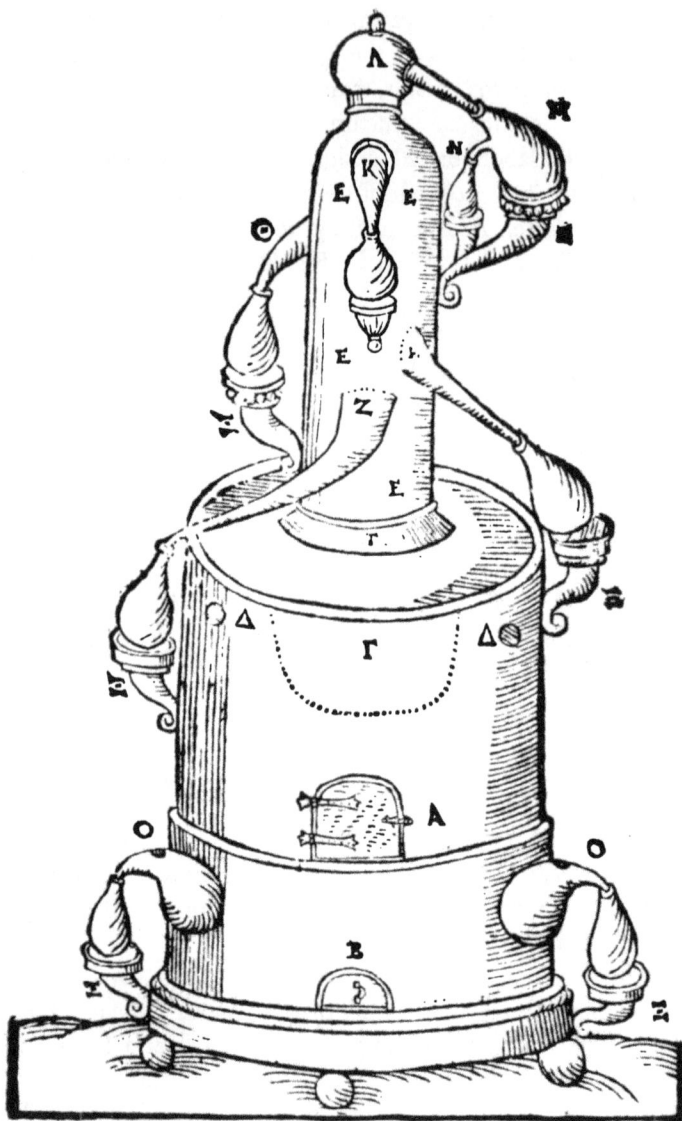

Fig. 80. Fractionating tower, LIBAVIUS

Fig. 81. Still with condenser and cooler, LIBAVIUS

Fig. 82. Still with two cooled streams, LIBAVIUS

Fig. 83. Still with condenser and double cooler, LIBAVIUS

from his work (Figs. 85 & 86) which hardly differ from those in other alchemical works prove this.

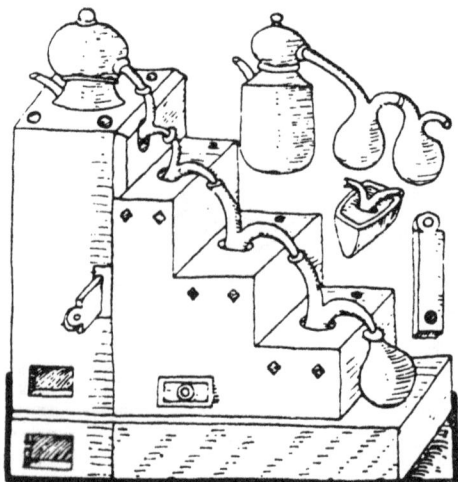

Fig. 84. Reboiling stills, LIBAVIUS

Another fervent Paracelsist and figure as bombastic as the Master himself was JOSEPH DU CHESNE, also called Quercetanus. He was born at Armagnac (1546), but visited German universites and obtained his degree at Basel. Then he became a burgher of Geneva (1584), where he was chosen as a member of the Council of 200 and sent on many diplomatic missions. In

Fig. 85. Alchemical plate from KHUNRATH's work

Fig. 86. Alchemical plate from KHUNRATH's work

1593 he was called to Paris as a court-physician to HENRY IV, where he died in 1609.

He is the first chemist expressily to mention Cohobatio side by side with Exaltatio, Exhalatio, Circulatio and Destillatio as one of the major operations of this kind. By cohobation these alchemists meant repeated distillations, each time returning the distillate to the residue ("faeces") to get the highest possible yield of light fractions or to "edulcoriere" (sweeten) or "elciere" (lixiviate) the "spirit" desired. These words are frequently used by subsequent generations of chemists until the end of the eighteenth century (499) (500).

Neither does the *Basilia chymica* of the Paracelsist OSWALD CROLL (147) contain anything new, it is not even illustrated but it is worth mentioning because it contains a table of symbols many of which relate to distillation and some of which have been inserted in our Table II. CROLL (1580-1609) was physician to the Prince CHRISTIAN of Anhalt-Bernburg.

ALEXIUS RUSCELLO PEDEMONTANUS was a rather myterious figure. Tradition has it that he was a noble Piedmontese who worked for more than 57 years as a physician and finally died at the age of 83 after he had written his book *On the secrets* (528) because somebody died of the "stone" who according to him could have been saved quite easily by some simple medicine. His book was very often reprinted and it was translated into German by JOH. JAC. WECKER and published at Basel in 1563. Some think that PEDEMONTANUS is identical with HIERONYMUS RUSCELLIS the astronomer, philosopher and theologian of Mount Cassino, who was three times general of the order of Benedictines and died at the age of 70 in 1604. On the other hand others believe that this book was written before 1550 as it does not yet use continuous cooling when distilling alcohol (rather a weak argument!). His methods and recipes are mainly derived from German authors on the subject.

In Holland distillation in its modern form was introduced by the *Constelyc Distilleerboec* of PHILLIP HERMANNI (296). It shows a strong influence of BRUNSCHWYGK's and GESNER's writings and it is said to reflect the "Many Years' Experience of Master Hermanni". The book was first printed by JAN ROELANTS at Antwerp (1552), then by SIMON COCK (1558) and JAN ROELANTS (1566) in the same town, but in 1622 an edition "now entirely corrected and improved" was published by BROER JANSZ. at Amsterdam. This edition too contains an appendix on the proper method of distilling wine. The writer warns against the use of anything but pure old wine with which one should fill two-thirds of the still. The alembic is then luted with a mixture of white of egg and flour and the fire lighted, which should be kept as small as possible as soon as the distillation begins. The same book also gives the earliest recipe in the Dutch language of aqua juniperi or Hollands gin (jenever) which liqueur is recommended for all kinds of diseases. Two figures illustrating this oldest Dutch book on the art are given in Figs. 87 & 88.

Fig. 87 en 88. Illustrations from HERMANNI's *Distilleerboec*

"Dye maniere hoe men den Ghebranden wyn maken sal metten onderwysinghe der instrumenten die men daetoe hebben oft besigen moet.

Als ghi nu wilt gaen maken den ghebranden wyn, soo en sult ghi niet doen ghelyck vele bedrieghers doen, die daertoe nemen veelderhande dinghen daer si den wyn mede verderven ghelyck als wyndroesem oft gist van bier ende diergelycken onreinich- heit, dwelck (hoe wel dat de ghebranden wyn die also gemaeckt is somtyts oock sterck sy) so ist nochtans een groot bedroch ende en behoort niet ghedaen te worden. Maer ghi sult nemen tot uwen wercke wyn die lanck oft oncleer gheworden is, maer noch- tans niet suur, want hoe sueter hys is, hoe crachtiger ende beteren wyn gheven sal ende oock meer wyns gheven sal. Ende hoe wel dattet beter ware dat men goeden, oprechten welrieckenden wyn daertoe name, soo hebbe ic u nochtans dat om den grooten cost te vrmyden toeghelaten ende het geet oock goeden wyn.

Van den voorgeschreven wyn sult ghi uwen ketel de twee deel volghieten en het derdedeel ledich laten blyven, dan sult ghy uwen helm daer weder opsetten, ende dan sult ghi nemen wit van het ei ende cloppent tot niet dan water en worde. Dit water sult ghi nemen ende maken een papken daeraf tarwenbloemen te samen onder een geclopt; hiermede sult ghi doecxkens die int papken voorgeschreven genet zyn om den tote van den helm, daer hi in de coperne pype comet te legghen ende des geluckx oock aan dat benedenste einde van de pyp, daer u receptakelaen oft onder staet, dat ist vat daer den ghebranden wyn in ontfangen wort, die door de pype comt geloopen...

Om dat receptakel sult ghi oock doecxkens, als voorgeschreven is, omslaen ende oock wel dicht toemaken datter ooc geen locht uut en mach.

Als die voeghen daer de instrumenten in een comen aldus dichtgestoppet syn, soo sult ghi vier maken in den oven, totdat ghi siet dat het begint terdeghe te branden. Dan sult ghi den oven stoppen, eerst het gat daer men de assche treckt uit ende dan het deurcken, daer men de barminghe door in den oven werpt.

Die bovenste lochtgat sal men open laten, totdat men siet oft hoort dat beghint te droppen, dan sal men de twee noch stoppen op datter vier niet te heet en si, want hoe langsamer ghebrant wort, hoebeter dat is... Daerom sult ghi hem met cleinder hitte aldus distilleeren totdat den besten wyn over is; dat suldi aldus probeeren, ghi somtyts u receptakel wegnemen ende proeven tghene datter druipt oftet noch sterck is......

Dat C ende XXI capittel: Van geneverbessenwater Aqua juniperi

De bessen van geneverhout sal men stooten ende met wyn besprenghen. Dit is een seer goet water teghen veel siecktten die van couden ghecomen syn ghelyc als verstoptheit der leveren ende der milten, vercoude ende vervuilde magen, this goet teghen alle fenyn uutwendich ende inwendich ende teghen de pestilentie, men sal den siecken des morghens nuchtern gheven van desen water een glas vol te drincken, soo en sal hem geen fenyn moghen schaden, ende oft iemand gebeten waar van een fenynjch dier, dien sal men van den water te drincken gheven ende wasschen hem de wonde met desen water......

The next important Dutch author on this subject is CASPAR JANSZ. COOLHAES (1536-1615), a minister at Leiden, who was unfrocked by the synod of Haarlem because of his heteredoxy. He made and distilled waters in his house at the Rapenburch, Leyden, where he learnt from his friend, professor HEURNIUS (1581) "the free art to draw the spirit with the help of the forces of fire and appropriate instruments from all spices, roots, herbs, flowers and seeds, and also from Rhinewine, Spanish wine, French and other wines which can all be used as medicines against different diseases, internal and external". His profession was considered very unseemingly for a minister and COOLHAES was probably the reason for the law enacted by the synod of Brielle forbidding that a member of the clergy should distill brandy and the like. COOLHAES' first book was published at Leyden (1588), at least written at his house on the Rapenburch. It describes 51 waters and 17 oils with a rather lenghty introduction declaring the fermentation and distillation of corn a sin (137). In 1600 COOLHAES moved to Amsterdam where he published his second book (138) in his house in the Warmoesstraat near the St. Olofspoort, called "In den vergulden Mortier" (In the gilded mortar). This *Booklet on waters* is a short edition of the earlier one, containing only 32 pages of the original 111, it has become a purely pharmacological pamphlet giving just a short preface condemning corn-spirit.

For the sake of a good survey of the literature on our subject we will now mention a few of the later writers on distillation, though the early seventeenth century is particularly poor in original scientists working on our subject.

ANDREA BACCIO's book *On wine* (33) is of some importance. The author was born at Ancona about 1550, though he says in his book, that his birthplace is Milan. He taught botany and pharmacology at Rome, was always heavily in debt and so had to give up his plans of travelling to other countries and stay at Rome from 1567 until his death (1600). His history of wine is very detailed as are his remarks on the distillation of wine but his book contains no plates.

JOHANNES RHENANUS was a fervent propagandist for the new distillation methods. This Hessian doctor and Paracelsist was born at Cassel. The title of his thesis is *De chymia technia, in qua totius operationis chymicae methoda practica clara ad oculus ponitur* (Marburg 1610) and in his *Aureus Tractatus* (504) he also deals with distillation and its effects on the "spirits" and "tinctures" which Nature had embodied in plants and minerals.

ANGELO SALA (ANGLI SALAE VICENTINI) was a far better chemist whom both BOERHAAVE and HALLER praise. He was born at Vicenza and obtained his M.D. In 1609 we find him at Winterthur, between 1613 and 1617 at the Hague, then he becomes court physician to count GÜNTHER of Oldenburg (1620). In 1625 he joins the duke of Mecklenburg in the same function and accompanies him on his flight before WALLENSTEIN's army, living at Bernburg, Lübeck (1628-1630), etc. He died at Bützow on October 2nd 1637. In his *Hydrelaeologia* (544), also reprinted in his *Opera Omnia* (545), we find the usual recipes for the distillation of brandies, essences, etc. He gives some details on the then common manufacture of corn spirit. Everywhere in northern Europe, so he says, brandy is made. One takes any grain suited to the brewer, grinds it not too finely, mixing it with tepid water in a vessel to a paste-like consistency. The yeast of beer is added and fermentation starts. This is an expert's job. The distillation of the alcohol that follows, however, is very simple. The fermentation is "an intimate movement of elementary principles which tend to group themselves in a different order to give birth to a new substance". SALA is a skilful manipulator and one of the first to get fairly constant temperatures in his ash-, sand- or waterbath and furnace. Thereby he is able to make fairly exact guesses at the alcohol content of various beers and other alcoholic beverages. According to him there was already an important alcohol-industry making corn-spirit near Magdeburg and Wernigerode (Harz) before the Thirty Years' War, that is to say before 1618, though local documents do not date further back than 1623.

He knows that it is important to be informed on the part of the plant which one should treat with water or alcohol. Each part is more suited to produce a specific principle by distillation for instance "a smell" or "a taste". Alcohol penetrates better to the "principles of scent" (essential oils) and water dissolves the "bitter principles" better.

In his essay on De natura spiritus vitrioli he asks himself whether the "oil of vitriol" distilled from copper- or iron-vitriol (sulphate) is the same. He proves this and says that the distillate is different from a sulphurous vapour, because "it has taken something from the air". He seems therefore to have had an inkling of the conversion of sulphur dioxide to sulphur trioxide by oxygen, and by air.

MARTIN SCHMUCK, born at Leipzig in the second half of the sixteenth century, was town physician of Hersbruck near Nürnberg, where he died in 1640. In his book (562) he gives details on the preparation and rectification of the spirit of wine which are worth mentioning (pars 2, dec. 2, § 1, p. 42).

The important Art of Distillation written by an English physician JOHN FRENCH contains nothing out of the way but describes the different methods very clearly. FRENCH too was a Paracelsist who used the word "distillation" in the sense of "filtration" also.

Two Frenchmen are worth mentioning as typical examples of their generation. The first is JEAN BROUAT (BREVOTIUS), a doctor and chemist who lived in the seventeenth century. The only fact known about his life is that he travelled in Holland. Many analyses of food-stuffs convinced him that they all contained a percentage of the "alcoholic principle". Thence he waxes very enthusiastic about well rectified alcohol and advises its moderate use against many diseases. His book (93) is really a rhapsody on alcohol, which according to him was invented by the Arabian chemists. This book is still written in the ancient alchemist style, which is clear if we quote a few titles of paragraphs:

"Why the Water of Life (eau-de-vie) bears its name and why there are two kinds, the Water of Life and the Water of Death".

"That the Water of Life is the general essence of all plants and akin to ether".

"Why the philosophers call the Water of Life heaven".
"Of the circulation and conservation of the quintessence".
"How to draw the tinctures with the spirit of wine and how to separate their soul".

The excellent methods described by a doctor ARNAUD (30), a native of Lyon, about whom nothing further is known, are very practical. Though he calles distillation "wet sublimation" his descriptions of furnaces, lutes, vessels and methods are simple and clear.

The same is true of the works of NICOLAS LE FÈVRE, who was born in the Ardennes in the early seventeenth century. He studied at the Calvinist university of Sedan and became "démonstrateur" of VALLOT, then professor of chemistry "au Jardin du Roy" at Paris. CHARLES II called him to London and entrusted him with the management of the chemical laboratory of St. James. Here he found a second native country, dying in London in 1674. He earnestly tried to design better distilling apparatus (354) but did not succeed. He was the first to use the word "alcoholize" and "alcohol" in chemical books in the sense of "making fine powders", which meaning was already quite common in medical and pharmacological handbooks. Distilling essential oils is a careful business, "one should not smell them at 300 to 400 paces from a chemist's shop", he exclaimed.

These less important figures slowly lead us up to the Honourable ROBERT BOYLE, the pioneer of the new chemistry that arose in the wake of the new natural philosophy of NEWTON and his school. He was born at Lismore (Ireland) on January 25th 1627 and went to Eton at the age of eight. In 1638 he went travelling abroad meeting GALILEI in 1641. From his return in 1644 his life was devoted to study. He was soon a prominent member of the Invisible College from which body the Royal Society was destined to grow of which BOYLE was a president in 1680. He went to live at Oxford (1654), then moved to London (1668) where he died on December 30th 1691.

BOYLE did not study distillation as a separate subject but in the course of his work he discovered many facts which were to be of singular importance in later days. Thus for instance he notices that effect of pressure on the boiling point of liquids such as water, wine, etc. and states too that the effervescence stops by the internal cooling of the liquid if the access of heat is not sufficient to maintain the boiling. He designed an apparatus for laboratory vacuum distillation shown in Fig. 89. Part II, sect. 1 of his *Experiments* (88) deals with the "production of vinous spirits". Here he devotes considerable space to the question whether the distillation products are preformed or whether they are formed during the heating and distillation only. He denies the latter conclusion but he also proves that the grape itself does not contain alcohol (87), which is only formed after the addition of malt and during fermentation, which seems to stop in vacuo! This was an important step in combating the idea that volatile compounds had to be "detached and formed" from the original substance by prolonged heating in the form of digestion or circulation in pelicans and other similar vessels or by cohobation, another cause of the loss of much distillate.

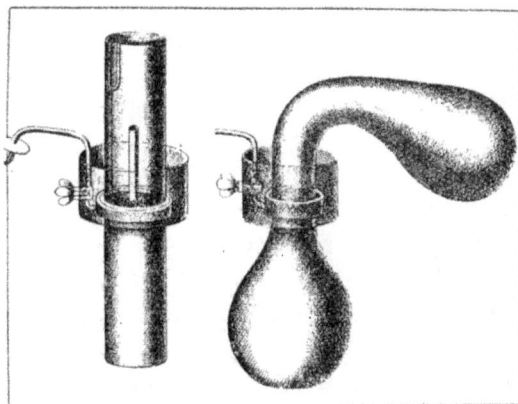

Fig. 89. BOYLE's apparatus for vacuum distillation

His correct and meticulous experiments can not be better illustrated than by giving some details on his work on the distillation of wood. In his *Sceptical Chymist* (85) we read: "Wood distilled in a retort yields far other heterogenities and is resolved into oil, spirit, vinegar, water and charcoal. The degree of fire by which the analysis is attempted is of no small importance. For a milde balneum will never sever unfermented blood but into phlegme and caput mortuum, the latter whereof pressed by a good fire in a retort yields a spirit, an oil or two, and a volatile salte besides another carput mortuum". And again in the next chapter: "I had long observed that by the distillation of divers woods, both in ordinary and some unusuall sorts of vessels, the copious spirit that came over had, besides a strong taste, an acidity like vinegar, wherefor I suspected that though the sowrish liquor be lookt upon by chimists as barely the spirit of it, and therefore as one element or principle, yet may be divisible in two different substances. Having distilled a quantity of boxwood per se and slowly rectified the sowrish spirit, the better to free it from oyle and phlegme, I cast into this rectifyed liquor a convenient quantity of powdered coral, expecting the acid part of the liquor would corrode the coral and so retained by it, that the other part of the liquor which was not of an acid nature nor fit to fasten upon the corals would be permitted to ascend. Nor was I deceived in my expectations for having gently abstracted the liquor from the corals there came over a spirit of strong smell and of a taste very piercing but without any sourness." He then proceeds to distinguish further characteristics of both the acetic acid (his "acetum radicatum") and the methyl alcohol (his "spiritus anolymus, adiaphorous spirit") and states that these compounds could possibly be separated by careful and slow distillation!

To illustrate the conflicts of this period between the new type of chemist and the older alchemist, who still survived, we will devote a few lines to a famous contemporary of BOYLE, JEAN BAPTIST VAN HELMONT. He was born at Brussels in 1577 and studied philosophy at Louvain, teaching it later at the College of the Jesuits and meanwhile obtaining his M.D. degree at Louvain in 1599. Then he travelled through Switzerland, Italy, France and England to settle finally at Vilvorde in 1605 where he died on December 30th 1644. This discoverer of carbon dioxide, which he calls "gas sylvester", was called by some the "Faust of the seventeenth century". He noticed (293) that during fermentation his "gas sylvester" escaped from the liquids and therefore he calls "the mother of the transmutation which divides the bodies into very small particles". He observes the same phenomenon with grapes, malt, etc.

He is no friend of distillates to be used as medicines for the "art of firing" is usually handled by amateurish persons. "Therefore distillates are in truth nothing but the sweat of the herbes but not their blood, and, would God that they did not also adulterate them because of their lust of money". "So everything that is distilled is a new creation brought forth by the fire and not pre-existing in the herb itself". Here, in the case of VAN HELMONT, much reasoning and little facts, whilst BOYLE develops his thesis from the experiments which he discloses to every intelligent reader.

If we now turn to the *essential oil industries* we see that the methods by which "summers distillation was left a liquid prisoner pent up in walls of glass" as SHAKESPEARE expressed it (Sonnet V), profited by the new discoveries and inventions, but there is relatively little change in the apparatus as compared with the great increase of knowledge and the field covered by the new essential oils from the colonial products which began to stream towards Europe in this period. The perfume industry always remained an industry on a small scale which was that of the laboratory or the apothecary. Still we saw that scientists like CORDUS, GESNER, RYFF, KHUNRATH and PORTA collected more details on essential oils and the subject was also studied by BRUNSCHWYGK, ULSTADT, LONICER, MATTHIOLUS, RUBEUS, CAMERARIUS, QUERCETANUS and DARIOT.

Apart from these we must mention JOHANNES WINTER OF ANDERNACH (GUINTHERUS ANDERNACENSIS) born in 1487, who taught Greek in Louvain and Strassburg and then obtained his M.D. and became court-physician at Paris. He had, however, to fly to Metz and Strassburg as he became a Protestant, in the latter town he died in 1574. In his book he describes the essential oils of cinnamon, nutmeg, pepper and spices of the genus Lilliacea. He and CORDUS were the first to make a proper distinction between fatty and essential oils.

BESSON prepared the essential oils from the peelings of oranges' and lemons (77) and G. COSTEAUS of Lodi, a physician and teacher at Turin, later professor at Bologna until his death in 1603, wrote a careful study on the preparation of these oils (145) in which he advises the use of sand- or water-baths and cooling the spout of the alembic.

Apart from distillation the other older methods were still used in the preparation of essential oils, viz. extraction, pressing and extraction with fats ("enfleurage"). The same centres still remained important, the monasteries, Venice, Nürnberg, France (Languedoc, Provence and Bourgogne) but others like the Netherlands came to play a part, where perfumes were prepared with essences from exotic plants and spices, a trade which grew more and more important as the number of ships sailing to America and the East increased.

On the other hand many products came to be distilled on an industrial scale as they profited from the growth of the mining and metallurgical industry in the early sixteenth century, which was in its turn stimulated by rising capitalism. Hence many *technological handbooks* were published in which the manufacture of mercury, mineral acids, sulphur and similar compounds by distillation on a large scale were described.

One of the earliest technologists was PEDER MANSSON (1460-1534) who travelled in Middle Europe and learned a good deal from the rising Italian and German technology. His book (332) shows strong influ-

ence from GEBER. In it he describes the manufacture of mercury by
destillatio per descensum in the "botus barbatus". Acids are made
by distillation of alumn and nitre in an alembic.

By far the most important author on technology of this period was
GEORG AGRICOLA whose *De Re Metallica* (12) is a mine of information on technological distillation methods of the period. GEORG
BAUER was born at Chemnitz on March 24th 1490 and studied philology, medicine and physics at Leipzig. Then he became physician at
Joachimstal in the mines of the Fuggers. In 1530 he moved to Chemnitz, where he became lord-mayor. However, he remained a Roman

Fig. 90. Mercury distilled per descensum, AGRICOLA

Catholic and therefore he had to resign. At his death on November
21st 1555 he was buried at Zeitz because no permission was given to
bury him in his native town which he had served so well. AGRICOLA's
book is well illustrated and he always compares the practice of his days
with that of the classics showing that little change had been made
until his own generation.

Mercury is mostly distilled directly from the ore by destillatio per
descensum (Fig. 90), often in batches of 700 pairs of flasks at the
same time. Another method is the destillatio par ascensum in
alembics with long spouts, usually combined in pairs, delivering the

distilled mercury into the same receiver (Fig. 91). A third method is
the sublimation of mercury from the ores in a large chamber in which

Fig. 91. Mercury distilled with alembics, AGRICOLA

Fig. 92. Mercury refined by chamber process, AGRICOLA

the mercury condenses on branches and drips to the floor to be collected (Fig. 92). Distillation of the ore covered with two fingers of

sand in a botus barbatus (Fig. 93) delivers the mercury on the sand, from which it is then washed.

In the Tenth Book we are informed on the manufacture of "Scheidewasser" (aqua regia) which is made by distilling a mixture of vitriol, sodium chloride, alumn, nitre and sometimes pulverised bricks. The glass cucurbits and alembics rest on an iron plate placed in the furnace, but usually the cucurbit is either placed in a cupel or covered with a protective layer of mud (Fig. 94). The receivers rest on straw rings.

Fig. 93. Sublimation of mercury from its ores, AGRICOLA

Cucurbit and alembic are joined by linen strips covered with white of egg mixed with flour and starch. Similar apparatus is used for the recovery of silver from the aqua regia (Fig. 95).

In the Twelfth Book AGRICOLA describes the distillation of sulphur from mixtures of ores and sulphur. The ore is distilled in pots with alembics, pairs of which deliver into one receiver with a hole at the bottom, allowing the sulphur to flow into forms (Fig. 96). Another method is the destillatio per descensum which is also used for the production of bitumen from bituminous ores.

The handbook of LAZARUS ERCKER (1574) shows the same apparatus for the manufacture of nitric acid and aqua regia (Figs. 97 & 98).

Again BIRINGUCCIO has almost the same methods for the manufacture and refining of mercury and sulphur. Nitric acid is prepared

Fig. 94. Distilling nitric acid, AGRICOLA

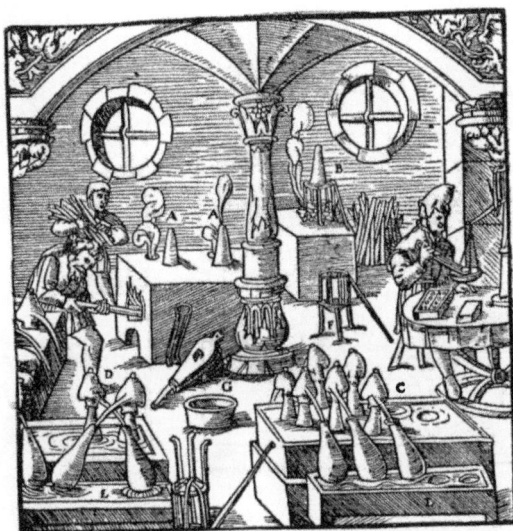

Fig. 95. Redistilling spent aqua regia, AGRICOLA

Fig. 96. Refining sulphur by distillation, AGRICOLA

Fig. 97 & 98. Manufacture of nitric acid, ERCKER

from a mixture of one part of nitre and three parts of alumn. BIRINGUCCIO remarks that in the case of the acids the "waters" are useless as they contain nothing but the phlegm, but the "spirits" should be recovered. After dissolving the silver from the crude gold, a useful nitric acid can be recovered by distilling the liquid containing the silver. His Book IV, caput 5 is devoted entirely to the construction of furnaces, cucurbits, etc. to be used in the manufacture of nitric acid, which is summarised in twelve rules.

Most of the sulphuric acid in those days was distilled from vitriol and GESNER gives a good description of this manufacture of "oyl of vitriol" (249): "Take four pounds of Vitriol of Rome, dry it in an earthen vessel till it waxes red, after when it is better put in a bely of glasse diligently defenced with clay... and first distill with a soft fire, encreasing the degree of fire by little and little, until whyte fumes begin to issue out at the nose of the bely. Then set a great receiving vessel fenced with clay and make a fire with wood continuing for the space of twelve hours and a length shall issue out red drops and heaven. When the receiver beginneth to bee clear, the matter is finished, wherefore then cease that the vessel may be couled."

The manufacture of sulphuric acid was still on a laboratory scale even when it developed into a sideline of some mining industries. This is clear not only from the above description of GESNER, but also from that given by PARACELSUS, CORDUS, PORTA and DARIOT. But at the same time it was well known that sulphuric acid could be prepared by burning sulphur with excess air and in contact with water. SALA mentions that this method "per campanum" is used by the apothecaries and LIBAVIUS, who gives pictures of the apparatus used (Fig. 99) proved the identity of this "oyl of sulphur" with the "oyl of vitriol" both in his *Alchymia* and in his *Aquarum mineralium*. GESNER gives a good description of the "bell-process" (251): "Take a vessel of glas... not much unlyke to a little bell, daubed with clay, hang it the space of a cubit from the grounde, by a wyer of bras or iron, under ye which thou shalt set a basen of glas of a great compas, with a pot of earthe turned upside downe. Moreover the bottom of the pot shall hold an iron plate four fingers broode, made red hot, whereupon the Brimstone may be brent. Whyles this is brent, new shal be added upont. Thereupon it shal cum to pass that by the smoke ascending the hanging vessel in short space shal destill drops down into the basen that stands under an oyll, which gathered diligently thou shalt serve in a phyall of glass".

Finally we must discuss the *evolution of the distilling apparatus* in this period, many of which we have already mentioned in detail when calling attention to the main authors on distillation.

As for the chemical furnaces used in distilling there are two main types, one for prolonged heating, generally with a soft fire (for which the supply funnel was constructed) called "Faule Heintz" or "Slow Harry" (Piger Henricus), the other the furnace for fierce

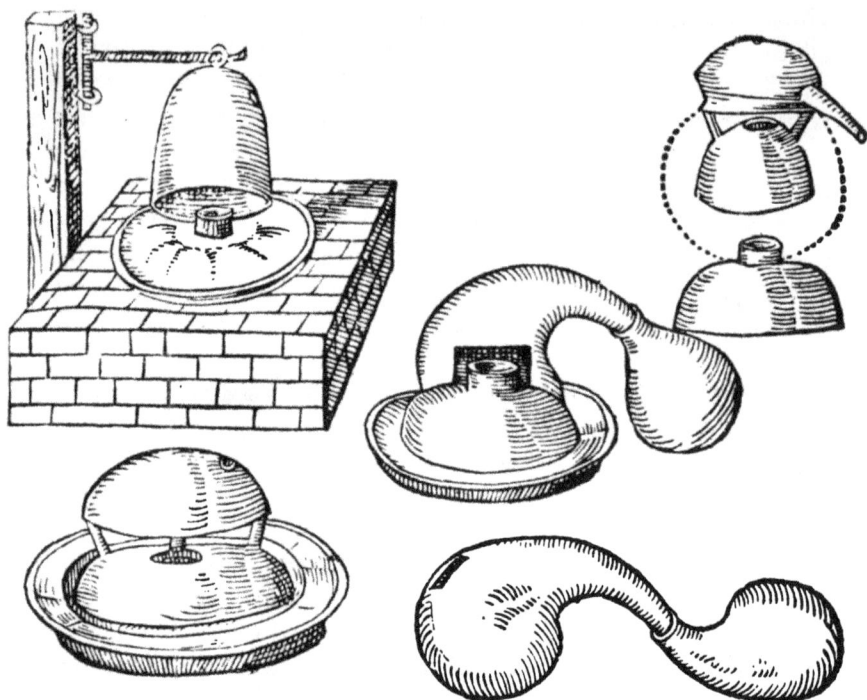

Fig. 99. Making sulphuric acid by the bell-process, LIBAVIUS

heating which was generally the same as the metallurgical furnace. Some scientists like LEONARDO DA VINCI tried their hand at the construction of proper furnaces for a better and more regular heating (597), but most of these attempts were fruitless as long as the temperatures were not properly measured. Therefore the scales of temperature designed by many remained purely academical. Generally four, six, eight or twelve "gradus ignis" were distinguished and VAN HELMONT even proposed sixteen grades. We hear of an "ignis lapidis philosophici", an "ignis lentus (blandus)" for a soft fire, an "ignis parabolicus" for a fire blown with bellows, an "ignis sapientum" obtained with manure and an "ignis suppressionis" used for the destillatio per descensum. But generally four grades of fire were distinguished:

1) a soft fire for a water-bath, just hot enough to allow one to touch the distilling vessel, such a heat as obtained by the sun or by "a sitting hen", which is usually required for digestion and the like operations;

2) a somewhat hotter fire used for the distillation from an alembic in a sand-bath;

3) a still hotter fire is used for the distillation of solids from retorts and for sublimations;

TABLE I

Element	Syriac MS 7th.- 9 th.cy.	Causa Causarum XI th.cy.	XV th.cy.	XVI th.cy.	XVII th.cy	1783	1808 Dalton	1814 Berzelius
SILVER (Selene)	(⌒	) ⌐	)	)	.)	)	⑤	Ág
GOLD (Helios)	♂ ♋	♉ ✻	⚆	⚭	⊙	⊙	⑧	Au
ARSENIC	-	-	∤	∤	⋂	•—•	Ⓐ	As
COPPER (Aphrodite,Phosphoros)	♀ ◆	—∘ ♀	♀	♀	♀	♀	©	Cu
IRON (Ares,Thurios)	♂ ♂ ♐ ⊷	⊷ ♂	∤	⟋	♂	♂	①	Fé
MERCURY (Hermes Stibbon)	⚷ ♉ ⟩∘⟨ ⟩∘⟨	—⟨ ♉	♃	♉	♉	☿	⊙	Hg
LEAD (Kronos)	♄ ♄♇ ♒	♄ ⟋ ⟲	♄	♄	♄	♄	ⓛ	Pb

The Evolution of some alchemical symbols (acc. to HOCH)

4) the highest degree of heat is obtained by direct contact with a full fire and used for the "spirites minerales", for instance for the manufacture of "oyl of vitriol" by calcination of vitriol.

Other methods of heating (destillatio per balneum roris, per vaporis, per cinerum, per limaturam ferri or martis, per ignem nudem, etc.) have been discussed already. Some even mention a "distillation with ice" (LIÉBAUT) by which they mean freezing a solution or mixture of liquids to solidify the "phlegm" and obtain the "oily parts".

No new vessels are designed in this period as a selection from GESNER shows (Fig. 100). We have already mentioned several places which seem to have specialised in the manufacture of proper laboratory ware such as the glass from Venice (Murano). Others mention earthenware from Waldenburg (Saxony) as very tight and fire-proof, AGRICOLA obtains his from Ipsen, GLAUBER from Almenrode (338). It is clear that specialists already design and make this apparatus for the chemist. For instance we find that the court-apothecary of Dresden buys his apparatus from GEORG STURM at Augsburg.

The apparatus is generally resting on or supported by rings of straw. Metal vessels were generally used in industrial work, the laboratory shifted over to glass vessels, as their quality improved. Their names remain unchanged (alcara, botia, botus barbatus, kymia, zucca, alembic, cucurbitae) (for long necked flasks such names as phiolae, buccia, ovum vitreum, ovum philosophicum, locus chimicus, urinales,

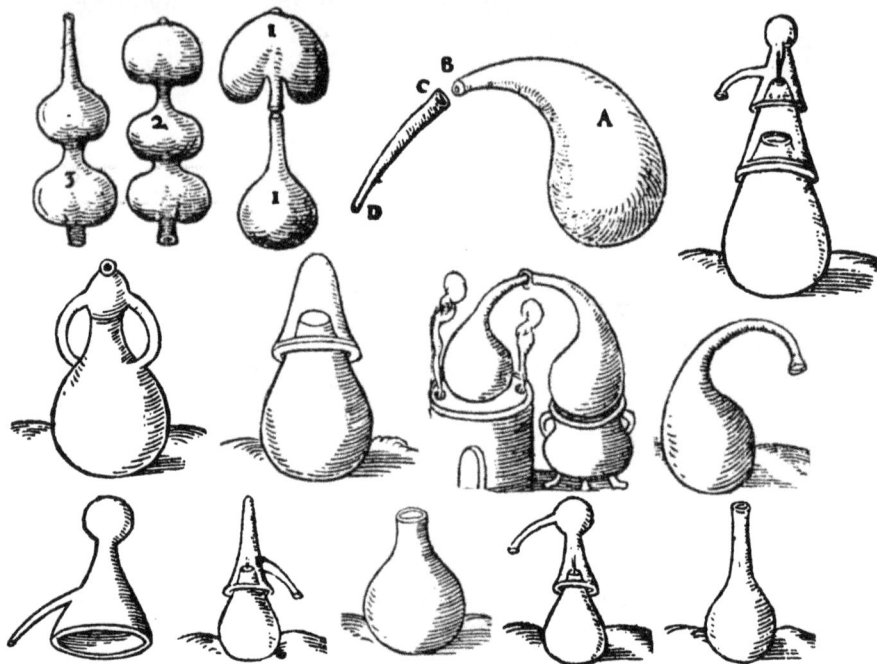

Fig. 100. Distilling vessels, GESNER

sublimatorium). In each country local names were formed and used (34) (558). Thus in England the *Book of Simples* (105) speaks of Horned Stills, Bagpipe Stills and Pelican Stills, etc.. Its author WILLIAM BULLEYN was born at Ely and studied at Cambridge. He travelled extensively in Europe, especially in Germany where he studied the new distillation methods. The brother of Lord HILTON falsely suspected him of the murder of this gentleman and therefore he suffered much persecution and even imprisonment. BULLEYN wrote a number of popular medical works creating many terms for chemical apparatus. He died in 1576.

The retort came into general use under the name of "retort"

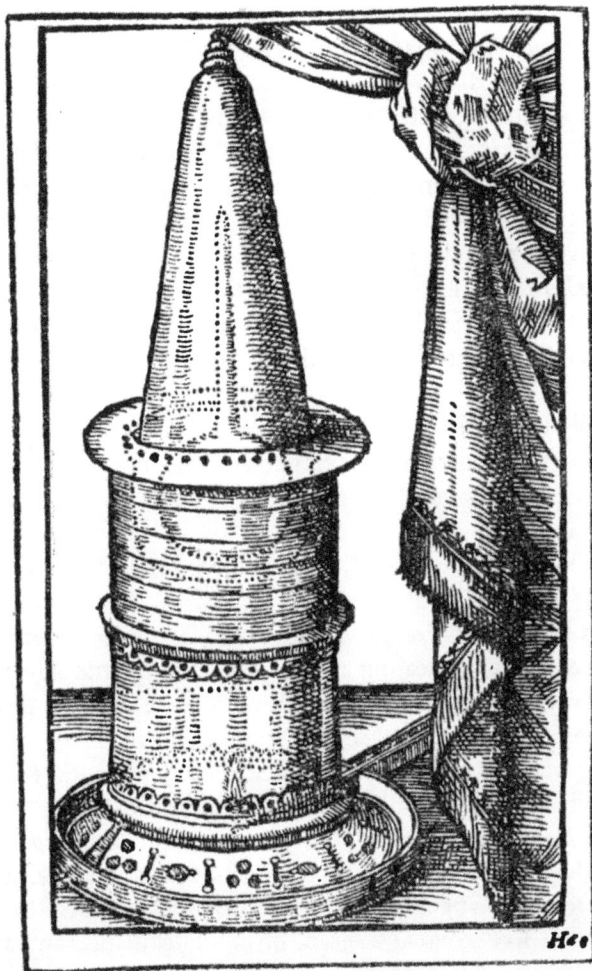

Fig. 101. Astral for prolonged heating
of liquids, KHUNRATH

"cornue", "cornemuse", "Dudelsack (Krummhals, Elefant, Storchen-schnabel)", "Bagpipe still", etc. It was especially suited for work at high temperatures when the lutes began to fail.

These lutes were still compounded from clay, alumn, lead oxide, red lead, white of egg, blood and flour. For fireproof lutes glass powder, iron filings, galena and animal hair were added. Much importance was ascribed to the recipe and these secret mixtures were often sold at high prices. GLAUBER sold his secret lute for two hundred guilders.

The operations remained the same as in the Middle Ages and the same importance was attached to their execution and duration. The faulty impression that volatile products were only formed by the application of heat persisted notwithstanding some experiments which proved the contrary. Therefore prolonged pre-heating of the material to be distilled was often deemed necessary, especially in the case of the production of perfumes and alcohol. Such heatings for which special furnaces were designed (as the astral of KHUNRATH, Fig. 101) often lasted for months. Generally in these cases the material was heated by the sun aided by mirrors and lenses or by manure, the putrefication of which was sometimes hastened by the injection of steam.

The symbols used by the later alchemists differed locally and often individually, but generally speaking there was not much change in the symbols for common things and operations or apparatus, as is shown by Table II (140).

Some new principles were, often unwittingly, introduced in distillation practice.

There is no sign of *preheating the intake* as some modern authors have claimed. The preheating to "loosen" the volatile spirits was a separate operation, the material was always cooled before distillation proper. Some designs, however, show an arrangement for preheating the cooling water used for cooling the still-head (Figs. 60 & 71), but it is not certain whether this is done intentionally.

There is certainly no sign of *filling the alembic during distillation,* which would have meant the first step towards continuous distillation.

There are some attempts at *steam distillation,* that is to say the introduction of live steam as a means of rapid distillation and avoiding superheating. The latter point is important in DARIOT's design for the distillation of gums and resins (Fig. 77) though the principle was here misapplied as the only effect might be a suction of the resinous vapours towards the cooler by the condensing steam. Still in the distillations of macerated herbs with water steam distillation was already unwittingly applied by the chemists of this period. The idea of heating by the introduction of steam was not new as will be seen from the design of a bathing-house (Fig. 102), but the logical application to distillation was still waiting for inventors who had more insight into evaporation and the heat balance of a distillation.

TABLE II

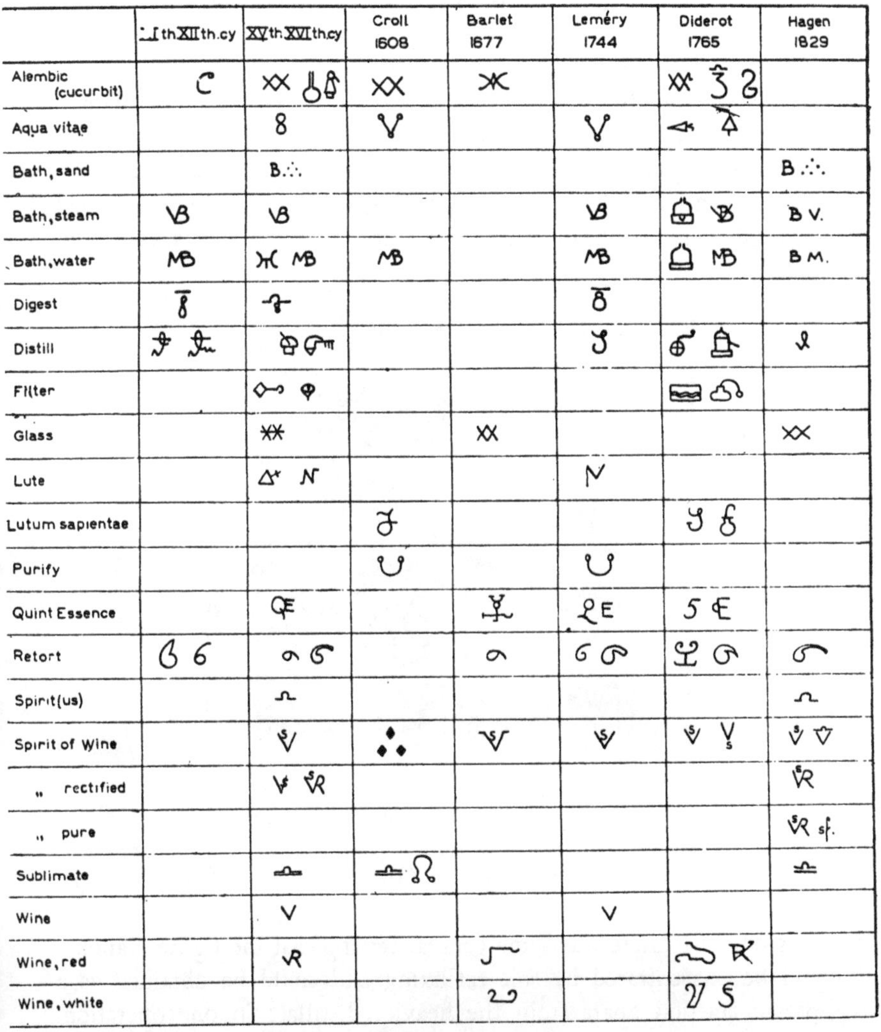

	˙₋I th XII th.cy	XV th XVI th.cy	Croll 1608	Barlet 1677	Leméry 1744	Diderot 1765	Hagen 1829
Alembic (cucurbit)	C	✕✕ 👁	✕✕	✕		✕✕ ⌃3 ₂	
Aqua vitae		8	V		V	◁ A	
Bath, sand		B∴					B∴
Bath, steam	ᴠᴮ	ᴠᴮ			ᴠᴮ	⌂ ᴠᴮ	Ɓ V.
Bath, water	ᴹᴮ	✕(ᴹᴮ	ᴧᴮ		ᴧᴮ	⌂ ᴹᴮ	Ɓ M.
Digest	8̄	⸮			8̄		
Distill	⸮ ⸮	⸮ ⸮			3	⸮ ⸮	⸮
Filter		�⟩ φ				⟨symbol⟩	
Glass		✕✕	✕✕				✕✕
Lute		Δᵗ N			N		
Lutum sapientae			⸮			⸮ ⸮	
Purify			∪		∪		
Quint Essence		Œ		⸮	⸮E	5 €	
Retort	𝒢 6	⌒ 𝒢		⌒	6 𝒢	⸮ 𝒢	𝒢
Spirit(us)		⌐					⌐
Spirit of Wine		⸮	∴	⸮	⸮	⸮ ⸮	⸮ ⸮
„ rectified		⸮ ⸮					ᴿ
„ pure							ᴿ sf.
Sublimate		⌐	⌐⟨symbol⟩				⌐
Wine		V			V		
Wine, red		ᴿ		⌒		⸮ ᴿ	
Wine, white				⸮		2/ S	

Some symbols of distillation through the ages

The two elements used in cooling the vapours were the still-head and the separate cooler, the latter becoming more prominent as experience grew. The *temperature of the still-head* was now generally adjusted by the application of cooling water, first discontinuous, later continuous cooling being applied. We see the evolution of the water-cooled still-head from the simple bladder attached to it (GESNER, LONICER) (Figs. 53 & 62) to discontinuous cooling (DARIOT, LONI-

CER) (Figs. 62 & 78) and continuous cooling (MATTHIOLUS, LIÉBAUT, BESSON, DARIOT) (Figs. 60, 70, 74, 76).

There were also some attempts at *refluxing* to obtain better fractionation, apart from the cooling of the still-head. Sometimes substances were introduced into the still-head to hold the phlegm back, such as sponges or the sulphur, advocated by ULSTADT or the charcoal which LOWITZ mentioned later. Refluxing was of course already obtained by cooling the still-head itself, but some scientists use a cooler to

Fig. 102. Heating a bathing-house with steam according to the Bellefortis MS (Fifteenth century)

condense part of the rising vapours (BIRINGUCCIO, LIÉBAUT) (Figs. 57 & 73).

It seems that there was some general feeling that the lighter vapours could be concentrated by this refluxing and could be obtained as a separate fraction apart from the heavier distillate in one operation. Usually they tried to obtain this dephlegmation by excessively long glass tubes emerging from the still and leading to the receiver. This introduced the rather uncontrolable air-cooling (LIBAVIUS, KHUNRATH). PORTA tried to make the distance between still-head and receiver shorter while maintaining the same air-cooled surface. This was the idea of the hydra which he introduced.

Fractionation was often obtained in a very simple way by making the still-head larger as in the case of the Rosenhut which increased its surface appreciably (Fig. 40), often two distillates are obtained from it (RYFF, LONICER) (Fig. 63). GESNER designed an apparatus to distill four fractions from wine and LIBAVIUS developed this idea and mentions a still for two fractions (Fig. 82) and even a tower for several streams (Fig. 80). It is doubtful whether any of these towers and stills achieved proper fractionation seeing the indifferent and difficult temperature control. The introduction of a condenser between still and receiver (PORTA, LONICER, LIBAVIUS) (Figs. 41, 68, 81, 83) was a step towards well-defined cuts.

The general effect of *proper cooling of the vapours* was well understood. We have already mentioned the cooling with the Moor's head or "caput Aethiopicus". Better cooling avoided distillation losses, which must have been considerable in older times in the case of volatile components, and at the same time it allowed for more rapid distillation. It was generally felt that the receiver should be far from the still or at least protected against the heat of the furnace into which the still was built. We saw that this was sometimes achieved by lenghty glass tubes wound in coils and bends and by introducing air cooling. But the introduction of the water-cooled coil or worm was a far better means to obtain the ends desired. This is what was generally called the "serpena" or "serpentina", the "canalis refrigatorii cum multiplici gyro, serpentini nominati" of LIBAVIUS. We see its gradual evolution from the tube crossing the barrel of water horizontally (GESNER) (Fig. 50), diagonally (GESNER, MATTHIOLUS, LONICER, BESSON) (Figs. 51, 61, 64, 75) (sometimes splitting the stream in two to obtain more efficient cooling, Fig. 64) to a horizontal coil (BIRINGUCCIO) (Fig. 57) and later the vertical coil (GESNER, ULSTADT, LONICER) (Figs. 49, 56, 65, 66). The position which the coil should have relative to the still is not yet properly understood, for often it is placed too high and the whole effect would be nothing but a severe reflux action (Figs. 56 & 67). Another very general mistake is the wrong introduction of fresh water to the condenser or still-head. It is then introduced at the highest point, because "the water becomes hot there much sooner than at other places of the cooling trough" (DARIOT). The idea of bringing the hottest vapours in contact with the coolest water is not yet understood.

Continuous cooling is first met in the case of the water-cooled still-head. There is some doubt about the date of its introduction though it is now generally agreed that it is not common before 1575. VON LIPPMANN (376) says that it was unknown to ALDEROTTI, SAVONA-ROLA, BIRINGUCCIO, RUSCELLI, PORTA, BRUNSCHWYGK and RYFF. According to him it is first clearly stated and understood in a small essay entitled *Del modo di distillare le acque da tutte le piante e como vi possino conservare i loro veri odori e sapori,* dated Venice, 1550, which shows the design given in Fig. 60. This tract was written by FRANCESCO CALCIOLAIO, an apothecary of Verona and a great friend of MATTHIOLUS. This was probably introduced by a mistake in MATTHIOLUS' essay *De ratione distillandi* which was published in his *Opera Omnia* (Basel, 1674). SPETER remarked that this figure does indeed occur in the Czech edition of MATTHIOLUS (1562) and the Latin edition of 1565. The essay mentioned by VON LIPPMANN was published by BARTOLEMEO DEGLI ALBERTI, who also published the *Discorsi* of *Matthiolus,* in the 1604 edition, but not in the seven earlier editions. It is more probable that MATTHIOLUS got the idea from JOAN. BAPTISTA MONTANUS (1498-1551) who wrote sixteen books on distillation which are still in manuscript form in the San Marco Library at Venice and which were never published. There is also the same fundamental idea in BESSON's work (Fig. 74). GESNER in his foreword to the German edition of BESSON states that the idea was formed somewhere in Saxony. Anyway the idea was in the air by that time and GESNER himself applied it (595). In LIÉBAUT's work we find proof that it is well understood by then (Fig. 70).

Though the forms of the condensers are manifold, there are only two types, the cooled still-head and the tube or coil cooled in some kind or trough. Sometimes the vessels for distillation or fractionation receive names different from those used in primary distillation. LIBAVIUS for instance calls the latter "fistulata" and the former "rostrata".

The principle of *reboiling* was used though probably not understood (GESNER, LIBAVIUS) (Figs. 54 & 84), this was of course an excellent means of obtaining good fractions.

The only man who really understood the idea of fractionation and its intimate connection with the boiling point was BOYLE. In discussing the results of his distillation of boxwood and the discovery of its different components he says: "All seems to argue that the acid portion of such distilled liquors is more ponderous or more fixed than the adiaphorous (neutral) spirit, which upon that account may be in great part separated from it by bare distillation if carefully performed". But then he never realised this idea.

The *separation of the water from the distillate* was achieved by three methods. The first was the use of a simple separating funnel, worked by keeping the thumb over the hole at the bottom and thus regulating the outflow of the water. The second method consisted in

Map II. Towns that play a part in this story between 1650-1850 A.D.

filtering the distillate through a wet filter and the third was the destillatio per filtrum, drawing off the excess water by means of a felt which played the part of a syphon. Sometimes we find whole batteries of such filtrations placed in a cascade arrangement.

Testing the quality of the distillate was still difficult, it was especially hampered by the lack of temperature control. In the case of alcohol there was the evaporation test on a piece of parchment, the fire test with a piece of linen and the test of observing a drop of oliveoil put into the alcohol, noting whether it sank or floated. Others tested it by smelling or pouring some drops on the hands and rubbing them together at the same time noting its smell (PORTA, GESNER).

Most of the principles enumerated could only become a succes when they were understood theoretically and applied with all the force of science. Technology was already the handmaid of pure science and had profited to some extent in the early sixteenth century but in later generations it was kept waiting at the' door. But the door remained shut, for the miraculous developments in the days of BOYLE and NEWTON kept the minds of the scientists busy in the realm of pure science and there was little time to be devoted to its practical applications. The cooperation which had yielded some promising results stopped and the field remained barren until the days of the French Revolution. Yet the few tentative results of development tried out by the practical distillers in the following period show that "there is some soule of goodnesse in things evil would men observingly distill it out" (*Henry V*, Act IV, scene 1, line 5).

CHAPTER SIX

FROM BOYLE TO LAVOISIER

„Die U ontbindt, die is Uw vriend.
Het grove wezen op den test
Geeft zynen geest of alderbest.
Zoo zoekt de Wysheid door het lyden
Naar liefdes eigenschap en wensch
Het geestelyk wezen uit den mens
Van 't grove deel tot haar te scheiden."

(JAN LUYKEN) (396)

The picture in JAN LUYKEN's *Spiegel van het menschelyk bedryf* accompanying this poem shows a "scheider" (distiller) equipped with the traditional cucurbit, alembic and phial. If this is the typical distiller's equipment in 1790, it goes to show that stagnation in the evolution of the distilling apparatus which set in in the seventeenth century, and even earlier. Of course this did not mean that there was also a stagnation in the technological applications of distillation. As a matter of fact the distilling trade was a most prosperous one in this period and the applications of distilling grew on every side. Slight as the qualitative evolution may be, the more important the quantitative development was.

In this period the industrial centre shifts definitely to Western Europe as Germany and Italy are exhausted by wars and the uncertain conditions make these countries less attractive to industry, which follows trade to the harbours of the Atlantic Ocean.

Science, especially physics and biology, took enormous strides from the days of Newton onwards and it opened fields for industry as soon as the scientists turned towards practical technology. But this was not yet the case in the seventeenth and early eighteenth centuries, when theoretical science and its development absorbed the best minds of Europe and only few had the courage to point out new technological possibilities. In Engeland we find only BOYLE, in France such men as LEFÈVRE, LEMERY, and GLASER, in Germany BECHER, GLAUBER and KUNCKEL, who, dissatisfied by the lack of interest of their own countrymen often did their most important work abroad. Upto the middle of the eighteenth century there is a remarkable lack of good technological works, which forms a striking contrast with the many

handbooks published in the sixteenth century. Then new generations
of men are born, who believe in the possibility of turning the new
science to practical account. Here France leads with such names as
MACQUER, RÉAUMUR, LAVOISIER, BAUMÉ, DEMACHY, CHAPTAL and
BERTHOLLET followed in the wake by such German scientists as TAUBE,
WIEGLEB, BECKMANN, FERBER, VON POPPE, HERMBSTÄDT and
WESTRUMB, less original perhaps, but none the less very energetic
propagandists of the new technology. England strange to say has
hardly any technological work to present until SAMUEL PARKES'
Chemical Essays of 1815, though theoretical science of this period can
boast of many Englishmen of international fame.

The books of science of the seventeenth and eighteenth centuries
are widely read and we can truly speak of an "Age of Enlightenment"
in which the bourgeoisie toyed with the new scientific results and apparatus and discussed them in club and drawing-room. But the seed was
also sown in other classes, especially in the days shortly before the
French Revolution when new possibilities were opened to create
the necessary skilled labour and training-schools like the Ecole Polytechnique were opened in France. The example of France was soon
followed in England, though there this technical education was still
mostly in private hands.

The Industrial Revolution of the later eigtheenth century came to
full bloom in England, which led for many generations.

One of the most important social changes of our period is the
gradual disappearance of the guilds, which form of organisation
remained here and there until the French Revolution, for instance in
the case of the distillers, to be swept away in the upheaval "distill'd
from limbecks foul as hell".

Even mineral acids which became technologically very important in
the new chemical industry were still made on a laboratory scale and
industries employing more than ten men were rare. The difficulties
for larger industries were mainly fuel supply and the lack of adequate
mechanical power. These had to wait for the advent of such industrial
fuel as coke and coal and the invention of the steam engine.

The typical form of industry of this period is no longer the home
industry or the laboratory but the "manufacture", a creation of COL-
BERT's age which is still far away from our factory. The products
passed but few hands and the "manufacture" is hardly more than an
organisation of many home-industries coupled by capital, such as the
linen industry of medieval Flanders had shown (338). But chemical

industry could not profit by a horizontal coupling of many small home industries or laboratories. Only larger apparatus, more power and larger castings could turn the wheel. The recasting of industry which COLBERT had tried failed between 1690 and 1730 notwithstanding the help of state protection and heavy import duties. This was not due to the death of COLBERT himself or the withdrawal of the duties, but mainly to the unstable and foolish policy of the French state and the development in other countries. In France and Germany the Industrial Revolution failed at its start because of the lack of monied interests and it had to wait for better conditions.

The state industries in these countries lingered on to make place for private enterprise. The "industrialists" of France were either states-men, government officials or merchants. In some industries such as breweries, etc. we find some concentration and here real factories arise but these are of no importance on the total industry of France.

The frequent European wars of this period did not hamper the new industries, for they often lacked sufficient supply of labour and prisons had to be opened for this purpose. Their main difficulty, apart from the lack of sufficient capital for their development, was the lack of skilled labour. Still the distilleries of France flourished in this period as there was little change in their apparatus and they remained more or less on the old footing.

The Low Countries had very flourishing distilleries. Here there was plenty capital for a proper Industrial Revolution, had not its develop-ment been seriously hampered by the Act of Navigation and COLBERT's protective measures which defeated all the energy of the Dutch in-dustrialists, who had to live by exporting as the inland market was far too small for industrial expansion. In fact the only country in which the Industrial Revolution found the right conditions in this period was England. Here plentiful supplies of raw materials com-bined with a liberal spirit and monied interests created ideal conditions and the eighteenth century shows the birth of English industrial leadership which continued for a considerable time into the next century.

The importance of this development for the history of distillation will be clear from the following pages, but we must first follow a trade, which shows a quantitative but not a qualitative technological expansion in this period, the *manufacture of alcoholic beverages*.

We must begin to remark that this period from BOYLE to LAVOISIER was one of enormous consumption of alcoholic beverages, of beer, wine and above all all kinds of brandies and liqueurs (337). Both Continental and English taverns bore signs telling the traveller that here he could have as much brandy as he desired and plenty of straw to sleep off his debauch for the sum of two pence. It was quite common in this period that the guests had to be borne home after a dinner and kings and clergymen led the dance. The pamphlets praising the virtues of alcohol as published in the sixteenth century and later bore fruit. Even in the seventeenth century doctors advise everyone to be drunk from time to time for the sake of their health. This question was thrice the subject of a prize-essay of the University of Paris and each time the reward was given to an essay (1643, 1658, 1665) advising the reader to get drunk at least twice a month "to strengthen the gastric juice".

The seventeenth and eighteenth centuries had not yet made their choice among the many new alcoholic beverages, there was no definite brand of brandy on the market as there was a standard quality of beer. From chemistry and technology the distilling trade derived some new suggestions. New types of apparatus were tried out, more especially wooden ones as we shall see. The generation of KRÜNITZ, NEUEN-HAHN and BECKMANN was diligent in trying out new base materials but in general alcohol was still manufactured from wine or corn.

The *Low Countries* possessed a flourishing distilling and perfume industry, stimulated in this period by good trade conditions and the existence of sufficient capital to be invested in industry, though hampered by some severe measures of foreign governments in its exports trade. For Holland the eighteenth century meant the loss of the White Sea trade and this meant compensation by a growing trade with the Mediterranean and southwestern Europe. Even in the most prosperous colonial period this European trade was by far more important to Holland. Its distilling trade could have profited from the new technical inventions, but as it was in the hands of many private persons with little capital it did not. The emigration of the Huguenots in large numbers meant the death of many Dutch guilds, but did not affect the guilds of distillers, who profited from the good trade conditions when the larger industrial enterprises in Holland suffered badly from the English Industrial Revolution and the import-duties which many countries raised on their products.

In the wake of this general decline the labour conditions grew worse and worse, then wages fell, housing conditions were awful and these social evils led to great unrest at the end of the eighteenth century. But the distilling trade formed one of the few exceptions.

The brewing trade of Holland declined steadily and severely in this period. Haarlem had only 8 small breweries left in 1740, in Amsterdam their number fell from 23 in 1650 to 4 in 1749, in Rotterdam from 12 in 1750 to 2 in 1792. Delft lost most of its breweries due to the bad quality of its water.

This decline was due to German competition, declining inland consumption and the fact that tea, coffee and above all brandy became more prominent. The provinces of South and North Holland were the only ones in which a brewing trade of some importance remained by 1800 (320).

But the rise of the distilling industry was not easy. The popular prejudice which preferred brandy from wine to corn spirit, was very strong and only strictly prohibitive import duties on German wines and other base materials and prohibition of French wines such as those of the years 1671 and 1672 made the rise of the industry possible. Before that date the distilling industry produced practically for local consumption only. The early attempts at export were met by prohibitive duties in France and these of course met with a counter- attack, as the French wine imports in Holland were still considerable at that time. The Dutch distilling industry gained importance by the end of the sixteenth century and corn spirit from the Low Countries came into favour with the customers round the White Sea notwithstanding the libels distributed by French competitors. The new trade brought prosperity to Weesp, Amsterdam, Delft and Dordrecht but did not yet affect its modern centre Schiedam very much. There was a slight decline of the industry in the last decade of the seventeenth century because of the shortage of corn and the lowering of the import-duties on French wines. But this interval was relatively short and the industry grew considerably during the eighteenth century. However, its centre moved. The industries of Haarlem and Alkmaar vanished, and in Amsterdam, Delft and Dordrecht it declined severely, leaving only Weesp and the new centres on the river Maas (Schiedam, Delfshaven and Rotterdam) to profit from the new tide. To combat French competition the Dutch distillers even founded branch industries in such towns as Dunkerque, Nieuwpoort and Liège (32). The protection in their own country made it possible to use all such base materials as bad wines, lees of wine, bad beer, prunes, raisins and all kinds of cereals as early as the beginning of the seventeenth century. Some claimed that the climatological conditions and the water of the neighbourhood of Schiedam were especially favourable to the production of an excellent brandy, but this contention has never been proved (185). Still it is certain that the growing industry profited from the possibilities of exporting their product to all foreign countries with which the Dutch traded. The production of gin rose to fourfold in the period from 1733 to 1792.

At first the exports to the countries along the Baltic were prominent, but this trade declined as home production in these regions, rich in corn, started. Hamburg and the Rhine-country took fairly large quantities, but the oversea trade was by far the most important. Large quantities of brandy were shipped to England and Scotland and the end of the eighteenth century found the United States of America as a large consumer of Dutch gin. Other important quantities went to France, Portugal and Spain, who mainly re-exported them to their colonies. The export to the Dutch colonies in Asia and America was fairly considerable.

The exporting trade was concentrated in Schiedam, Delfshaven and Rotterdam, all on the river Maas, where we find respectively 122, 22 and 22 distilleries in the year 1771. The old distilling centre of Weesp declined after 1690. The reasons for this concentration in Schiedam and neighbourhood may be found in the excellent shipping facilities and the far-sighted policy of the Schiedam local authorities, who seem to have tried to compensate for the decline of their herring fisheries by stimulating this new industry (290). Schiedam had the earliest distiller's guild of Holland (513) and the local authorities furthered this industry by exacting very little imports and paying its way by all kinds of measures. By 1700 there were no less than 77 stills for the production of raw gin owned by 34 distillers.

This was only rendered possible by protective measures such as the prohibitions on French wines and brandies of 1668 (also on raw German gin) which were repeated in 1689, though they had to be repealed after the wars in 1678; but it lasted for a short time only.

The distilleries were relatively small (Fig. 103) and the guild of distillers of Schiedam was not a large one, but in 1694 12 new members were admitted and in 1695 another ten. The development of the industry ran parallel with the number of corn-mills. Gradually there was also some specialisation. Some distillers made only raw gin, other distilled this into export brandy or liqueurs. Then there were also malt-mills which malted barley and milled other cereals. These malt-mills together with the draining-mills were the first to employ steam-engines in Holland. Thus GOLDBERG saw one of the earliest steam-driven malt-mills at Nieuwwerk (Rotterdam) in 1800. The primitive distilleries ran great fire-risks and the municipality of Schiedam took every possible measure to prevent serious fires. There was a strict inspection of the distilleries and their buildings and every addition was carefully scrutinised. The malt was largely imported from England and the fuel came from the eastern provinces of Holland (coal, wood, peat). The labour came partly from the provinces of Holland and Brabant, but a large number of Lutheran immigrants from Münster and Osnabrück worked at Schiedam too.

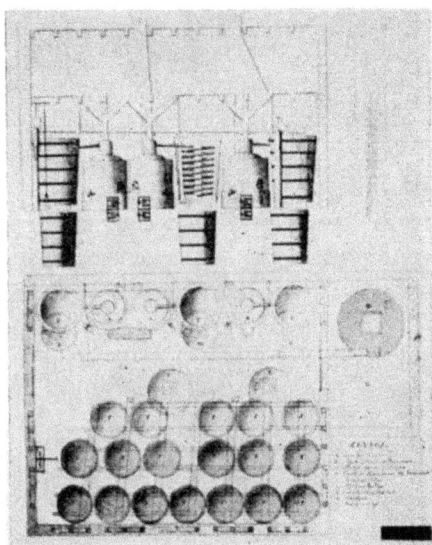

Fig. 103. Plan of the distillery "De Papegaay"
Delft, about 1790

The distilling trade developed steadily in the seventeenth century though there was a temporary slight decline after the peace of Ryswyk (1697) and again after that of Utrecht (1713). During the war against France a strict prohibition of French wines and brandies was still enforced (1672-1678), but after the war some more lenient measures had to be introduced on account of the peace-treaty. The Staten Generaal had to restrict the industry on account of the great shortage of cereals in 1698, a shortage all throughout Europe, to prevent famine, and many distilleries were closed in the years 1698 and 1699. In 1700 many of them received permission to reopen their business and a proposed restriction of 1709 never materialised. Another restriction was effected in 1771-1772.

Beer had long been one of the general beverages of army and fleet, but "jenever" or Dutch brandy appears as such only since 1692. Then it quickly ousted beer from its prominent place especially in the fleet. But as the consumption of gin rose it became subject to heavy taxes and by the end of the eighteenth century it belonged to the products that were traded forwards on the Amsterdam exchange. It was even an object of the gambling craze of that period.

Schiedam profited highly by its expanding distilling business and its large exports of gin and liqueurs (87) though it is strange that an observant technologist like BECKMANN does not mention the distilling trade in his dairy of travels through Holland in 1762 (52).

The great vogue of brandy and the high profits of the distilling business also showed themselves in the rise of prostitution and the growing number of women addicted to drink. In fact drunkenness was one of the chief vices of the proletariat of the manufacturing towns in the early nineteenth century and their chief drink was genever.

The French Revolution and the subsequent occupation of Holland by the French meant only a temporary decline to the industry, lasting until 1813. But after that date the profitable export trade was resumed. The yeast industry was subsequently concentrated in Delft, but Schiedam remained the most important distilling centre of Holland, still exporting gin and liqueurs in large quantities.

The apparatus changed very little in this period and it is characteristic of the lack of cooperation between science and technology that a prize-essay by ZILLESSEN (665) of 1783, an answer to a better design of stills asked for by the Hollandsche Maatschappij der Wetenschappen at Haarlem, has little to offer but an improvement of the way of heating the still. No change in the design of the still itself was proposed. ZILLESSEN proposed to use a smaller grate situated immediately behind the fire-door and to lead the flue gases round the still into the chimney. Indeed such a grate had some advantage if the firing was not too quick but as it had to be handled very carefully, this design was not adopted universally. ZILLESSEN himself doubted the adoption of his plans, for he says: "Labour in the distilleries are not Dutchmen but Germans and they care very little for the wellfare of the masters, especially if it makes their work even slightly more difficult".

The most important literature on distilleries in this period is nothing but a series of compilations of recipes for different waters and liqueurs, such as the *Secreetboeck* of CAROLUS BATTUM (49), a physician of the town of Dordrecht, who gives the old recipes and uses all the methods which we have already discussed in connection with the preceding period. Of the later books we mention that of PIETER VAN KEULEN (324). In this book, published in 1696, he mentions that the still should not be over-filled lest it boils over but anyway the distillate shall be redistilled at least three times to obtain the proper strength (for there was no proper fractionation in these stills). To make strong alcohol, which is tested with a piece of linen in the well-known way, it is sometimes even re-distilled six times, taking a strong wine as base material.

Another book of this type is the *Nieuw ontdekte Distilleerkonst* (81) (82) written by one J.K.B., who is certainly not identical with JACOB BOLS as some authors suggested. The title-page of the second edition is shown in Fig. 104. Here we see the typical glass used for genever ever since. B. mentions that wood or peat are the best fuels. The still and alembic are made of copper and tinned inside, though glass apparatus is the best, but this should be heated in a sand-bath and protected by a layer of salt, loam and dung. When rectifying wine one can obtain a stronger distillate by adding some salt

Fig. 104. Title page of *De Nieuw Ontdekte Distilleerkonst*

or Sal Tartar (which binds part of the water). It is also possible to distill in a cucurbit with a tin alembic and a tin worm-condenser. A mat in the still will prevent overheating of the contents. Typical is the lack of proper temperature control, for B. says "heat as long as it is possible to keep your hand on the end of the alembic, for then the distillate will run steadily and quietly". The fire is regulated by opening and shutting the fire-door and it is possible to cool the alembic with wet rags. The condensing coil is still cooled discontinuously. Apart from these few remarks on distilling in general the book contains a host of recipes.

This type of book was popular in Holland until the end of last century. For example we find a similar compilation by VAN DE WOLLENBERG in the early nineteenth century (666).

We are tempted to cite a poem praising the virtues of jenever

written by one HENNEBO in 1718 and often reprinted (295). HEN-
NEBO was the landlord of the "Gulden Vlies van Jason" near the
Princenhof in Amsterdam and later of the "Karseboom" in the Kalver-
straat, where his inn was famous for "his French soups, delicious
English Roast-beef and real Dutch chops and teals". HENNEBO was
the typical adventurer of his century, twice losing all his money and
twice regaining wealth. Here are a few lines from his long poem:

> „De Duitscher riep tot zyn confrater
> Herr, Broeder, had-er Wacholer water,
> Soo sauft einmahl aus deinen Fles.
> De Engelschman, die riep: Lord bles
> Here is Jenivi. En de Schotten
> Die zoopen als de Waterrotten.
> De Deen, de Onger, de Hussaer
> Elck zoop gelyck een Toovenaar.
> De Nederlander riep: Jenever,
> Jenever die heeft kuit en lever
> Die maekt ons moedig, vroolyk, bly,
> Weg kaas en brood, dat's snoepery."

Similar books of recipes for alcoholic beverages and stimulants
were produced in *England* by KENELM DIGBY in his *Closet* (180).
DIGBY was born at Buckingham and he was a very popular figure at
the court of CHARLES I and CHARLES II. He died in the wars against
the Turks in 1665. His collections of recipes (178) (179), many of
which are of a rather doubtful nature, were often reprinted.

Another collection of this type was THOMAS SHERLEY's *Curious
Distillatory* (575) which also contains some general notes on distilling.
SHERLEY was born at Westminster in 1638 and studied medicine at
Oxford and in France. He was physician to CHARLES II and wrote an
English essay on the preparation of the Stone. He died on August 5th
1678 when involved in a lawsuit over his father's heritage. His *Distil-
latory* is really a translation of ELSHOLT's *Distillatoria Curiosa*.

JOHN SHIRLEY's *Closet of Rarities* also contains recipes of "waters
and liqueurs" (576). SHIRLEY was the son of a clerk, born in London
on August 7th 1648, and did not finish his studies at Oxford but
travelled round posing as a doctor. He lived at Islington translating,
correcting and writing books, e.g. a *Compendium chirurgiae*. He died
on December 12th 1679.

The Distiller of London is a pamphlet of a more serious type compiled and set forth for the sole use of the Company of Distillers (183)
which Company was incorporated the year before (1638). We saw
that BAKER used the word "distiller" somewhat earlier and FENTON

when using the word "distiller" in 1577 refers to its more general meaning and speaks of a "distiller of water". Later it comes to mean more specially one who extracts alcoholic spirit by distillation. The above mentioned pamphlet is one of the earliest examples of the use of this word in the restricted sense.

In England too the distilling industry gained ground on the brewing industry in the seventeenth century. "Before brandy came over to England in such quantities as it doth now we drank good strong beer and ale and all laborious people used to drink a pot of ale or a flagon of strong beer" (*Harleian Misc.* VIII, 537). In London it took £ 2000 to £ 10,000 to build a brewery whilst a distillery required only £ 500 to £ 5000. The large Dutch imports of the early seventeenth century showed the big profits of the trade and the English distillers were not slow in taking up this hint. From the beginning of the eighteenth century the English distilling trade shows specialisation. The "malt-distillers" produced raw gin and "vie with the Brewery for Return of Money and Profit for most are large concerns". They delivered raw gin to the "rectifiers" who prepared the finished gins and liqueurs as in Holland. MANDEVILLE ridicules a pair of large distillers whose business was a large capitalistic concern involving malt-mills and pigbreeding, a combination which we also find in Holland.

It seems that the distilling business in England was largely concentrated in London as is seen from the following figures. Between September 10th 1784 and June 5th 1785 the Distillers of London treated 96,909 gallons and the Rectifiers 102,643 gallons, while for the rest of England these figures were 126,968 and 57,208 gallons respectively.

Similar figures could be quoted from *Germany* where Quedlinenburg and Nordhausen were the principal centres of the eighteenth century. The *Nordhäuser Chronik* mentions the first distillery in 1713. We also hear that the Hessian distillers were ordered to use coal in 1756 as there was a considerable shortage of wood, used by the metallurgical industry. In 1781 the first German brandy was exported to the United States of America.

In *France* the corporation of distillers found its privileges affirmed by a Royal decree of August 14th 1674, but by decree of May 15th 1676 the "limonadiers" and "marchands d'eau-de-vie" · were incorporated. For in 1673 there was still a guild of brandy merchants with abouth twenty members existing as a separate organisation. By this

decree of 1676 the entire distilling trade was combined into one corporation. It is strange to find the word "brandevin" used as slang by
soldiers in this period, though the official word always remained
"eau-de-vie". Different tariffs of 1664 and 1691 fixed the taxes on
drinks made from brandy, wine or water, such as liqueurs and syrups,
details of which can be gleaned from the publications of SAVARY
(553) and PAGANUCCI (462).

The above-mentioned corporaton became a good source of income
to the Government, it was dissolved in December 1704 and a new
corporation created with a fixed number of members (50 for Paris),
who had to pay the State a large entry-fee. Some years later this corporation was disbanded and the old corporation reinstituted (July
1705). But the game was so profitable, that it was tried again in 1706
when a new corporaton was planned with 500 members for Paris.
This, however, was a failure for the old corporation had to be reinstituted in November 1713 and all the distillers and retail-merchants
were combined in this corporation which was to last until the French
Revolution. Not before 1776 were the pharmaceutical chemists detached from this corporation and formed into a separate guild.

The French distilling trade flourished and was able to free itself
from Spanish competition in such centres as Barcelona and from that
of Dutch and Scottish imports. Since 1729 the distilling industry of
Languedoc had carried on a notable export trade, but the best liqueurs
were made at Montpellier, Nancy and Turin (462). Still in a wine-
producing country like France the consumption of brandies and
liqueurs was relatively slight when compared with that in other
countries. Thus we find LAVOISIER reporting that the yearly consumption of alcoholic beverages for Paris amounted to 250,000 muids
of wine, 20,000 muids of beer, 2000 muids of cider and only 800
muids of brandy. After 1762 the use of cereals or molasses for the
production of alcohol was strictly forbidden, in fact nothing but wine
remained free as a suitable base-material. Still we find the earliest
attempts to ferment sugar-beets and potatoes on an industrial scale in
the eighteenth century.

Both the closer study of fermentation and the production of alcohol
from potatoes were to influence the trend of the distilling apparatus.
The first increased the size of the apparatus and brought secondary
factors in its design to the fore, the latter was the source of new
designs able to cope with thick mashes. But as their effects belong to
the following century we shall deal with these developments in due
course.

It is now time to review the more *prominent chemists and scientists* of this period and hear their views on distillation and distilling apparatus. We shall see that the generations of the second half of the seventeenth century had nothing new to say, that the early eighteenth century produced a gap in the activity on this subject and that the second half of that century shows pronounced conflicts between the technologists suggesting new ideas and the scientists hanging on to old ones.

JOHANN SCHRÖDER was a typical specimen of the physicians of the late sixteenth century. He was born at Salzuffeln (Westphalia) in 1600 and after having obtained his M.D. degree he served in the Swedish army as a surgeon. Then he became town-physician of Frankfort, where he died on January 30th 1664. In his *Pharmacopoeia* (568) he defines distillation as the "luring of a liquid thinned by fire down into the attached receiver". He mentions the old apparatus and advises the use of glass or earthenware vessels instead of metal, thought tin may be used too. He distinguishes different degrees of heat as usual and uses the word alcohol both for a fine powder and for the spirit of wine, which is obtained by distilling with a cooling-coil, repeating the distillation if necessary.

His contemporary, ATHANASIUS KIRCHER, is a strange and rather fantastical figure. He was born at Geysa near Fulda on May 2nd 1602 and joined the order of the Jesuits in 1618. In 1631 he becomes professor of mathematics and philosophy but is forced to fly because of the Thirty Year's War, first to Avignon, thence to Rome where he taught mathematics and Semitic philology. Towards the end of his life he retired and died on October 30th 1680. His books are a strange mixture of abstruse theory and practical experiment. Though many of them touch our subject, the most important work is his *Mundus Subterraneous* (327), many editions of which are known. The first section of the eleventh book gives a table of all chemical operations according to his theories (Fig. 105), and an impression of the apparatus used by him can be gleaned from Figs. 106 and 107. The twelfth book, section IV discusses distillation, which "is nothing but the dissolving of the humid particles into vapour, elevating these until they touch the cold parts of the alembic and are reconverted into liquid by condensation".

He gives very elaborate schemes of destillatio per ascensum, per descensum, ad latus and the like, each divided into four types (distillation by fire, earth, water and air). The distillation by fire converts

Chymia verfatur circa

Operationi infervientia, quæ funt

Locus

Furnus

Apertus { Probatorius. / Ventofus.

Tectus { Simplex { Calcinatorius { Cæmentatorius. / Reverberatorius. / Afcenforius. { Siccus { Veficæ, / Catini arenarii. / Defcenforius. { Humidus: balneum. / Diffolutorius.

Compofitus. { Athannor. / Acediæ.

Vafa, quorum aliud igni

Admovetur,

Ex certa materia factum { Vitreum { Phialæ / Circulatorium { Pelecanus { Ovum Philofoph. / Dyota. { Ahenum.

Minerale. { Metallicum fervit { Subtiliationi { Vefica. / Fufioni { Infundibulum. / Pyramis.

Terrenum capiens { Materiam { Fuforia { Catillus cinereus. / Crucibulum. / Non fu- { Pyxis cæmenta-foria, { toria.

Aliud vas, { Catinus arenarius. / Tegula fornicata.

Non admovetur.

Pro artificum arbitrio. { Superiora, { Coctus. / Alembicus, { Roftratus. / Inferiora. { Cucurbita. / Retorta,

Continens { Receptaculum. / Concha.

Tranfmittens { Tritorium. / Separatorium.

Caufa adjuvans

Inftrumentum manuarium igni.

Admovendum { Semper { Gracile { Craticula / Baculus ferreus. / Latum { Spatula. / Forfex.

Ad arbitrium { Rutabulum. / Cochleare. / Circulus.

Non admovendum. { Ligneum: dyoptra.

Naturalis; radii folares. { Metallicum. { Tabula ferrea. / Mortariolum.

Calor.

Artificialis,

Simplex { Digerens, { Athannor. / Fimus.

Separans. { Senis, { Veficæ. / Cinerum. { Arenæ. / Impeditus. { Scobis ferri. / Fortis. { Liber, { Carbonum. / Flammæ.

Miftus. { Maris.
Balneum. { Roris. / Vaporofa.

Operationem ipfam, quæ eft

Solutio.

Calcinatio { Corrofio. { Immerfiva. { Humida, { Amalgamatio. / Præcipitatio. / Sicca, { Cæmentum. / Commixtio.

Ignitio. { Combuftio. { Vitrificatio. / Reverberatio. { Incineratio.

Diffolutio.

Subtiliatio. { Elevatio. { Sicca fublimatio. / Directa { Alembici. / Humida. { Veficæ. / Defcenfio. { Calida { Obliqua, { Retortam. / Frigida } per. { Latus. / Circulatio. { Deliquium.

Exaltatio. { Ablutio. { Filtratio. / Putrefactio. { Imbibitio. { Coratio. / Digeftio. { Extractio. { Cohobatio.

Liquefactio. { Simplex, { Cineritium, / Probatoria, per { Antimonium.

Coagulatio, { Frigida. / Calida.

C A.

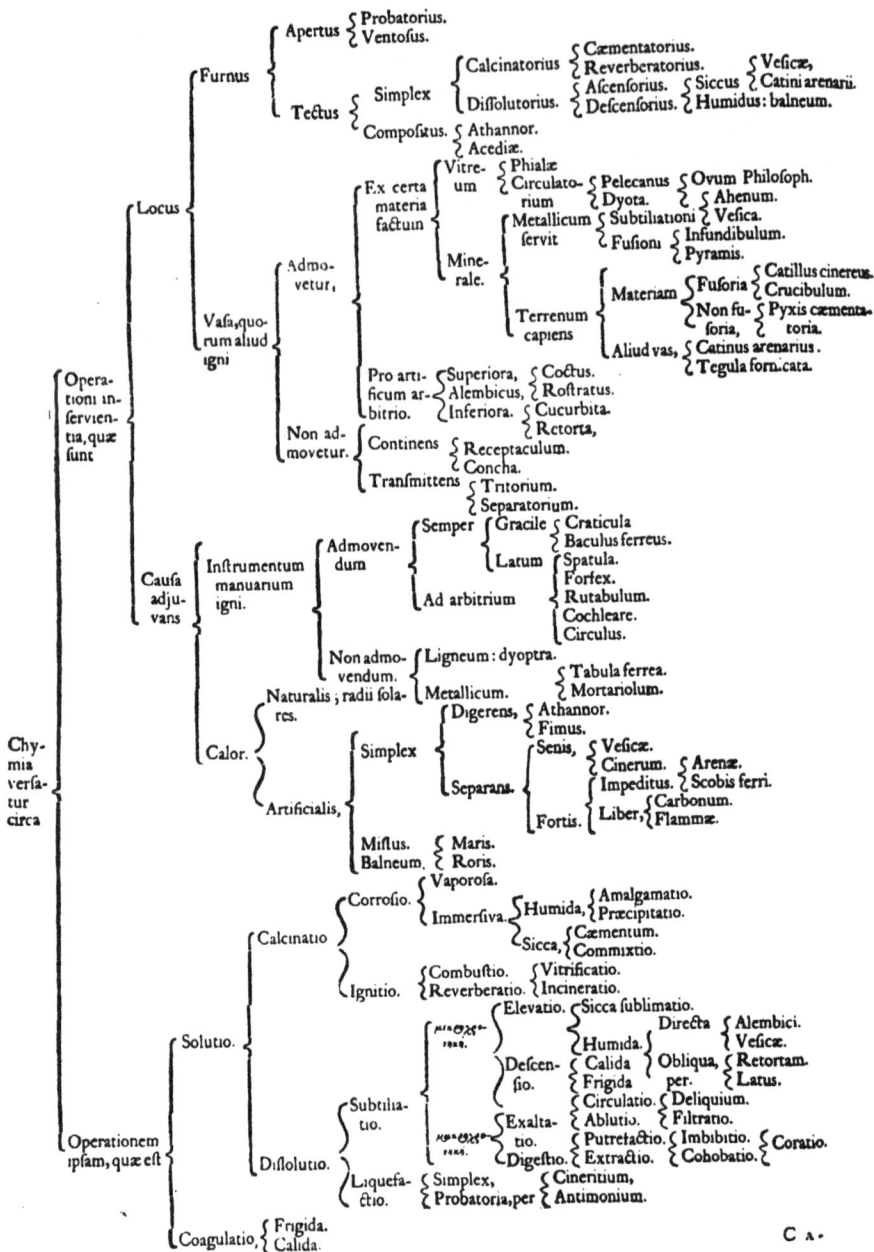

Fig. 105. The chemical operations according to ATHANASIUS KIRCHER

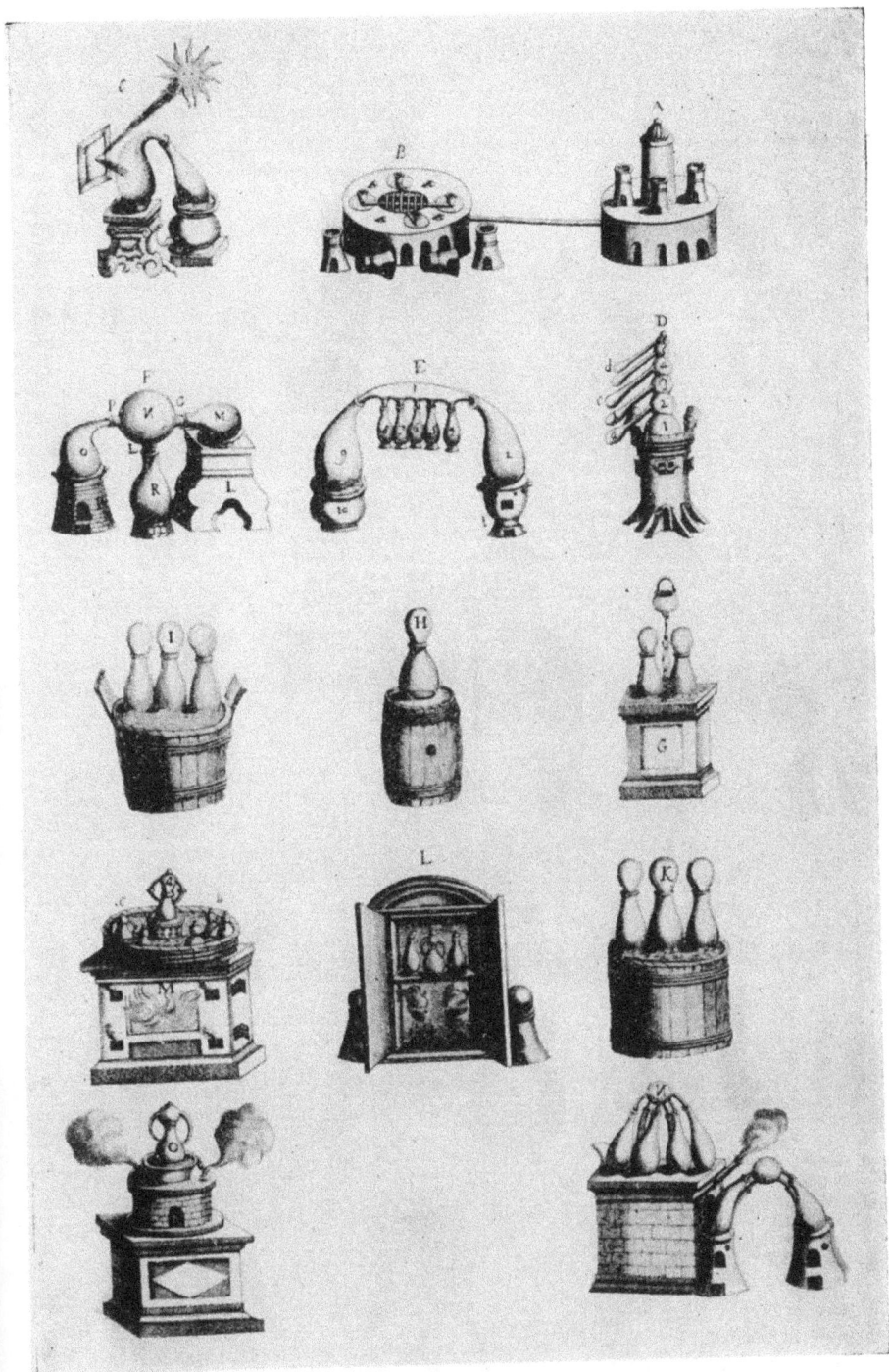

Fig. 106. Chemical apparatus used by ATH. KIRCHER

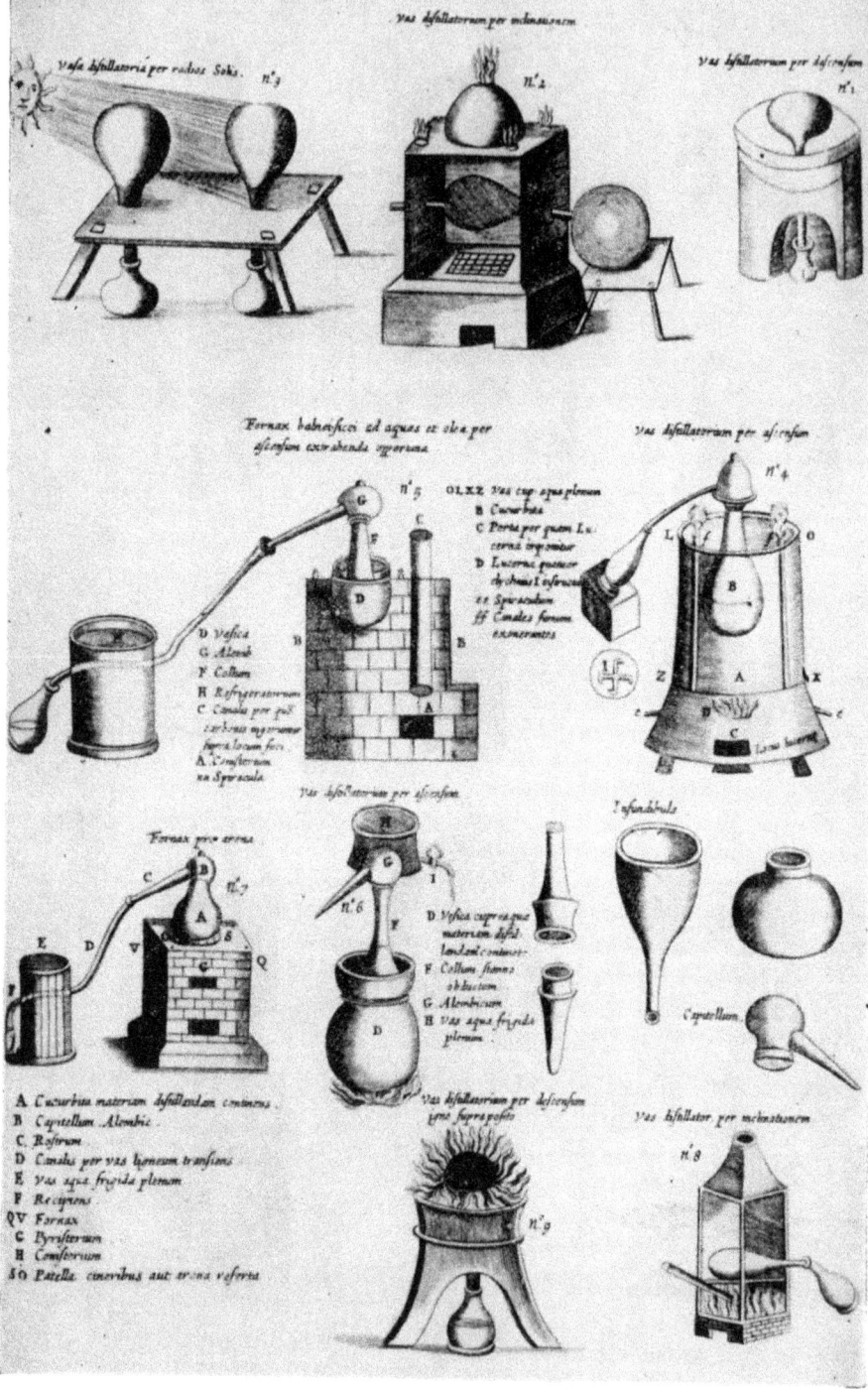

Fig. 107. Chemical apparatus used by ATH. KIRCHER

the "waters" or "spirits" in the earthy parts into oil by heat, while they lose their watery parts. Distillation by earth means simply using a sand-, ash- or dung-bath, that by water is achieved in a water-bath and that by air in an air-bath. Apart from this theorizing KIRCHER has very little to say on distillation.

JOHANN RUDOLF GLAUBER was far more practical. This physician and chemist was born at Karlstadt (Francony) and lived at Salzburg, Vienna, Frankfort and Cologne before he moved to Holland in 1648, where he died twenty years later at Amsterdam. He belonged to that group of alchemists who left the preparation of the Stone and the making of gold alone and turned to chemical technology. His contemporaries were so impressed by his books and his constructions of furnaces that they called him "the second Paracelsus".

Apart from a book showing the economic value of the production of alcohol for Germany, which book (265) is really but the first part of his *Teutschlands Wohlfahrt*, he wrote two books which deal with distillation (263) (264). In his book on the furnaces he mentions the possibility of distilling in apparatus that consists for the greater part of wood, an idea that was taken up at the end of the eighteenth century and that was widely discussed among technologists. Now it was forgotten for some time like many other good suggestions made by GLAUBER.

GLAUBER was very much impressed by the idea that steam could be used to convey heat and he has different projects using this principle. One of them is the production of steam in a copper retort heated by a furnace (Fig. 108), which steam is then injected into a cask of wine. The alcohol that distills off is cooled in a coil inserted in a tub. In this project the wine is at the same time base material to be distilled and the pre-heated liquid, which saves a considerable amount of heat.

In another figure (Fig. 109) we see his design for a better cooling vessel, in which the condensate is retained and the non-condensed gases are syphoned over from the water-cooled cooler into next vessel. There is the possibility of arranging rows of these vessels thus obtaining proper fractionation by dephlegmation.

His contemporary, OTTO TACHENIUS, was a practical apothecary, born at Herford (Westphalia) and trained by his father, from whom he had to fly because he had stolen money. His life is most adventurous, he lived at Lemgo, Kiel, Danzig and Königsbergen (1641) earning his pay as a chemist. In 1644 he moves to Italy and studies at Padua obtaining his degree. Then he moves to Venice where he sells

Fig. 108. Steam distillation according to GLAUBER

Fig. 109. GLAUBER's dephlegmators

"secret medicines" and finally dies in 1670. TACHENIUS has much practical experience. He knows how to test waters with silver nitrate and shows that proper distillation removes the chlorides and salts from water. Therefore one should always use distilled water in the laboratory. He also showed that waters distilled in copper vessels contain traces of copper which are an aperient. The laxative action of many waters is therefore due to the copper salts and not to the water itself. It disappears if the same waters are prepared in properly tinned vessels (612).

PHILLIP JAKOB SACHS (VON LEWENHAIMB) was a physician, born in Breslau on August 26th 1627. He studied in his native town, then at Leipzig, several Dutch universities, Paris, Montpellier and finally obtained his degree at Padua. In 1651 he returned to Breslau where he became town-physician two years before his death on January 7th 1672. His main work is a book (543) on the medical applications of wine and the spirit of wine in which he cites many authorities such as BACCIO, CAMERARIUS, CARDANUS, EVONYMUS, GLAUBER, KIRCHER, LIBAVIUS, LONICER, LULL, MATTHIOLUS, PORTA, QUERCETANUS, RUBEUS, RYFF, SALA, SAVONAROLA, SCHROEDER, VILLANOVA and ULSTADT. He identifies aqua ardens, aqua vitae and brandy (Branntwein) discussing the proper way of distilling it. Quinta essentia or alcohol is obtained by rectifying aqua ardens at least three times. He gives many synonyms of alcohol from the earlier alchemists such as NORTON, LULL, etc. and says that though most prescribe rectifying, only few judged it necessary to start with circulation now. The rest of his book is devoted to different recipes of alcoholates.

A book more especially devoted to distillation was written by the physician JOHANN SIGISMUND ELSHOLTZ (207) who was born in Frankfort in 1623 and studied both in his native town and at Wittenberg and Königsbergen. He travelled extensively in Holland, France and Italy taking his M.D. at Padua in 1653. He became court-physician to the Elector at Berlin where he died in 1688. ELSHOLTZ is convinced that the result of the distillation is determined by the type of cucurbit and furnace used. Several compounds as nitre are broken up when distilled giving coloured fumes. The essential oils have natural colours which most of the waters do not show after distillation. As he is convinced that good distillates should show the same colours as their base materials to show the same properties he distills a host of animal, vegetable and mineral compounds without reaching a general conclusion on the subject. His book was translated into English by SHERLEY (575).

JOHANN KUNCKEL, Baron LÖWENSTEIN, was another practical chemist whose views on fermentation we shall discuss in due course. He
was born at Rensburg in 1630, his father was an alchemist attached
to the court of Holstein. He studied chemistry and pharmacy at Lauen-
burg and was attached to the court of the Elector of Saxony. Then he
was professor of chemistry at Wittenberg when FRIEDERICH WILHELM
called him to Berlin in 1670 to his laboratory and the glass industry
which he intended to found there. For this work on glass technology
KUNCKEL is rightly famous, he introduced many new ideas, new
methods into the glass blowing technique which made glass into a
proper laboratory material. He was so successful in business that he
was able to buy a large estate. In 1693 he was called to Stockholm by
CHARLES XI who created him a Baron and a Bergbaurat. He discovered
the element phosporus and combated the wrong chemical theories of
VAN HELMONT and PARACELSUS. He died at Stockholm in 1702.

The proper handling of glass meant a revolution in laboratory
technique and the possibility of designing new apparatus. Still a
chemist like JOHANN CONRAD BARKHAUSEN (or BARCHUSEN as his
Dutch editors generally call him) did not profit by the new technique.
BARKHAUSEN was born on March 16th 1666 at Horn (Lippe-Det-
mold) and studied at Berlin, Vienna, Mainz, etc. between 1683 and
1693. Then he goes with the Venetian army to Turkey but in 1694 we
find him at Utrecht he teaches chemistry. In 1698 he gets a post
as tutor at the University of Utrecht and in 1703 he is nominated
professor, teaching at Utrecht until his death in 1723. In his works
(42) (43) he says that "in distilling the exhalations that are apt to
be raised by fire are expelled from the instruments to be condensed
in the vessel superimposed on the heated vessel".

The destillatio recta or per ascensum is especially suited for the
distillation of "waters" and "spirits" from fermented masses, the
destillatio obliqua or ad latus is performed with the retort and direct
fire, it is suited for solids which give heavy vapours such as mineral
acids, nitric acid, etc. The destillatio per descensum is used in the
case of compounds giving off vapours and other humid parts, such
as some minerals and all kinds of wood. Sublimation is the "elevation
of solids by heat in the dry form", it is used for mercury, sulphur,
antimony, etc. His apparatus are the same as those of CHARAS and
other French chemists.

This is also the case with CASPAR NEUMANN, apothecary and
intimate friend of Prince FRIEDERICH WILHELM I. In his works (454)

455) he minutely describes the proper ways of sealing glass vessels "hermetically" by the new methods described and invented by KUNCKEL.

Neither did the Dutch chemists of this generation produce anything strikingly new. ANTONIUS DE HEYDE (HEIDE), born at Philippine in 1640 was educated by a relation who was a surgeon and a member of the guild of Middelburg. In 1667 we find him at Leyden as a student and on November 1st 1668 he obtained his M.D. degree. Then he worked as a physician at Middelburg and wrote his celebrated work on the mussel, describing for the first time the movements of the vibrissae. Under the influence of JEAN DE LABADIE he left Middelburg and we find him at Amsterdam in 1680 where he died, presumably in 1695. His work on pharmacology (291) also translated into German (292) complains that few understood proper distillation and therefore produced waters which had lost all the virtues of the original herbs. One should preferably take wine or alcohol to macerate the herbs and distill them. When properly distilled with water the oil from the herbs can be obtained in a pure state with a separating funnel. He mentions that he has seen applications of the destillatio per descensum in the tar industry of Bohemia "where they use small baking ovens open on both sides. Therein long logs of wood are placed, the ends of which project from both sides. If one of the sides is fired a green liquid drips from the other ends of the logs and this is used in Bohemia and Austria as a lubricant for waggons and carts". His main aim is the proper distillation and preparation of simplicia and composita which he describes and performs quite painstakingly without introducing any new ideas.

Nor had the famous HERMAN BOERHAAVE anything to add to the existing distilling apparatus. BOERHAAVE was born at Voorhout on December 31st 1668 and obtained his Ph. D. at Leyden (1690) and his M.D. at Harderwyk (1693). He taught theoretical medicine at Leyden from 1701, obtaining a post as professor of medicine and botany in 1709. From 1714 he taught practical medicine and after 1718 chemistry. He died at Leyden on September 23rd 1738. BOERHAAVE, who achieved international fame as a physician and a chemist, was an excellent tutor but his work is not as original as his fame would make us believe. In his chemical work (78) (79) he devotes much space to distillation and he describes a kind of alembic, which like the fantastical forms devised by LEFEVRE and BARCHUSEN has a large-cooled surface. He says that "one should take a conical condenser in the form of a sugar-loaf, if the parts which one should elevate are easily loosened" (Fig. 110). BOERHAAVE is the man who introduced the use of the word alcohol for the spirit of wine only. According to him the volatile oils consists of "a coarse resinous part insoluble in water" and a subtile "spiritus rector" (esprit recteur). The latter is essential for the smell and the taste. By action of air and the evaporation of the spirit the oil becomes thick and resinous. In this way he tries to explain the polymerisation of these oils.

The French produced some excellent chemists in this period most of them being Calvinists, who wrote some famous textbooks used all over Europe. One of them was MOYSE CHARAS, an "apothicaire-artiste du Roy en son jardin royal des plantes", a chemist, who later also became a physician. After the repeal of the Edict of Nantes he fled to England (1688) where he obtained his M.D. Then he worked at Amsterdam to be called to Madrid where he became court-physician to CHARLES II. The Inquisition forced him to recant, then he returned to LOUIS XIV of France who made him "membre de l'Académie". He died in 1698 at the age of 80. According to his *Pharmacopée Royale* (126) distillation is "the elevation of the watery, or mixed saline parts and spirits followed by a descent". Apart from the old tripartite division of the destillatio he also distinguishes a destillatio per deliquium, which is nothing but the deliquescence of salts. Rectification is the "new purification or exaltation of the essential parts", but according to CHARAS this also includes the calcination of salts, dissolution, filtration of paints, etc.

This work dedicated to COLBERT contains a few chapters devoted to distillation. The first fraction obtained in distillation is called "water", the second a spirit called "mercury", then an oil called "sulphur", a "salt" may distill over but usually remains partly in the residue called "earth". When talking of the distillates of herbs he calls special attention to absinthe but he also prepares the usual waters from all kinds of substances even from "skulls".

The production of alcohol from wine is still considered a matter of much patience, which goes to show that it was not yet prepared in great quantities in France by 1680. The yield is about one sixth of the wine taken. CHARAS believes that the tartar content is directly connected with the yield of alcohol. He is however right in telling us that sponges hung in the alembic do not help in distilling strong alcohol. A new test for the strength of the alcohol is developed. A few drops are poured on gunpowder which will not burn if the alcohol still contains appreciable amounts of water. When rectifying alcohol one should distill off three quarters of the contents of the still and repeat this four or five times. This is the only way of obtaining

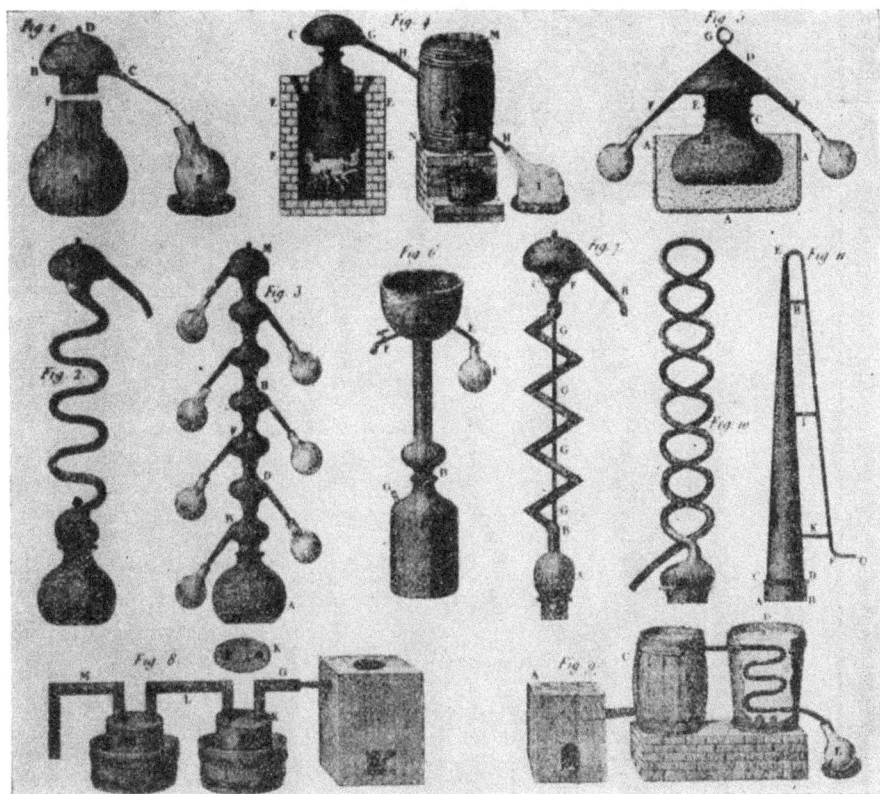

Fig. 110. Distilling apparatus from different periods, LENORMAND

    fig. 1. The alembic of the ancients
    fig. 2 & 3. PORTA's still
    fig. 4, 5, 6, & 7. Apparatus designed by LE FÈVRE
    fig. 8 & 9. GLAUBER's apparatus
    fig. 10. BARCHUSEN's condenser
    fig. 11. BOERHAAVE's condenser

strong alcohol with the primitive apparatus employed in those days (Figs. 111, 112, 113). It will be seen from CHARAS, designs that he too built very high air-cooled coils on the alembic, using a Moor's head in the shape of a heart to collect the volatile distillate (Fig. 113).

Similar remarks are made by CHRISTOFLE GLASER, a good chemist and "démonstrateur au Jardin des Plantes", who was pharmaceutical

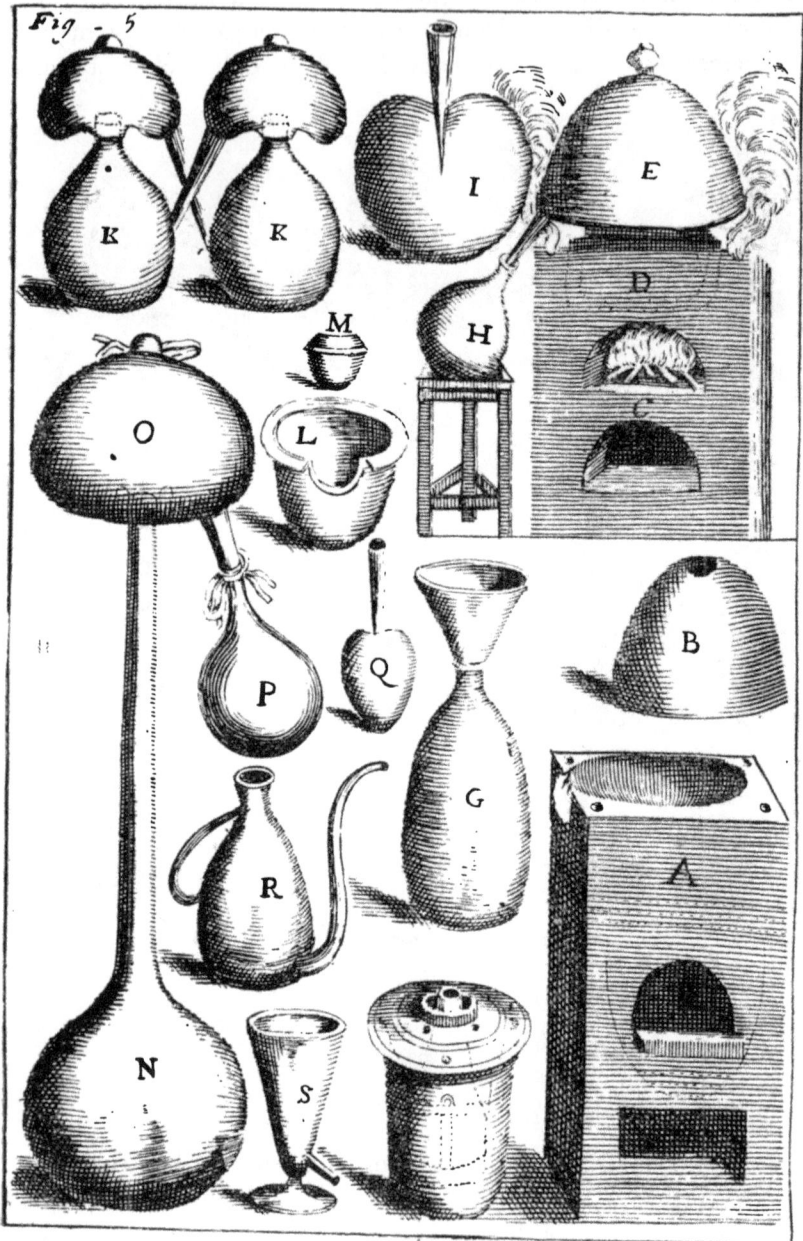

Fig. 111. Distilling apparatus, CHARAS

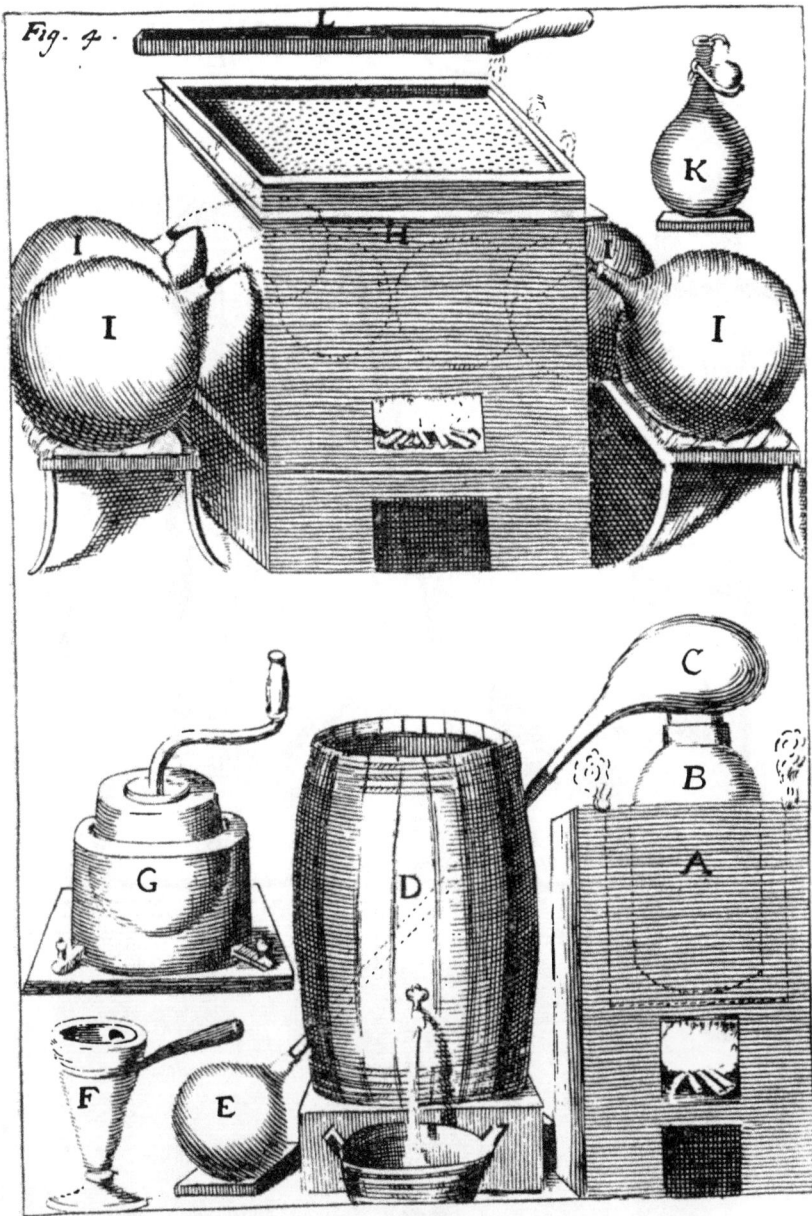

Fig. 112. Distilling apparatus, CHARAS

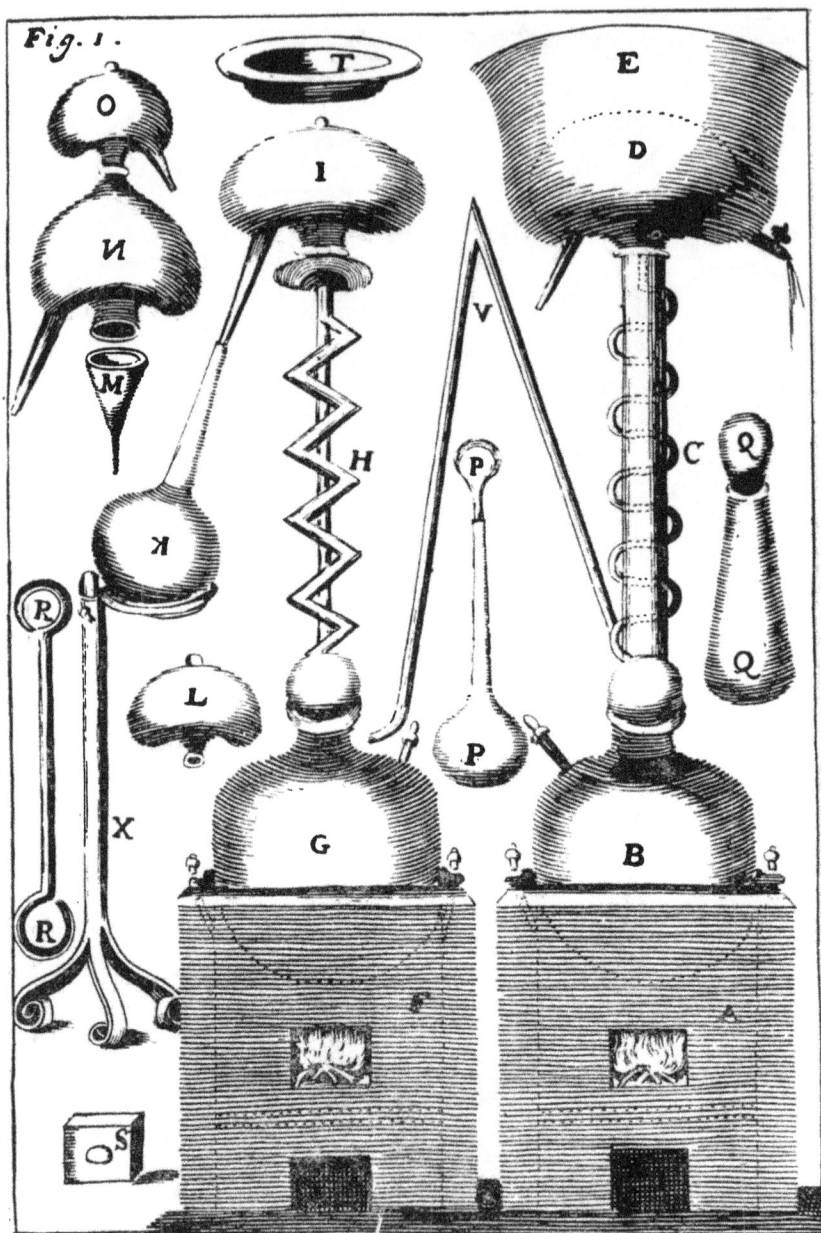

Fig. 113. Distilling apparatus, CHARAS

chemist to King Louis XIV and the Duke of Orléans. He was involved in the criminal case of the marquise DE BRINVILLIERS (1667)
and though acquitted had to leave the country and return to his native
town Basel, where he died. The figures in his book exactly like those
of CHARAS and need no further discussion. In his works (261) we
find the word "alcoholiser" used for "pulverising", but also for the
"exaltation of any spirit or essence which is stripped of its phlegm and
all other impurities, hence the name alcohol of wine". He frequently
advises cohobition, that is "distill several times, each time adding the
distillate to the residue in the cucurbit and redistilling again". Rectifying is "redistilling the spirits to render them more subtile and to
exalt their virtues".

Some impressions of the contemporary courses in chemistry can be
obtained from the pictures (Figs. 114, 115 & 116) from the book
(44) by ANNIBAL BARLET, a French physician who taught alchemy in
Paris in the days of CHARAS. He designed a distilling tower like that
of LIBAVIUS, which he calls a "cosmic furnace", but which is far
more primitive than the original. BARLET mentions, correctly, that
distillation should not be interrupted. Volatile spirits should always
be distilled in alembics with a worm-cooler and heated on a slow fire.
For the distillation of roots and soft fruit he advises the use of an
ash-bath, but for leaves, flowers and seeds one should use a water-
bath. When distilling fruit or "waters" the still should not be filled
more than two thirds. The "spirit of wine" which is nothing but a
sulphurous liquor very subtile because of its heavenly nature" when
well distilled should give no "sweat" on evaporation. It should be
distilled "giving a fluid stream by the gaseous elevation with the
help of mixed watery and sulphurous parts" and redistilled at the
same temperature. The first redistillation shall take off six pints
from twelve, the second five from six, the third four from five, etc.

The last of this generation was NICHOLAS LEMERY, whose Cours de
Chymie (358) was reprinted no less than 23 times. This chemist who
reformed the pharmacology of his time was born at Rouen on November 17th 1645 and studied pharmacy and chemistry under GLASER at
Paris (1666) and afterwards for three years at Montpellier. In 1672
he was back in Paris as apothecary and teacher of chemistry. As he was
a Calvinist he had to fly to England in 1683 but he returned to Paris
in 1684 and renounced his faith in 1686. He died at Paris on June
19th 1715 after having created a new nomenclature of pharmacological
products. Some of the symbols used by him are included in Table II.

Des Vegetaux.　　　　2.Figure.

Fig. 114. Views of a laboratory from BARLET's book

Des Vegetaux. 3·FiG·

Fig. 115. Views of a laboratory from BARLET's book

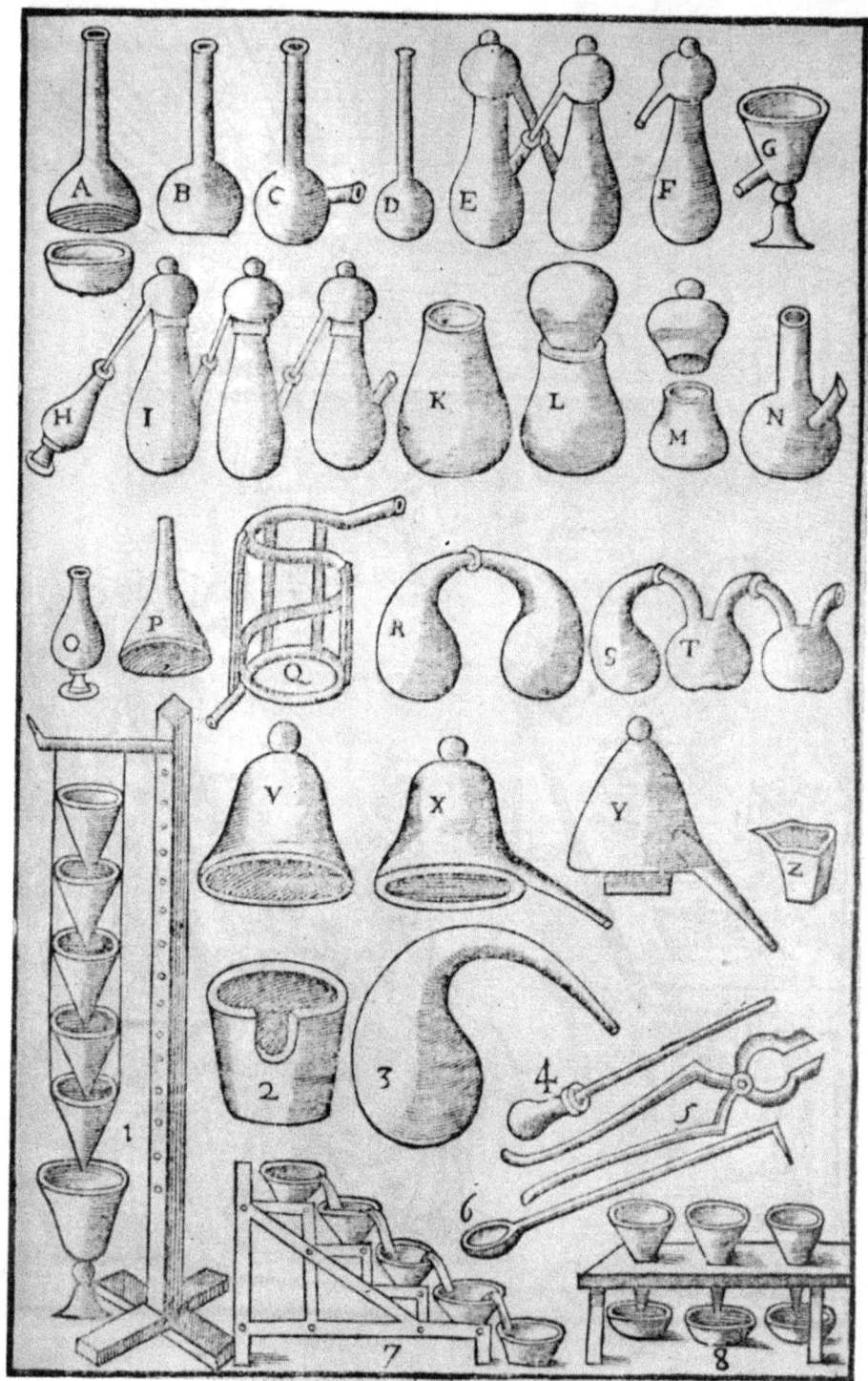

Fig. 116. Chemical apparatus, BARLET

In his book he discusses in detail the different vessels used in chemical operations. He still uses the Moor's head (Tête de More) and mentions a "rosaire", an alembic especially suited to the distillation of rose-oil. Mineral acids are distilled in retorts made of earthenware which he calls "cuine", while the glass retorts retain their old name "cornue". The receivers have the name of "matras". He still mentions that cohobation is used "to open up the bodies" and he repeats the old theories on distillation (Fig. 117).

The early eighteenth century shows a considerable gap in technological literature, but the second half of that century produced several theoretical and practical chemists worth discussing notably in France. The earliest of these is PIERRE JEAN MACQUER. This member of the Royal Academy of Sciences, member of the Faculté de médecine and professor of pharmacy was born at Paris on October 9th 1718. He first studied medicine but then applied himself solely to chemistry under ROUELLE. Already in 1745 we find him a "membre de l'Académie", then professor of chemistry "au jardin du Roy" and controller of the porcelain factories of Sèvres. He died on February 15th 1784.

Of his different works (397) (398) (399) his *Dictionary* was the most influential one. Here he describes the "florentine bottle" used in the distillation of essential oils to draw off regularly the excess water. Distillation is "the operation in which one separates and collects the liquid and volatile parts of a body by means of a proper degree of heat". Both destillatio per ascensum and per descensum are used, but the latter is rarely suitable according to MACQUER. Bodies with a high boiling point are distilled in a retort, ad latus, especially mineral acids, sal ammoniac, etc. If the condensate is solid one should call the operation "sublimation". Redistilling is called rectifying and "dephlegmation is nothing but dehydrating".

"Alcohol" still occurs in the sense of a very fine powder and of spirit of wine. The latter is obtained by simple distillation of wine and better designs of stills were the objects of competition in MACQUER's days as we shall see. He mentions that the taste of brandy, especially the difference in taste between brandy from wine and from corn, is a function of the "size of its oily parts", as was proved by GERH. GYSBERT TEN HAAF in the *Haarlemmer Verhandelingen* XIX, p. 189. As distillation is usually conducted too quickly and the distillate does not drip as it should one must always rectify the crude alcohol.

When discussing the form and size of the distilling apparatus he

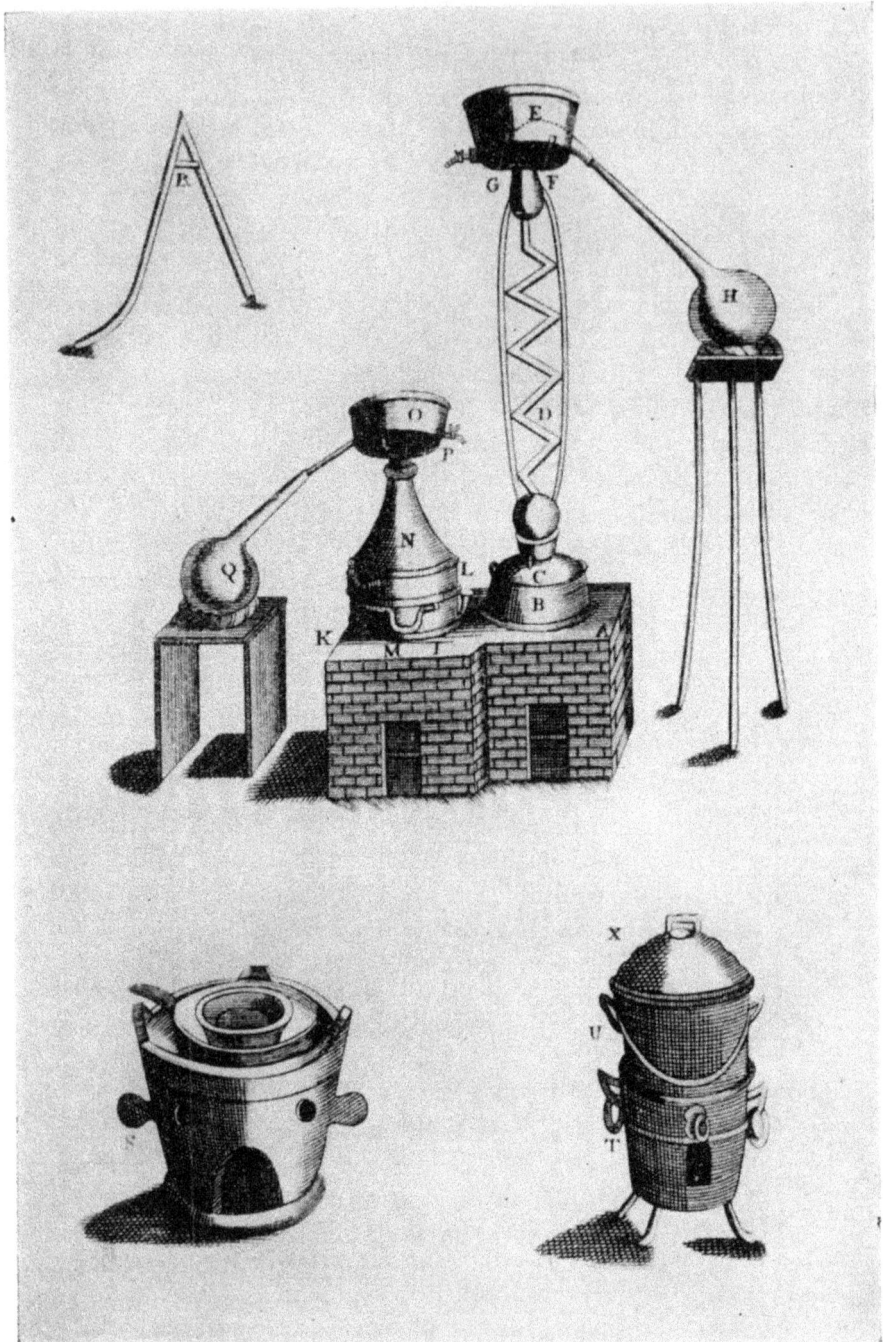

Fig. 117. Apparatus designed by LEMERY

remarks that a still should not be too deep or too wide, for the velocity of distillation is proportional to the heated surface. The spout of the alembic should slant towards the receiver. He describes in detail the Moor's head cooler, though he actually says that "some prefer a worm-cooler". In that case one should not interrupt the operation. Acid or alkaline liquids should be distilled in retorts. Alcohol burns up without leaving a trace, this is a good test. BOERHAAVE mentions that alcohol forms pure water when burning. The distilling vessels should be well proof against the action of fire and chemicals. The best test is to melt lead in them. Metal vessels, especially copper and iron ones, give poisonous distillates. One should use glazed or preferably glass ones. His description of the furnaces and other details is very precise and practical. It should be noted that he mentions the normal globular retort and a pear-shaped retort, which he calls "English retort" (cornue angloise). It is also clear that KUNCKEL's new glass technique is by now well developed for MACQUER makes frequent use of ground glass stoppers.

ANTOINE BAUMÉ was not only an apothecary but also a very technologically minded person. He was born at Senlis in 1728 and spent many years as an apothecary's assistent first at Compiègne, then at Paris (1745). He studied pharmacy and obtained his Master's degree "summa cum laude" in 1752. His remarkable pharmaceutical work opened him the doors of the Academy des Sciences (1773). As he was ruined by the French Revolution he had to take up his old profession again, which he had left for scientific work since 1780. In 1785 he became Pensionnaire de l'Académie and in 1796 of the Institut de France. He was one of the last adherents of the theory of the "four elements" and called fire "an essential liquid matter, principle of fluidity of other bodies and always moving".

In his pharmaceutical handbook (53), the ninth edition of which was printed in 1818, he discusses the distillation of wine, essential oils, etc. in different chapters. The laboratory stills should be preferably made of silver, tinned copper, tin, glass or earthenware. He too puts all his trust in the water-cooled Moor's head (Fig. 118), though he admits that the worm-cooler often gives distillates which are better in taste and smell. Of particular interest to us is the prize-essay (54) which won him the first prize in the competition of the Société Libre d'Emulation pour l'encouragement des Arts, Métiers et Inventions utiles, établie à Paris, which asked in June 1777: "What are the most advantageous forms of the stills, furnaces and all the instruments used

in the work of the large distilleries?" BAUMÉ received the first prize
of 1200 livres, the abbé MOLINE the second prize of 600 livres. The
original publication was severely criticised by the abbé ROZIER and
PARMENTIER, so BAUMÉ changed his still design shortly afterwards.
This original essay was published in the *Journal de Physique* (54), its

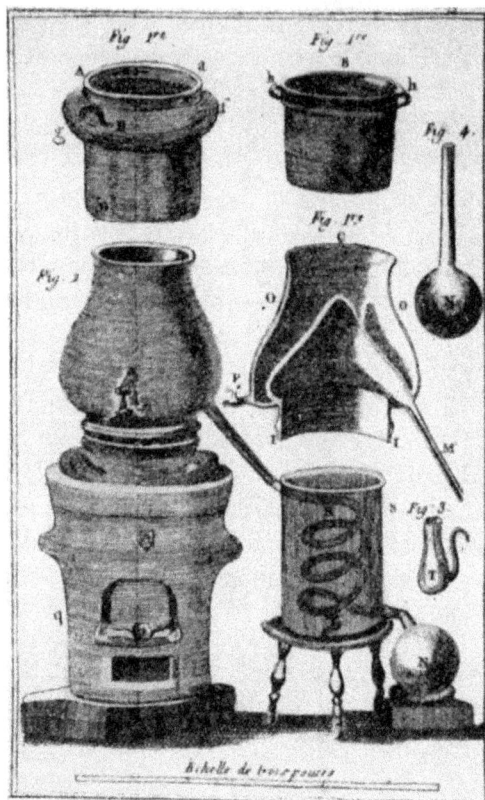

Fig. 118. BAUMÉ's laboratory still

contents were strictly limited by the questions asked by the Society
which were:
1. Is the actual form of the stills and their capital the best?
2. Which would be the most advantageous form of stills?
3. Which should be the relations between the fire-door, the furnace
   and the flue conducting the smoke to the chimney and what is the
   best place of the chimney to consume less wood and keep the heat
   longer in the furnace?

4. Is it more advantageous to heat several stills on one fire or one still of a larger surface and the same conditions as the smaller ones together and in which case does one obtain the largest yield of brandy?
5. Show the advantages of burning coal and what should the form of the furnace and still be in that case?
6. Describe the improvement of the coolers, receivers, etc. which are used in the distilleries.
7. What quality should the wood or coal have that is most profitable at about the same price?

BAUMÉ designed an elliptical still with a host of smaller alembics instead of one still-head with a fairly narrow spout. He therefore both enlarged the heating surface and the outlet of the vapours. These ideas were later adopted by MILLAR and the Scottish distillers. BAUMÉ gave six different forms of this type of still, but later went back to the cylindrical still after ROZIER's and PARMENTIER's remarks. He proves that the chimney should be placed as far as possible from the grate and that one large still is better and more efficient than many smaller ones. Coals want a slightly different grate, than that normally used for wood. His coolers and receivers are of the well-known type. Coal is too expensive in France like charcoal, therefore wood is the only practical proposition (Fig. 119). To avoid overheating metal gauze is better than the usual mat.

The abbé MOLINE (434) received the second prize for a still shown in Fig. 120 being a combination of four cucurbits with watercooled Moor's heads built into one furnace. He too decides that stills should not be too deep, but as the evaporation occurs at the surface and the vapours should be drawn away, both the surface and draught are essential factors. The fire or better still the flue gases should heat the entire still and the heat should not be lost through the walls of the furnace. When using the more expensive coal less grate surface is required, but in France good dry wood is more economic.

The above mentioned abbé ROZIER was a great authority on all questions connected with agriculture. He was born at Lyon in 1734 and became a priest though he was very much interested in agriculture. BOURGELET called him to Lyon as a professor of the local university, but after some years we find him directing several scientific publications. The king of Poland tried to entice him to his country as an agricultural expert but when he declined the king used his influence at the French court and got ROZIER the place of prieur commanditaire

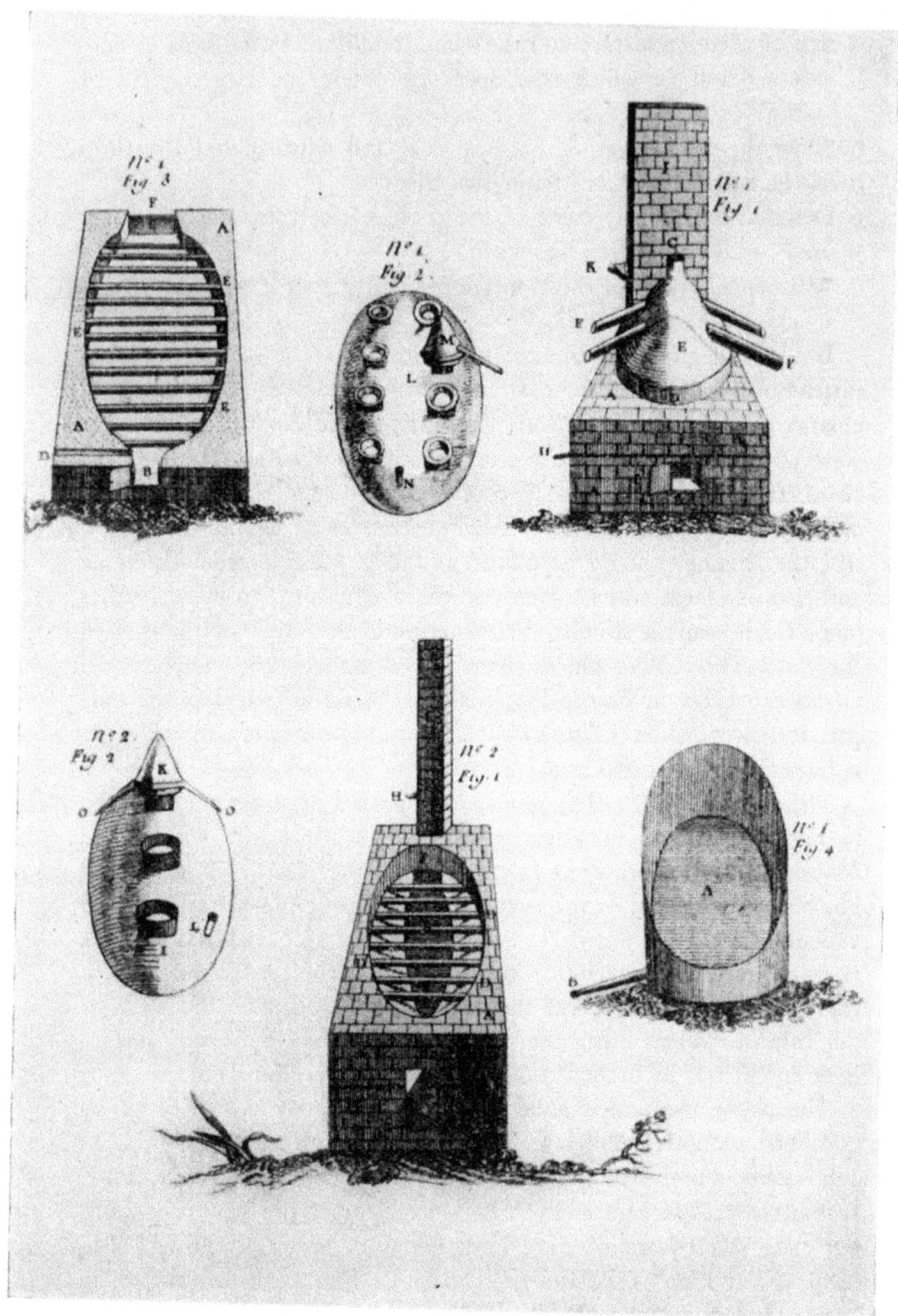

Fig. 119. BAUMÉ's still from his prize-essay

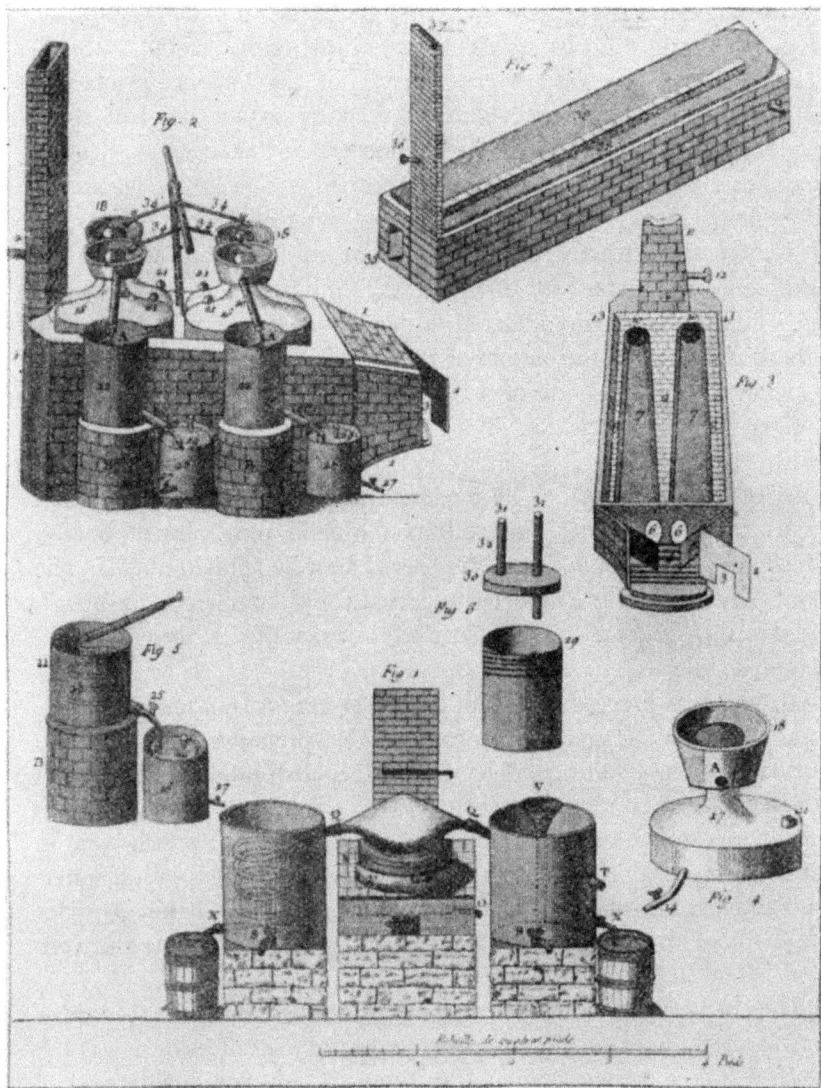

Fig. 120. The stills of BAUMÉ and MOLINE (figs. 2-3)

of Nanteuil-le-Haudoin. After the French Revolution broke out he was priest at Lyon where he was killed in his bed by a bomb in 1793. He did not finish his great work (523) for the tenth part was published in 1798 only and two supplements not earlier than 1800.

In this work we find a good survey of the knowledge on our subject at the end of the eighteenth century as ROZIER had some excellent collaborators like the distillation expert PARMENTIER. ROZIER mentions that the stills are generally heated with coal or wood the latter being already quite scarce in some parts of France. The stills of BAUMÉ and MOLINE are criticised. For crude alcohol and brandy distillation in a water-bath or by direct heating are preferable. In the case of lees of wine the material is put in the still in a basket and a wire-gauze is put on the bottom of the still. Puking of the still contents is avoided by the addition of a few drops of olive-oil. More attention is given to the construction of the furnaces than to the design of the stills.

Distillation was defined by the Academy of Lyon as "an operation by which one separates and collects the fluid principles of bodies, which are volatile at different degrees of heat by applying heat". The word "distillateur" is now often superseded by "bouilleur", "brûleur". The receiver, generally called "bassiot" in France, is dubbed "buguet" in the distilleries.

ROZIER then proceeds to tell of ARGAND's pre-heater and its virtues, these will be discussed in due course. This preheater is also recommendable when distilling lees of wine. When distilling wine the capital is often placed on the still when the wine is already boiling and its distillate is tested with a hydrometer, several forms of which were in use in France by this time. The best brandy is obtained with a water-bath and cooling with cold water in a worm-condenser. If oily particles are distilled over, the distillate turns turbid ("bronzer" is the verb used by ROZIER) and the taste is bad.

The chapters on wine written by CHAPTAL mention that the alembic was formerly built as high as possible, but in that case there was too much reflux and the empyreumatic products gave the distillate a bad taste. One should therefore insulate the alembic and should not use it as a cooling element but heat the entire mass of the liquid as quickly as possible, draw off the vapours quickly and condense them effectively and rapidly. The bottom of the still should be slightly convex and the spout should make an angle of 75° with the alembic. Thus the brothers ARGAND constructed their stills and obtained excellent brandy. On the other hand the tastes of public differ widely and the peoples of northern Europe like their gin better when it is raw. The French prefer a softer brandy. The distillate is tested with the drop of olive-oil test or by the gunpowder test. The first test probably originated in Spain and was introduced by BORIE and POUJET OF CETTE in the Languedoc trade.

Before compiling these chapters and volumes ROZIER had already competed for the prize of the Société d'Agriculture offered in 1766 for the best answer to the question: "What is the most economical way of distilling brandy from wine as regards the quality and quantity of brandy obtained and the costs?" Apart from the prize-essay of ROZIER, which is nothing but a general survey of the question (522), the society at Limoges awarded prizes to the essays of DE VANNE, an apothecary of Besancon, and MEUNIER, an "ingénieur des Ponts et Chaussées". DE VANNE first defines fermentation, effervescense and ebullition and proceeds to design a still with a wooden or copper stirring apparatus that can be turned from the outside. This still is used for the primary distillation, but rectification remains necessary. MEUNIER is not interested in theory for as he says "in the silence of the study one can discuss theories, but only in practice one becomes an artist". When distilling wine one has to distill three times obtaining brandy, spirit of wine and alcohol respectively. He then describes the still in general use in the distilleries of Saintonge and Augoumois and mentions that generally people are far too careless with this material containing so many volatile compounds. The strength of the distillate is tested by tasting, weighing with the hydrometer and a burning test. He then proceeds to describe a "parabolic furnace" which is of no importance to us. His stills are of the usual kinds, with slight, unimportant modifications.

POLICARPE PONCELET's book (490) on the chemistry of taste is a typical French book which states that the different flavours are vibrations acting with more or less force on the tongue. There is therefore a music of flavour like one of the ear and he proceeds to construct scales of flavour which should be studied by the distiller. PONCELET was an agriculturist born at Verdun and a contemporary of ROZIER.

ROZIER found a competitor in JEAN FRANÇOIS DEMACHY whose work on the art of distilling (167) had considerable influence outside France as it was, for instance, translated into German (165). DE-MACHY, a pharmacological chemist and director of the military hospital of Saint Denis, was born at Paris in 1728. Later he became director of the central dispensary of the French army. He was a fierce opponent of LAVOISIER and the new chemistry. His book is a large compilation of facts on the distilling trade. He mentions trials of the fermentation of rice, cane-sugar, dates, milk, beans, coconuts, figs, manna, molasses and all kinds of fruit, stating that none of these trials had yet been an economic succes. The first part of his book deals with the manufac-ture of mineral acids. The second part discusses the manufacture of "fluid chemicals" such as spirit of wine from wine and molasses, and essential oils. The third part is concerned with the manufacture of solid chemicals such as borax, magnesia, etc. The second part of the volume is called Le Distillateur-liquoriste and gives details on the manufacture of alcoholic beverages, liqueurs and syrups but also on the retail-trade and the way to manage a restaurant or café. The third book deals with the manufacture of vinegar and its distillation.

The earliest work of the eighteenth century devoted entirely to distillation was written by DEJEAN (169), it is dedicated to the "maîtres limonadiers-distillateurs de toutes sortes de liqueurs, esprits de vin, huiles, essences de la Ville et Faubourg de Paris". He mentions that practically only the destillatio per ascensum is used and that there were more than 14 kinds of alembics in use. The stills are made of glass, earthenware, tin or tinned copper. The general cooler is the copper water-cooled Moor's head, though some use worm-coolers. Though direct heating is often used a water-bath has several advantages, especially when safety is imperative. The worm-cooler is seldom used because then distillation would take too long (sic!). DEJEAN goes into details on the safety measures to be taken and the construction of the furnaces which should be adapted to the kind of still used. The rest of his book is filled with recipes of different liqueurs, etc.

Many interesting articles on the distillation industry are found in the famous Encyclopédie of DIDEROT and D'ALEMBERT (175) which is of course a compilation, but which contains many bibliographical references and some interesting historical details.

Real progress was achieved by the introduction of the preheater by AIMÉ ARGAND. This practical scientist, better known as the inventor of the Argand burner, was a physician of Geneva who came to France with his brother. In 1783 he crossed to England where he in-vented his lamp and got into touch with the famous manufacturer Boulton through a Mr. Parker. He patented his lamp in 1784 (British patent No. 1425 of the twelfth March, 1784) and on May 1st announced his intention to commit to Boulton's care the manu-

facture of most parts of the lamp. The very ease of its manufacture
was his undoing. For the market was flooded with infringements, the
tide of which ARGAND vainly attempted to stem. In 1787 for instance
we find him in Paris to defend his rights against QUINQET and LANGE.
His opponents challenged his patent on the ground of prior disclosure
and succeeded in getting it declared invalid, a crushing blow to ARGAND
and even to Boulton who lost some money in it, but the latter did
make some of these lamps, at least in 1797. ARGAND, however,
impoverished by his lawsuits devoted himself to alchemy to try and
win back his fortune by making gold, but he became more and more
a victim to bouts of melancholy and died poor in 1803 in London.
His lamp is of some interest to us for it was the reason why petroleum
or "rock oil" was distilled in the sixties thus giving birth to a new
flourishing application of distillation, the petroleum industry, when
the older colza-oil became too expensive.

ARGAND's fate is that of many a great inventor by whom the world
has benefited immensely, for he made many other practical inventions.
CHAPTAL, for instance, mentions that he greatly improved the furnace
of the distillery where he and his brother worked in France. In 1780
he had been called to France by a M. JOUBERT who had built a
distillery on his estate of Valignac near Colombiers between Nîmes
and Montpellier. Here ARGAND tried out his idea of using the wine
to be distilled as the cooling liquid in the worm-cooler belonging to
the still. This pre-heater proved an immediate success, it saved large
quantities of fuel. ARGAND was able to distill in 6 hours and 18
minutes 92 veltes of wine with 44 livres of coal obtaining 18 veltes
brandy "preuve d'Hollande". His competitors in the time of 5 hours
and 42 minutes distilled only 50 veltes of wine with 60 livres of coal
obtaining only 5 veltes of brandy of the same quality. The succes was
so striking that his pre-heater (Fig. 142) was introduced in a period
remarkably shorter than that of most other improvements of the
distilling trade.

This pre-heater was equipped with stirrers in Germany, where it
was introduced soon after, because of the thick washes treated there
and also in the United States of America early in the nineteenth
century.

A still for the distillation of sea-water on board ships (Fig. 121) was
specially designed by POISONNIER in 1770, it was also adopted for
the distillation of alcohol from wine (489). We shall discuss the
merits of the Poisonnier cooler in due course.

The last important figure of the eighteenth century distillers and the promotor of the new still designs in the early nineteenth century was J. A. C. CHAPTAL comte DE CHANTELOUP, who was born at Nogaret in 1756 and who studied medicine at Montpellier in 1777

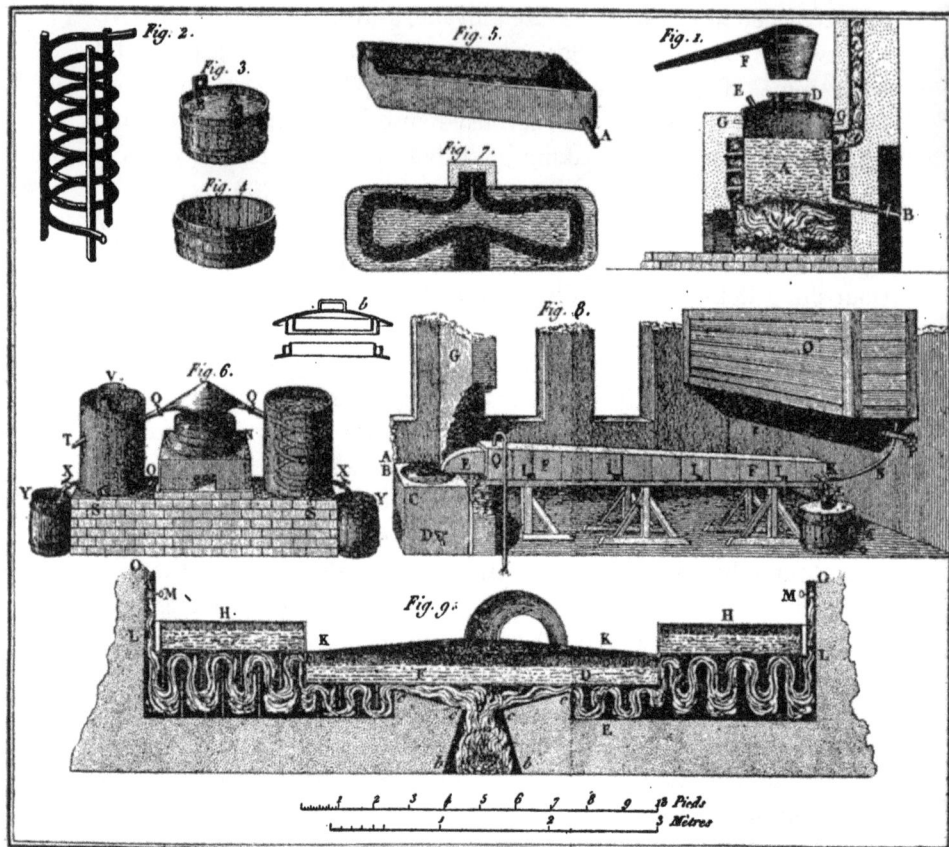

Fig. 121. Eighteenth century distilling apparatus, LENORMAND

fig. 1-4. CHAPTAL's still
fig. 5, 7. BAUMÉ's still
fig. 6. MOLINE's design

fig. 8. POISONNIER's still for sea-water
fig. 9. Furnace designed by CURAUDAU

moving to Paris to study chemistry. He became professor of chemistry at Montpellier in 1781 and soon achieved European fame because of his articles in encyclopaedic works like that of ROZIER, etc. This renowned technologist became director of the works of GRENELLE in 1793, teaching chemistry at the Ecole Polytechnique, soon becoming "membre de l'Institut".

From the 18th Brumaire until 1804 he was Minister of the Interior ·
and showed a strong initiative in stimulating French industry and
trade. Soon he was a senator and in 1806 he became comte DE CHAN-
TELOUP, member of the Academy of Sciences in 1816 and also member
of the Chambre des Pairs (the French House of Lords) which he
remained until his death in 1832. CHAPTAL, whom Napoleon often
consulted personally, was the man who created the modern French
chemical industry. He wrote many books, for instance a large work on
viniculture in collaboration with ROZIER and PARMENTIER (127), a
special work on the preparation of wines (with some interesting plates
in the second and third editions only) (122) and several general works
on technology and economy (123) (124) (125) which contain
chapters on the distilling trade. Many of CHAPTAL's works were translated and profoundly influenced the distilling art in other countries.

CHAPTAL mentions that the purpose of distillation is separation and
that it remained the only means of analysing plants and vegetable
products until the end of the eighteenth century. He gives many
details of the developments in the beginning of the nineteenth century
until the advent of CELLIER-BLUMENTHAL's distilling column for which
we refer to chapter VIII. It is interesting to note that CHAPTAL says
that the invention of distillation is not due to the Arabian chemists
but to VILLANOVA. A simple apparatus designed by him for laboratory
use in shown in Fig. 122. At the end of the eighteenth century there
had been little progress as CHAPTAL remarks, for fractionation and condensation were still bad and there were no proper means of regulating
the fire. These were all achievements of the early nineteenth century.
Only in CHAPTAL's time were filtration, sublimation and all the
other operations formerly included in "distillation" properly distinguished by all the chemists. This better nomenclature was no less a help
towards more rational distilling than the application of all the new
physical principles.

In *Germany* there were of course many chemists who, like some of
the French, went on repeating the old methods without trying to
improve them. This is for instance the case with G. H. BURGHART
(1705-1776), an apothecary, who took his M.D. degree at Frankfort
(1730). In his *Distillierkunst* (106) by which word he means pharmaceutical chemistry, he goes on enumerating the old forms of
distilling and heating as if the times were not changing. According
to him distillation is "an operation by which one separates by means
of heat the more watery, spirituous, volatile parts from the gross, earthy

and fixed ones in the somewhat cooler alembic or receiver in the form of smoke or vapour, where they collect in the form of a liquid". He admits that rectification is often the same as dephlegmation, which is nothing but the separation of superfluous humidity. The rest of the 400 pages of this book and the 550 of the *Neue Zusätze* contain nothing but the normal pharmaceutical recipes of the period.

A similar book is that written by ARTHUR CONRAD ERNSTING (1709-1768), a physician. We find here the same definition of distil-

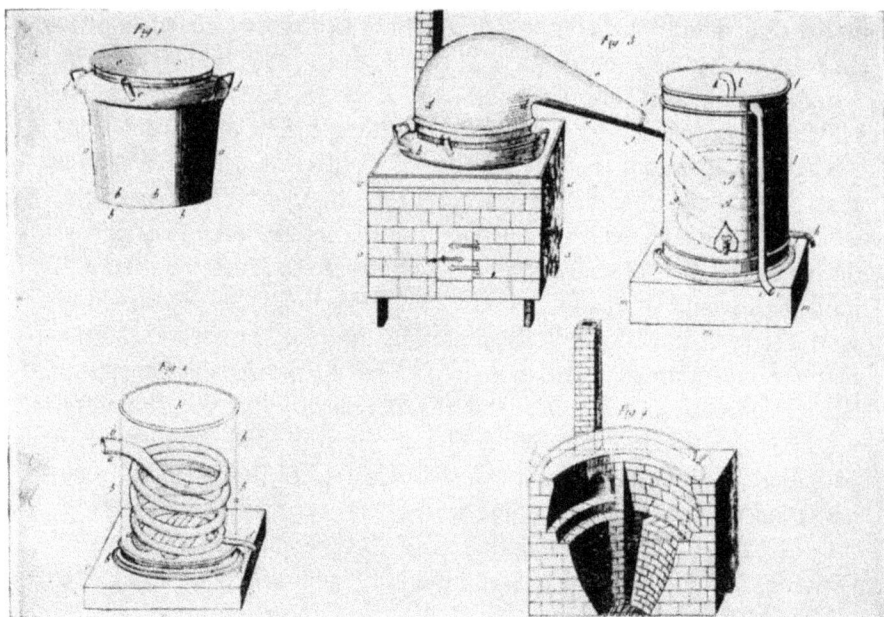

Fig. 122. CHAPTAL's laboratory still

lation. ERNSTING (211) cites a popular method of destillatio per descensum still in use in his days. For he says, that the old women still make rose-oil by covering a pot with a cloth, filling this with rose-leaves and heating the cover with a hot stone. The "rose-water" then sweats through the cloth into the pot. He also mentions destillatio per deliquium or destillatio per descensum frigida, as he calls it, which again is nothing but the deliquescence of salts.

C. G. HAGEN (1749-1829), professor of physics and natural history at Königsbergen took distillation to be a kind of evaporation, but in the latter case the "vapours are not collected". By dry distillation he meant the distillation of such products as the empyreumatic oils, mineral acids and volatile salts. Both rectification and dephlegmation are, according to him, a kind of concentration (279).

J. F. WESTRUMB, the technologist, was born at Nörthen on December 2nd 1751 and after completing his studies he devoted himself to scientific work under the influence of the chemist KLAPROTH and the botanist EHRHARDT. He became quite famous, but the decline of the phlogiston theory, of which he was a fanatical adherent, embittered his life and he died on December 31st 1819 in the midst of writing pamphlets to vindicate his beloved phlogiston as the last of its faithful. In his book (653) he classes distillation with sublimation and evaporation saying: "All three are founded on the principle that heat lessens the cohaesion of the bodies and slackens the bonds between their particles, so that the lighter ones can rise and fly away. The liquids are transformed by heat into vapours, the solids into smoke (halitus) and both are separated from the fixed and fireproof parts with which they were formerly mixed or to which they were bound."

"Heat spreads out in every direction and gladly enters into equilibrium or in other words a heated body gives off heat to a cold one which touches it until both have the same degree of heat. If therefore a body loosened into vapour or smoke touches a cold body, its heat-matter is attracted thereby, it separates itself from the vapour which becomes visible as a mist. Slowly they return to their old condition and take their specific form again.

Apart from the liquid or dry distillation we have also the abstractio in which liquids are drawn over solids."

The rising German distilling industry, however, produced several practical technologists, whose work was more important than the theorizing of those mentioned above. Among the most important of these technologists we must count JOHANN BECKMANN, born at Hoya (Hannover) on June 4th 1739. He studied theology and natural science at Göttingen and went to St. Petersburg as a teacher. Thence he went to Sweden (1765) where he befriended LINNAEUS and in 1766 returned to Göttingen to teach philosophy to which courses in the economics of agriculture were added in 1770. BECKMANN, who was the founder of the history of science and technology, died in 1811. He wrote many books of major importance (57) (58) in which we read chapters on the improvement of hops, the manufacture of good wines, the story of the adulteration of wines, the story of beer, etc. In his handbook of technology (56) he mentions many details on the story of distillation, believing that the Arabian chemists had first made alcohol from wine and taught it to later generations. His work was most valuable to the distillery of his period especially his economic views and essays, but it gave no new suggestions as to distillation methods.

This is also the case with the works of J. H. M. VON POPPE, born at Göttingen on January 16th 1776 as the son of an engineer attached to the university. He earned the money for his studies by writing compilations on different subjects, finishing his studies in 1803 and taking a post as a teacher at Frankfort (1804-1818) until the university of Tübingen called him there to teach mathematics and technology. He died at Tübingen on February 21st 1854. His books (491) (492) are very valuable from the historical point of view and should be consulted by everyone writing on the history of technology.

Of more practical importance is the work of C. C. A. NEUENHAHN (450) (451) who was born at Nordhausen in 1745 and died in that town on July 9th 1807. NEUENHAHN, who was a member of many agricultural societies, had much practical experience and made several practical suggestions on the subject. One of his papers contains severe crticisms on the shape of the alembic used in Germany. This was the old alembic of the generation of BRUNSCHWYGK, without a collecting drain or rim in the interior and with a spout emerging from the side. The French distillers had returned to the ancient Coptic alembic, they fitted their heartshaped alembics with an inner drain (See Fig. 118) because they swore by the water-cooled Moor's head and therefore really collected their condensate in the alembic itself. But the Germans relied more and more on the worm-cooler for the condensing and their large copper alembic, if not properly insulated acted as a severe refluxing element, which in the case of brandy might lead to all sorts of contamination by empyreumatic oils. On the other hand the practical Dutch distillers kept the alembic fairly small and took off the distillate from its top by a semi-circular copper tube leading to the worm-cooler (92). This idea was not imitated in Germany but the suggestions of NEUENHAHN had the effect that the spout of the alembic was widened and the alembic itself kept as small as possible (59).

The ancient distilling apparatus were not very safe and NEUENHAHN, BORDOWIG (of Brandenburg) and HERMBSTÄDT suggested measures to prevent explosions and cracking open of the alembic. Both RECHBACH (of Dresden) and BRAUMÜLLER (90) advised conducting the distillation in sand-baths which would avoid the formation of empyreumatic oils.

The chaplain LAUBENDER suggested adding some olive-oil to the contents of the still to avoid puking (1798). This was of course quite an old recipe, which, however, did not always work. NEUENHAHN severely criticised the suggestion of the brothers GRAVENHORST who wanted to use Glauber salt, which did certainly not stimulate the yield of alcohol as they thought, though it held some water. Others suggested the use of some stirring apparatus in the still, as used by the Scottish distillers (429).

Another suggestion of NEUENHAHN was the more economical use of the heat of the flue gases in the distilling furnace, which gases he applied with much succes for the malting of his corn (1794).

A question on which much ink was spilt was the bad taste of the corn spirit and several authors of this period suggested the use of wooden stills of which NEUENHAHN disapproved (452). This bad taste was partly due to bad rectification but also to the presence of traces of copper in the distillate against which many authors like PARÉ, GLAUBER, GASS and RIEM had already warned. PLOUQUET (487) added that often the tin covering the copper still contained lead and antimony which entered into the distillate and when the tin lining was damaged even oxide of copper could be traced in the brandy. RIEM (512) took up the original proposal of the mechanic Gass (1766) and built a small copper furnace into a wooden cask which served as still. The alembic was attached to the lid of the cask and it was claimed that the brandy obtained in this manner had an excellent taste. In fact the Esthonian peasants always distilled in such primitive apparatus. EBBESEN (203) claimed that this saved at least 40 % of fuel as did NEUMANN (456) and LAMPADIUS (344). GÖTT-LING (of Jena) and FISCHER (of Berlin) worked on such distilling apparatus, but the idea was never realised in distilling practice.

The important work on refining alcohol by LOWITZ (389) (390) falls in this period. THEODOR LOWITZ was born at Göttingen in 1757 and he became professor of chemistry and director of the court-dispensary at St. Petersburg. He died in 1804. In 1785 he discovered the refining action of carbon which he made by driving off the volatile compontents of coal by dry distillation and carefully powdering the coke. The powder thus prepared kept well if shut off from the air. NYSROM suggested refining the alcohol with water and sulphuric acid before redistilling it and in England colcothar was put into the still to obtain a water-white distillate. But LOWITZ showed that it was much simpler to add some of his coal-powder to the still or better still to filter the alcohol over this powder or mix it with the powder, which would settle out in the tanks or casks. His invention must have been made somewhat earlier as he had already read a paper on the subject at a meeting at Erfurt on August 8th 1784.

By the end of the eighteenth century Russia and Sweden became more prominent in the world of science, though the most famous "Russians" of the period are still Germans or other foreigners called to St. Petersburg by the Emperor of Russia. One of these was J. G. MODEL (430) (431) (432) who wrote several books and essays on the production of alcohol from corn and who gave many details on the primitive methods still used in Russia and Siberia, a subject on which GMELIN had also written. MODEL gives very clear information on the point that all parts of a distillery should be adjusted to each other, and that no distillery but one designed scientifically could work without a loss. The industry of north-eastern Europe was already considerable by this time. In Sweden no less than 300,000 tons of rye were used yearly for the production of alcohol. One bushel of malt yielded 24 pounds of brandy, a figure which amounted to 32 pounds in the case of rye and to 40 pounds in that of wheat. MODEL does not favour a cooling coil, which is difficult to clean. The still should be heated regularly on all sides and the fire should never go out. The still should not be over-filled and a wire-gauze should be put on the bottom to prevent overheating. It is better from the point of view of safety that the cooler should be placed in a separate room.

Another "Russian" author, the influential minister VON CANCRIN (114), gves a plan of a well-conceived distillery. The first distillate ("Rauhbrennen") is called "Lutter" or "Lauter", the second distillate ("Klarbrennen") is the final product. Both MODEL and he insist on the utmost cleanliness in the distillery and the malting house, they seem to have understood its importance for proper fermentation. Contrary to MODEL, VON CANCRIN is a great friend of the worm-condenser which should be continuously cooled, not the discontinuous replenishing of the cooling water which MODEL suggested.

In *England* too we find that apart from the practical distillers there are several authors, who merely compile well-known data. One of these was BENJAMIN MARTIN, born at Worplesdon in 1704, an enthusiastic Newtonian, who gave up a teaching career to move to Fleet Street in 1741, where he became famous as a maker of spectacles and microscopes. He died at London on February 9th 1782 after having written on many subjects, especially on mathematics. He also produced a *Sure Guide for Distillers* (412) which contains the usual descriptions of distilling methods, like the works of GEORGE ADAMS (1750-1795) (10), the mathematical instrumentmaker to GEORGE III.

It should be noted that CHAMBERS *Cyclopaedia,* one of the earliest in England (120) mentions that distillation "is twofold: 1° per ascensum, by ascent and 2° per descensum, by descent, when the matter which is to be distilled is produced below the fire."

A notable advance in distilling apparatus was achieved by the Scottish distillers, who sold large quantities of spirit to London previous to the year 1786. Then the license charged on the capacity of the stills mentioned before was fixed at £. 1/10/- per gallon capacity for the Lowlands and £. 1-.- for the Highlands, by which measure it was believed the London distillers would be protected against the severe Scottish competition. Of course this measure solicited frauds on a large scale as soon as this act of July 1786 became known and the extra duty on imported Scottish spirit of 1788 stimulated Scotch ingenuity to fight against an ever-increasing import duty (508). The Lowlands stills now quickly acquired a flat form which enabled quick distillation and rapid recharging (508). This type of still (Fig. 123) was probably introduced by a firm from Leith which planned its design according to the principles discovered by JOHN PAYNE who had taken out a patent in 1736. PAYNE (476) rightly worked on the idea that "the quantity rarified is always proportional to the surface of the fluid in the vessel that contains it and not to the depth". He therefore enlarged the vessel's surface and diminished the depth by making the evaporating vessel longways instead of round as one and a half or two or more diameters in length to one in breadth. "To raise an ebullition whereby the surface of the fluid is enlarged without encreasing the fire, I have contrived a cilendar or tube, the lower end of which is to be placed near the bottom of the still under the liquor, with several small pipes or the like openings anext to it, to admitt in and out the fluid freely and the upper end of this cilendar to come up through the still, and the top to be open, by which, when the vapour that is confined in the void head of the boyler is strongly compressed, it will by its pressure raise the fluid in the tube and at every discharge of the fluid descend again in the small pipes and so alternatively there will be a flux and reflux", but he also suggests a stirrer, which is usually found in this type of still (Fig. 123). PAYNE also suggested a flat condenser instead of the usual coil.

The design was improved by MILLAR, who in 1795 constructed a

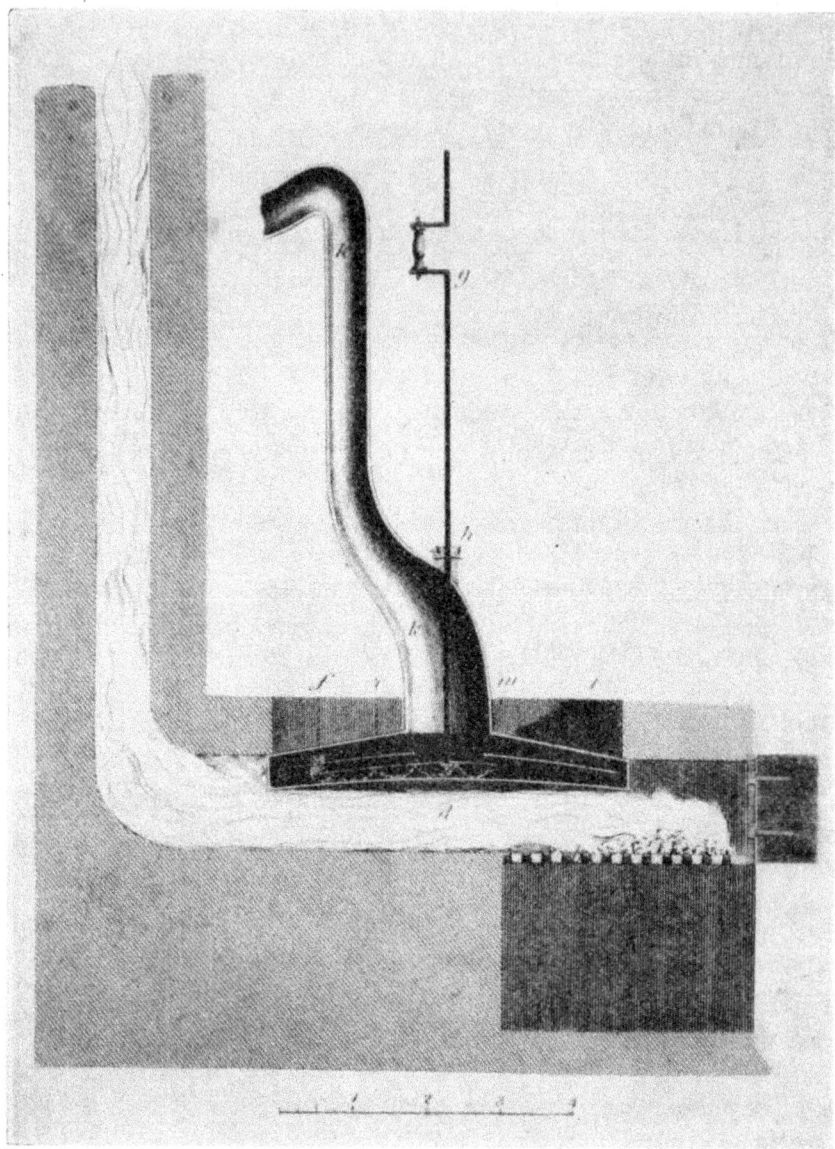

Fig. 123. A precursor of MILLAR's still

still shown in Figs. 124 and 125. It will be seen that he was inspired by PAYNE and BAUMÉ and that he expands the body of the alembic or upper part of the still by introducing hollow ribs to facilitate the escape of the vapours. A stirrer both destroyed the foam and kept the bottom clean.

With this new still MILLAR was able to distill over four hundred and eighty charges in twenty-four hours (509) as the Commission installed by the House of Commons reported (508) (509). The still is only 2½″ deep working with 22 gallons of wash, though the later types are somewhat smaller. Still it seems that even 40 gallon stills were built which is no wonder when we remember that the capacity duty rose steadily up to £ 162.— per gallon capacity. It is clear that this method of rapid distillation produced a very badly rectified alcohol. Apart from this evil the number of illicit stills rose rapidly until the town of Edinburgh claimed at least fifty illicit stills on every licensed one. This evil was not easily abolished, even when the House of Commons dropped the capacity duty on the advice of the Investigation Committee and introduced the normal excise duty in 1815. SCARIS-BRICK mentions that 6000 illicit stills were seized in 1811 alone and that even in 1822 3,000,000 gallons of legal spirit were produced in Scotland against 10,000,000 gallons of illicit brandy or whisky.

A typical eighteenth century question, which kept many prominent minds busy was that of the *"earth" formed in distilled water.* BOER-HAAVE had believed himself able to prove that a glass vessel filled with distilled water showed "earth" particles after a long period of shaking. MARGGRAF (1709-1782), a Berlin professor, tried to disprove this. Distilled water could not possible contain or form particles, so he thought correctly. He bound bottles full of water to the sails of a mindmill, but the result was indeterminate. Then he took similar bottles with distilled water and had them shaken during eight conse-cutive days without interruption by a group of assistants. Indeed he found a fine powder formed in the bottles, which looked like finely powdered glass (411), but he had not the courage to sustain this obvious conclusion. The final proof was brought by LAVOISIER, who in an essay of 1770 proved that the weight of the powder formed was exactly that which the glass bottles had lost. Therefore the "earth" formed was simply powdered glass formed by mechanical corrosion of the vessel. The conclusion was that properly distilled water is pure and could not contain nor form "earth".

Another question which was solved in the eighteenth century was

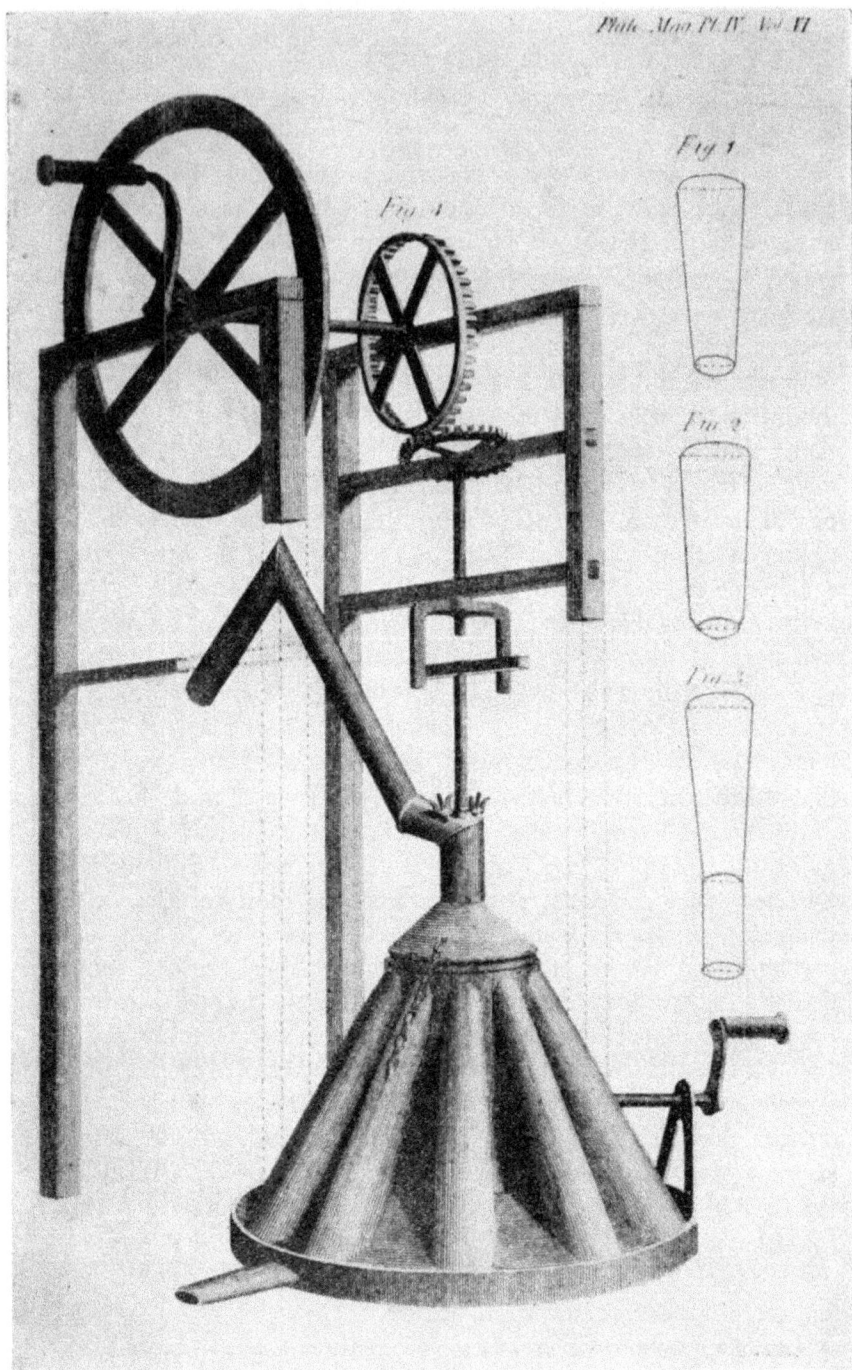

Fig. 124. MILLAR's still

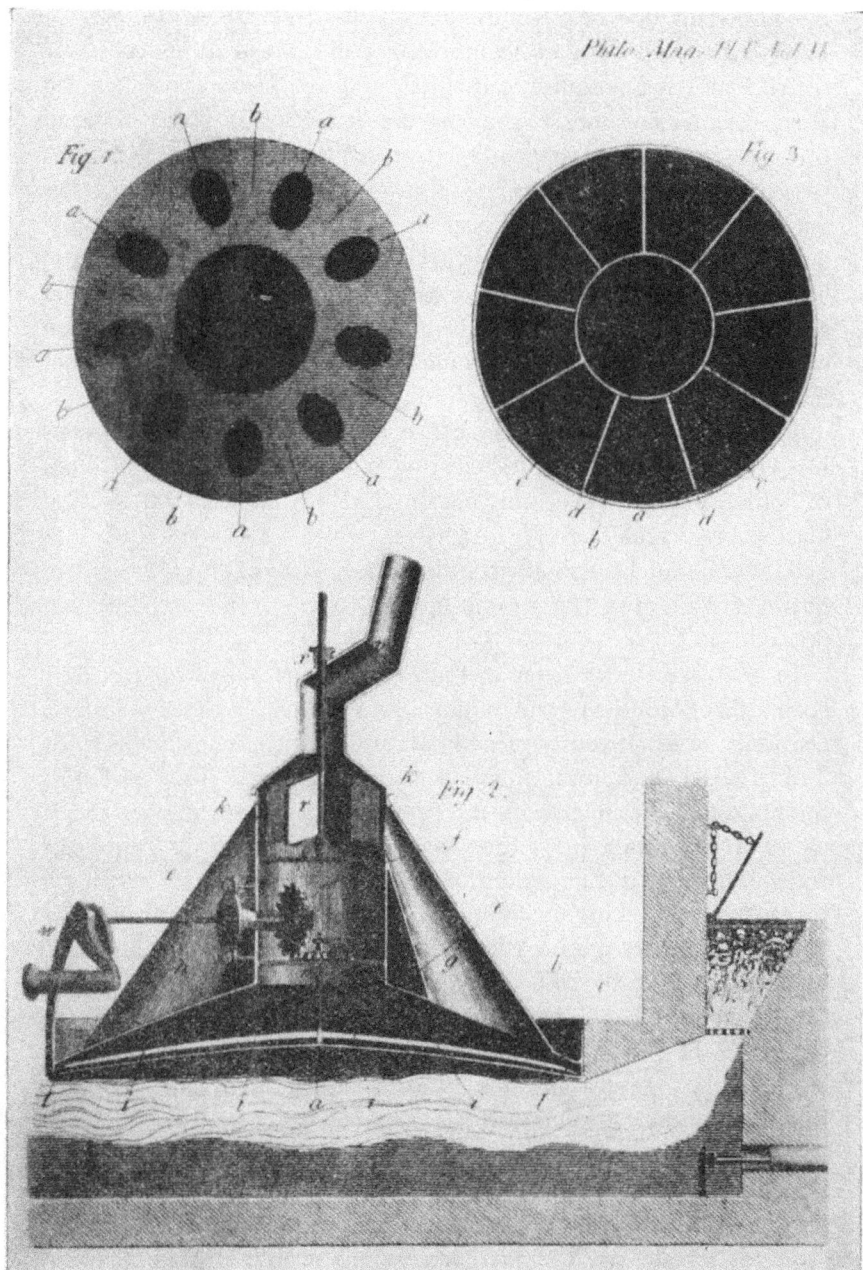

Fig. 125. MILLAR's still

the important one of a handy and cheap method of *distilling sea-water*. As the sea voyages became longer and longer in the course of the sixteenth and seventeenth centures, and the ships no longer sailed along the coast but took to the open sea, it became an urgent thing to become independant on board a ship of the water-supplies of the coasts and islands on the route. Especially the long voyages on the Pacific and other oceans where sailing ships might not be able to throw out their anchor for many a week, this was a most important question. The usual supplies of fresh water in wooden casks were often spoiled before the casks were opened and we need not wonder that the solution of this question occupied the minds of the Royal Society and other scientific bodies.

JOHANNES A GADESDEN (also called JOHANES ANGELICUS) mentions in 1516 that sea-water is freshened by laying a linen cloth over a pot of boiling water and squeezing out the cloth, in fact he advises nothing but the "woolcondenser" of the ancients, which we met in the writings of GESNER. But he also advises distillation, absorption of the salt in bowls of white wax (on account of ARISTOTLE's text?) and filtration through sand.

In 1670 we find a letter in the *Philosophical Transactions* (285) stating that "Monsieur Hauton hath now declared his secret of making Sea water sweet. It consists first in a praecipitation, made with Oyl of Tartar, which he knows to draw with small charges. Next he distills the sea water, in which work the furnace takes but little room and is so made that with a very little wood of coal he can distill 24 pots of water in a day, for the cooling of which, he hath this new invention, that instead of making the Worm pass through a Vessell full of water (as is the ordinary practice) he maketh it pass through one hole made on purpose out of the ship and so enter again through another. So that the Water of the Sea performeth the cooling Part, by which means he saveth the room which the common Refrigerium would take up; also the labour of charging the water, when the worm hath heated it. But then thirdly he joins to the two precedent Operations, Filtration thereby perfectly correcting the malignity of the water. This Filtration is made by a peculiar earth, which he mixeth and stirrs with the distilled water and at length suffers to settle at the Bottom."

In 1675 WILLIAM WALCOT got a patent for distilling sea-water, in 1683 FITZ-GERALD, son of the Earl of Kildare and relation to BOYLE got a grant of patent for a similar method. WALCOT's grant of patent by the Staten Generaal of the Netherlands was superseded (1684).

Still neither produced water wholesome enough for those who drank it in the long run. As STEPHEN HALES put it "no distillation works if the water is not purified first". This problem was solved by MARTIN LISTER (383) who remarked in 1684 in the *Philosophical Transactions:* "Now that the Sea water is made fresh by the Breath of Plants growing in it I have elsewhere demonstrated thus: I took a long glass bottle and having filled it pretty full with sea water, taken up at Scarborough, I put therin common sea weed (Alga Marina) fresh and new gathered, some with the roots naked and some growing on or adhering to Stones: the glass Bodies being full, I put thereon a head with a Beck and adapted a glass receiver thereto, all without any Lute or closing the joints; from these plants did distill dayly (tho' in small quantity) a fresh, very sweet and potable water, which had no Empegreuma or unpleasant taste as those distilled by fire necessarily have. I urge this experiment as the most natural, most easy and most safe way of having sweet water from the sea and which may be of greater use than perhaps some are apt first to fancy, even to supply the Necessity of Navigators."

The whole subject was studied in detail by STEPHEN HALES (280) (281). HALES was born at Beckesbourne (Kent) on September 7th 1677 and he studied theology but also mathematics and botany at Cambridge. He was vicar at Teddington from 1710 onwards and in 1717 became Fellow of the Royal Society. His important works, mentioned above, were published in 1727 and 1733. In the latter year he took his doctor's degree in theology at Oxford. The Paris Academy elected him a member in 1751 and he died at Teddington on January 4th 1761.

HALES often studied and discussed the sea water problem, not only in the above mentioned works but also in his *Philosophical Experiments* (282). There he gives two important papers, read before the Royal Society on "an account of some attempts to make distilled sea water wholesome" and "some considerations about means to preserve Fresh Water sweet", including a good history of the problem. The volume was quite logically dedicated to the Lords Commissioners of the Admirality. According to HALES distilling remained the most simple way; though HAUTON's proposal might be the scientifically correct one, it is too complicated for practical use. It is true that redistilled Oil and Salt of Tartar, Calx of Bones, Chalk, etc. have good effects on sweetening sea water. If however the sea water is placed in a cask of fresh water and left to putrify, redistillation will give excellent sweet water. The distillate can be tested for the presence of "spirit of salt" (chlorides) with silver salts. HALES adds that it is now usual for East-Indiamen to have stills and worm-tubes fitted for distillation, but the copper of these is not generally tinned.

In 1756 HALES published a small book which contains three essays, the first of which (283) is an extension of his remark in the *Philosophical Transactions,* vol. XLIX, No. 54, p. 312. On account of an instrument devised by BAILY for the steam engine of the type then in vogue, HALES announced the idea of introducing compressed air through a pipe or box with small holes into the liquid to be distilled. In this way the distillation might be quickened. LITTLEWOOD had already showed him a similar apparatus that made bad and undrinkable water potable.

Trials showed him that the distillation velocity was trebled. But he also intended to apply the invention of a German SCHMETOW (who in 1718 took a patent on the heating of a large quantity of water with little fuel) and use a coil built into a furnace through which the water was led to be heated. In fact, this seems the ancestor of our modern pipe-still. He would then have a small, economical apparatus with a relatively large output. It would of course be very economical first to treat the sea water with lime. HALES was so convinced of the succes of his proposal, that he adds that "It is to be hoped that this invention will not be used for the production of the spirit of wine".

Further important work on this subject was done by Lord PHIPPS in 1774 and in the meantime the subject was also tackled by French scientists of whom PIERRE ISAAC POISSONNIER had the most lasting success. POISSONNIER was born at Dijon on July 5th 1720 and studied pharmacy at Paris taking his degree in 1743. He was professor of chemistry at the College de France from 1746-1777, but in the meantime he was also inspector of the medical service of the French army from 1754 onwards, court-physician to the Empress ELIZABETH of Russia (1758-1760), inspector-general of hygiene for the colonies (1764-1791) and returning to France he became director of the Ecole Centrale (1795). He died at Paris on September 15th 1798.

POISSONNIER invented a good distilling apparatus for sea-water (Fig. 121) which differed little from that of IRVINE, who was allowed an annuity for his invention by Parliament. The POISSONNIER apparatus consisted of the usual still connected with a cooler made of two tubes, the distillate running through the inner, the water between inner and outer tube. The water ran in counter-current to the distillate and was fed continuously from a reservoir placed high on a support, the supply being regulated with a tap. The entire cooler was insulated by a wooden box built around it. A Commission of the Académie, consisting of TURGOT, TRUDAINE, MAQUER and LAVOISIER, reported very favourably on it and the apparatus was used by the French navy from 1770 upto about 1799 when it was displaced by better and more modern stills (489).

A field lost to distillation in the course of the eighteenth century was the *production of sulphuric acid.* The introduction of the "bell-process" and its improved form, the "lead-chamber process" made the ancient way of distilling vitriol inefficient and obsolete. The early history of the manufacture of sulphuric acid is given by SAMUEL PARKES (464) and more lately by DICKINSON (173).

We have seen that the name "vitriol" crops up in the thirteenth century and was probably given to the natural sulphates because of their glassy crystals. In the period which we are discussing there were still two ways of producing acid, firstly by dry distillation of vitriols and other admixtures, by which one obtained "oil of vitriol" (oleum vitrioli, oleum sulphuris, spiritus aluminosum), generally a strong acid with sulphur trioxide dissolved; and secondly the bell process, burning sulphur under a glass bell in the presence of water, the identity of the "Oyl of sulphur", obtained by the bell-process and the "oyl of vitriol" was proved by LIBÁVIUS, SALA and BOYLE. LE FEBVRE improved on the alembic form of the bell given to it by LIBÁVIUS and similar apparatus are depicted by CHARAS (fig. 126).

LE FEBVRE (355) gives the following description of the making of "spirit of sulphur": "Use a grey earthenware pan holding some water and an iron trefoot placed on the bottom. A glazed dish holding brimstone is placed thereon and the sulphur is set on fire with a red-hot iron (a horse-shoe is most proper!). A glass bell is then placed over the pan. When the brimstone is spent another dishful must be ready to displace it" and "The Artist may place as many earthen pans and bells under a chimney as it can hold to advance the more his work... above all times chuse that of the two Aequinoxes, vernal and autumnal, to work this Spirit". The latter remark is typical of the period and recurs in later works.

According to some authors like PAPIN and STAHL the oil of sulphur was better than the spirit of vitriol though more expensive. This was probably due to the fact that it was purer, though in general the acid from vitriol was stronger and need not be concentrated by subsequent

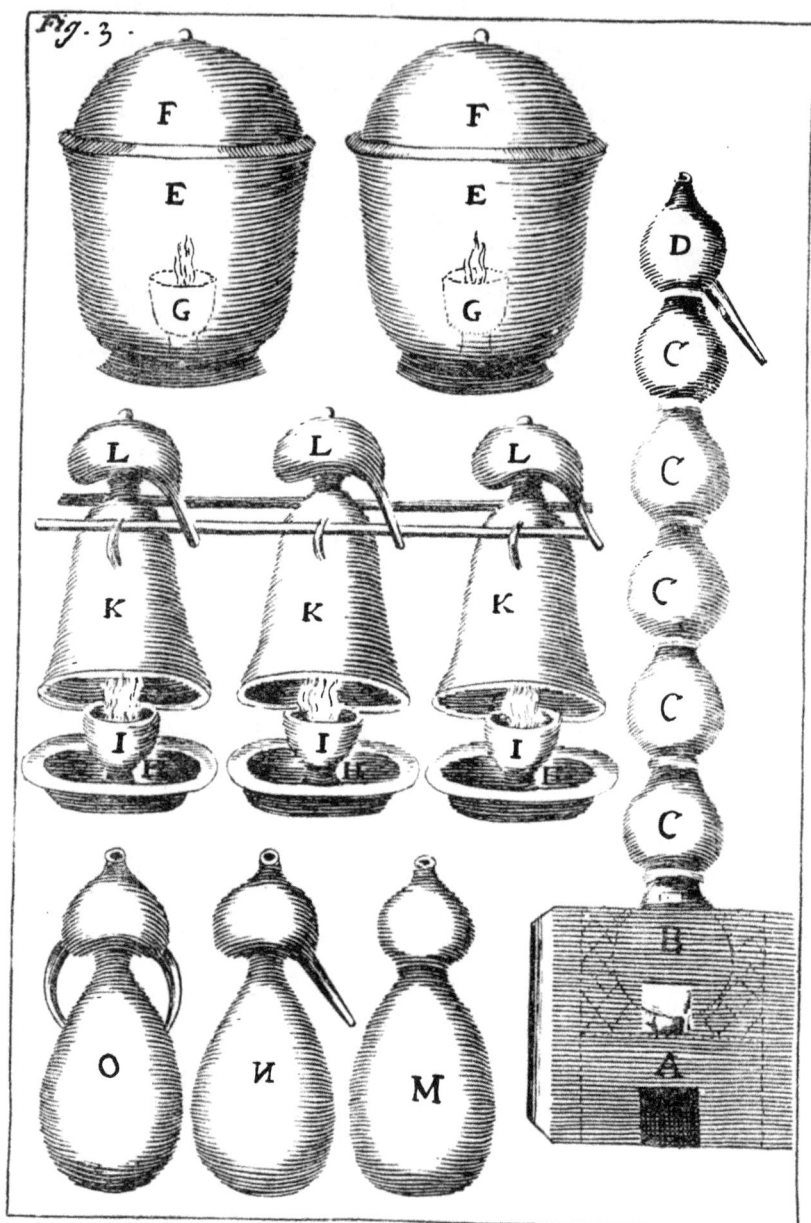

Fig. 126. The bell process according to CHARAS

distillation like the bell acid. On the other hand the "bell" acid could be concentrated to any desired strength. However, it still remained a laboratory product, LEMERY for instance ignites a mixture of four pounds of sulphur and four ounces of nitre in a closed pot with a little water on the bottom (Fig. 126).

In mean time there was a strong urge to increase the production of sulphuric acid, especially in the eighteenth century when new applications of this acid in technology were discovered, which required mass production. For in this period sulphuric acid was used in the tinning and gilding business, for the separation of silver and gold in metallurgy, for engraving on copper, for the manufacture of brass and for the treatment of felt for hats, etc. 1744 BARTH discovered the sulfonation of indigo which meant a new field for the acid in the dyeing of woollen fabrics. A further stimulant was the bleaching industry when HOME (of Edinburgh) discovered that sulphuric acid bleached better, cheaper and quicker than the customary buttermilk.

The development of the bell proces as a competitor of the vitriol process was more or less a nationalistic affair. For the distillation of vitriol was concentrated in Saxony, Nordhausen being an important producer of strong and "fuming" acid. In the eighteenth century Silesia and Bohemia grew to be important producers too. But the vitriol industry was of course seriously limited, for the acid was a dangerous thing to handle and the retorts were necessarily small and often combined in "galley"-ovens as the nearest approach to mass production. The lack of proper acid-proof material prevented the development of larger stills. On the other hand the bell-process was an invention of English and French scientists, it contained the germs of something greater, as was widely felt, and the eighteenth century showered all its energy on its development.

The addition of nitre to the burning sulphur was an important step in the right direction. It is not exactly known who was the inventor. GLAUBER mentions it (264) in 1651. DIGBY (178) does not refer to it, though in HARTMAN's edition of DIGBY's work (1682) there is some discussion on the excellent effect of the formation of steam by the floating dish with the burning sulphur. LEMERY mentions the addition of nitre in the ninth edition of his *Cours* (1697) but not yet in the fifth (1685). DICKINSON believes therefore that this invention must have originated on the continent about 1690, but as GLAUBER mentions it much earlier, the question should be reopened. At any rate the manufacture both with and without nitre is known to JOH. CHR.

BERNHARD (1755) and to ROBERT DOSSIE, the London apothecary and skillful chemist, who was a friend of Dr. JOHNSON and whom BOS-WELL mentions in 1758. DIGBY refers to the process in his *Elaboratory laid open* (191). The idea is also often ascribed to DREBBEL but his papers contain no such invention.

Two people were to stimulate the use of nitre in an improved bell process, JOSHUA WARD and JOHN ROEBUCK. JOSHUA WARD (1685-1761) was a quack doctor famous for his "drop and pill", the efficacy or otherwise of which excited heated controversy. WARD tried to enter Parliament illegally in 1717, but was obliged to flee the country, returning from St. Germain in 1733 when he was pardoned and waxing fat on a credulous public with potions and pills concoted abroad (463). He possibly used his acid for his nostrums, but anyway he established a laboratory in Twickenham as early as 1736 and annoyed the residents with the stench, moving to Richmond in 1740. With his partner JOHN WHITE he took out a patent (No. 644 of June 23rd 1749). Stress was laid in it on the proportion of nitre to brimstone and the amount of moisture in the air. As to the last factor, this was shown to be negligible by LEMERY. WARD and WHITE's process was operated as follows: Glass globes as large as could then be blown with safety, usually 40 to 50 gallons (26-28" diameter) were placed in a double row on a sand bed. A few pounds of water were placed in each, a stoneware pot introduced through the neck and on this pot an iron saucer previously made red-hot. Therein the nitre-sulphur mixture was placed and ignited with a red-hot iron, after which a wooden stopper was placed in the neck of the globe. As there was a row of globes the workman returning to the first would find it in a stage to refill it. This led to discarding the old bell method as the acid now soon cost as much per pound as it formerly cost per ounce, viz. 1/6 to 2/6.

The improvements introduced by JOHN ROEBUCK were of still greater importance. JOHN ROEBUCK (1718-1794) was the first partner of JAMES WATT, when the latter designed his first steam engine. He had studied chemistry and medicine at Edinburgh and taken his doctor's degree at Leyden (1743). He was one of the earliest industrial chemists. DICKINSON remarks (174) that he found sulphuric acid a laboratory product made in glass and advanced its manufacture to the industrial scale, with great changes in technique, from retorts to leaden chambers. ROEBUCK had a partner called SAMUEL GARBETT (1715-1803) with whom he had a chemical laboratory producing acid and corrosive sublimate, but both acted as consulting chemists, pro-bably the first in the world.

The "lead chamber process" was probably developed in this laboratory in Steelhouse Lane, Birmingham in 1746. WARD probably took his patent when he heard of it but ROEBUCK and GARBETT took no action as the price of their acid was not more than a quarter of that made by the older "bell process". In 1749 ROEBUCK established his works for the production of acid at Preston Pans, Midlothian near Edinburgh. Probably one reason for choosing a site in Scotland was because he had applied for a patent in Scotland, but not for an English one, for his process. This patent was granted on August 9th 1771 but annulled in January 1772. The chambers had a floor space of about six foot square covered with a little water and soon as many as thirty chambers are found in one factory, each 6 by 4 feet and 8½ feet high. A manuscript of a Sheffield chemist (1771-1790) published by GUTMAN (274) gives details of these early lead chambers.

The chamber process was an immediate succes, the price of sulphuric acid fell to sixpence a pound and we find a quick succession of works all over England. In 1772 the first factory was erected in Battersea, London; in 1766 one was built by HOLKER at Rouen, France. The first factory in Germany dates from 1812. Soon the process was taken up in Holland, France (especially near Paris and Montpellier) and America, but still England had practically a monopoly for a long time and chamber acid was usually referred to as "English acid" or "English oil of sulphur". The Frenchman JARS, when describing his journeys to England in 1758, 1765, 1766 and 1767 (313) tells of a sulphuric acid factory at Battersea and the manufacture of sulphuric acid "by the bell". Here more than 100 glass globes were used (about 2½ feet diameter) buried to the depth of 6" in sand. Spring stoppers acted as safety valves. The acid made was very weak and had to be concentrated by distillation in glass retorts as in other factories. Glass globes were also used for storing and transport. The process was still to a large extent dependent on the art of blowing such large vessels. The carboy padded with straw in a wicker work or hoop iron basket is the modern counterpart in shape and size of the old globe.

However, the lead chamber process was still intermittent. Further improvements were due to DE LA FOLIE, who in 1774 suggested a separate furnace for the burning of the sulphur and passing the fumes into the chamber where the nitre and steam were to be added (*Observations sur la Physique*, vol. IV, 1774, p. 333). The idea of injecting steam instead of water occured to him in 1777. However his ideas were lost but they were re-introduced at St. Rollox in the years 1813-1814. In mean time the price of the acid had already sunk from £ 32.— per Ton in 1798 to £ 18.15.— in 1806. The continuous chamber process was now within reach but technical and financial difficulties had to be overcome. It was realised in the first decade of the nineteenth century, when CLÉMENT and DÉSORMES suggested the continuous process and other scientists like GAY-LUSSAC worked on it. Finally THOMAS HILLS of Bromley-le-Bow introduced pyrites instead of brimstone (1818). But he admixed coal and this gave difficulties, so the process was later re-introduced by Messrs. Perret et Fils of Chessy, France.

By the beginning of the nineteenth century, therefore, the vitriol process was pushed entirely to the background and was used only to produce extra strong, "fuming" acid. The Bell process was left out and distillation figured only in the chamber process as a finishing operation to concentrate the acid if this should be necessary. It was still carried out in glass retorts until proper cast-iron retorts came into use. For the further development of this industry we must refer the reader to DICKINSON (173). We may note that England produced 3000 tons of acid by 1815.

There were few changes in the *manufacture of nitric and hydrochloric acids* in this period. Their production received a certain impetus when it was discovered that they were of great use in dyeing goods scarlet with cochineal (1630) (588). Nitric acid was still mostly produced by apothecaries and small distillers in the old earthenware retorts. Holland and Flanders were important producers, one "factory" producing no less than 18,000-20,000 pounds per year. The introduction of iron apparatus, partly due to GLAUBER, was important, especially for the manufacturing centres of Liège, Ostend, Bruges, Lille, Roubaix and Lukavic which suffered much from the Dutch competition. DEMACHY tells us (168) how the nitre and vitriol mixed with clay are distilled in retorts, sometimes using alumn instead of vitriol. The nitric acid was refined by re-distilling it in glass retorts, throwing away the phlegm distilling over before the "brown vapours" are emitted. Much of the acid used for the separation of silver from gold was refined by redistillation. GLAUBER calls it "spiritus nitri fumans" whence the later French "acide nitreux" (and since 1787 "acide nitrique").

The hydrochloric acid was made in LIBAVIUS' time by distilling common salt with vitriol. GLAUBER improved the manufacture of this expensive acid, the "spiritus salis" and propagates it as a panacea, even as a substitute for vinegar. To obtain strong acid he heats zinc-chloride with sand in an iron retort. The French name "acide muria-tique" dates of 1787. Hydrochloric acid became cheap as soon as the Leblanc soda process was introduced.

Both nitric and hydrochloric acid had very little effect on the development of distilling apparatus, since simple retorts were suf-ficient for their manufacture or concentration.

There were no changes in the methods of manufacture of *essential oils and perfumes* as practised in the great centres Italy, France and Holland. Some specialities were manufactured elsewhere, for instance in Cologne. In Languedoc and Provence flowers and other materials were still distilled with water or alcohol or treated with fats (en-fleurage) in the old way to obtain essential oils. In Holland the essential oils from spices and other Oriental products were specialities, this branch of industry profited by the local experience of the distil-leries. But on the whole the essential oil trade remained very conser-vative and the apparatus changed very little, even up to the present day.

It took quite a long time before distillation was applied to the *tar industry*. Before the eighteenth century only wood tar was produced as a by-product of the charcoal pits. The ancient charcoal industry of Macedonia and Bruttium (southern Italy) had gradually spread to Western Europe (563) and the Black Forest, Austria and Sweden had become its centres. The ancient "charcoal pits" superseded by the heaps or "meilers". The tar was of course easily but inefficiently obtained from these meilers and generally applied as such. Some scientists like BOYLE and GLAUBER were interested in the composition of tar. We have discussed how BOYLE discovered methyl alcohol and acetic acid in wood tar.

Towards the end of the seventeenth century AXTH describes the preparation of tar from pine trees and the manufacture of resins and pitch from this tar by evaporation. Similar handbooks by FUNCK (222) and DICHAEUS (172) show that the manufacture of tar and charcoal from wood had not risen above the level of the primitive industry described by THEOPHRAST and PLINY. The meiler was gradually dis-placed by a brick chamber-oven, but the tar was still collected with the acid water in a simple drain running from the oven. The distillation of this raw tar is described by WIESENHAVERN (658), MENEANDER (423) and BECKMANN (55), it was achieved in copper stills. Here the tar was mixed with water and heated, distillates like "pine-oil" were recovered and the tar obtained as a residue and concentrated to the right consistency over an open fire if necessary.

Wood-tar pitch was of course of importance to shipbuilders and as Sweden strove to obtain a kind of monopoly the English tried to get their pitch from America, whilst the French bought theirs from Russia and Sweden. In the eighteenth and nineteenth centuries many types of carbonisation furnaces were designed (328). Generally the furnace was used as a cucurbit and the raw distillate, especially that containing acetic acid was recovered from its mouth. Both BOERHAAVE and GLAUBER worked on the development of these furnaces. Proper dry distillation of wood in well-built furnaces had to wait until 1796 when the first installation was built in England. The early nineteenth century learnt to make cast iron retorts which improved the manufacture of raw tar considerably. Strangely enough the proper fractionation of wood tar had to wait for the solution of the same problem for coal tar, both depending again on the invention and use of coal gas and wood gas for lighting purposes.

The manufacture of coal tar from coal had been attempted by JULIUS VON BRAUNSCHWEIG-LÜNEBURG and STUMPELT in Germany in the seventeenth century. In England DUDLEY had tried to treat coal in the same way as wood to obtain tar, using the patent taken by THORNBOROUGH, dean of York in 1590. The German chemist JOHANN JOACHIM BECHER had obtained coke and tar from peat and coal both at trials at the Hague and in England in the presence of the court of Windsor and of BOYLE. Together with HENRY SERLE he took out a patent (British patent No. 214 of August 19th, 1681) called "A New Way of Makeing Pitch and Tarre out of Pitt Coale never found before or used by any other", but this patent does not contain any scrap of information about the way in which he did this. The patent taken out by MARTIN EELE, THOMAS HANCOCK and WILLIAM PORTLOCK (285) in 1694 does not seem to refer to a dry distillation of coal but to some treatment of a bituminous substance, but here again the patent does not contain any description of the process or the base material. In mean-time CLAYTON had demonstrated to BOYLE the possibility of making a gas that burnt with a luminous flame (1691).

The tar industry received a great impetus when DARBY discovered that coke was a splendid fuel, better and harder than charcoal, and that it was an excellent substitute for the charcoal, which grew more and more expensive, in the blast furnace (1735). Further trials by JARS in France and by HEINRICH in Germany confirmed this conclusion. This meant a total revolution of metallurgy. Coke became the principal fuel in England very soon and in the Sarre basin it was already common by 1770. The use of coke was not only stimulated by the rising new metallurgy but also by the development of the steam engine. On this question the reader is referred to the article on *The charcoal-iron industry* by H. GWYNN JONES in the *Journal of the Royal Society of Arts*, vol. 88, 1939, pp. 41-57.

In the meantime various experiments were made to find some useful outlet for the increasing amounts of tar. In 1746 HENRY HASKINS took a patent (287) for a "New Method of Extracting a Spirit or Oyl out of Tar and by the same Process produce the finest of Pitch". This method consisted in distilling the tar in a still with a worm-condenser, obtaining distillates such as a "pale acid phlegm", "a fetid, volatile spirit or oyle" and lastly a "black, glutinous oyle", and a caput mortuum consisting of good pitch. A few decades later ARCHIBALD, Earl of DUNDONALD took out a patent (British patent No. 1291 of August 20th, 1781) for "A Method of Extracting or Making Tar, Pitch, Essential oils, Volatile Alkali, Mineral Acids, Salts and Cinders from Pit Coal" which consisted of burning the coal with little air in a closed vessel external heating. The tar escaping from the retort is then condensed in several condensing vessels by the injection of steam. The difficulty for all these inventions was that apart from the pitch the industry had very little use for the distillates (see also K. R. WEBB, *The story of the distillation of coal, Chemistry and Industry*, vol. 62, 1943, pp. 158-159).

The situation for the wood tar and coal tar industries improved considerably when WILLIAM MURDOCH, continuing PHILLIPE LEBON's trials on dry distillation of all kinds of wood and coal, solved the question of producing light gas in the years 1792-1807 (See G. ATKINS, *A short history of gaslight, Repertory of Patent-Inventions,* August 1826, p. 84). The production of still larger quantities of tar in the new gas-industry invited further research and in 1815 ACCUM proved the value of solvent naphtha as a paint-thinner, etc. In 1820 REICHENBACH studied the wood tar from the new brick kilns and the coal-tar and discovered paraffin wax and other useful fractions. In 1838 BETHELL showed the preservative properties of creosote oil. Tar was already appreciated as a good fuel but the proper development of the distillation of tar had to wait until 1856 when PERKIN discovered "mauvein", a dye which he had made from a coal-tar fraction, aniline. But the wood-tar profited little from these developments as it declined with the rise of the coal-tar industry. When, however, the distillation methods were introduced in these industries, distillation was already fully developed in the alcohol industry. Our conclusion is, therefore, that the tar industries did little to stimulate development of distillation methods, though they did profit enormously from distillation once the presence of valuable industrial products was shown in tar.

The same can be said of the *petroleum industry*, the rise of which dates from Colonel DRAKE's succesful drillings in Pennsylvania in 1859. Of course the Arabian distillers had produced notable quantities

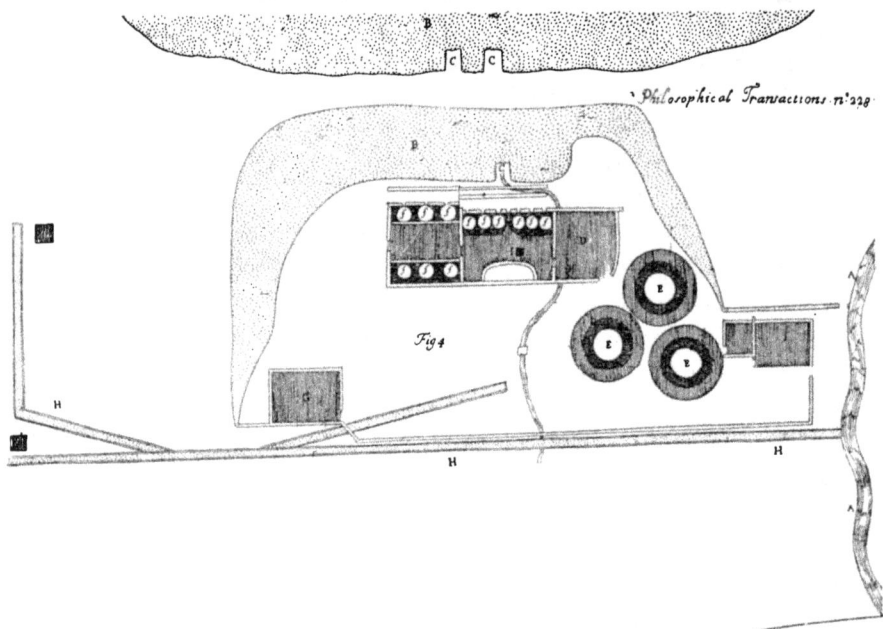

Fig. 127. The "tar oil" works near Brosely. G indicates the stillhouse, fff are the kettles in which the crushed rock is boiled with water, EEE being the crushers and mills

of light tops from petroleum for use in Greek fire or for cleaning silk and from time to time we hear of the distillation of petroleum in later centuries. EELE, HANCOCK and PORTLOCK took their patent (285) to treat bituminous rocks, as we hear from other sources. They ground the rock, boiled it in copper kettles with water and skimmed off the floating oil. The oil collected was then "evaporated to Pitch". A distillate was obtained "like Turpentine or Oil and tried by divers Persons in Aches and Pains with much benefit". It was later sold as Betton's British Oil. The works (Fig. 127) were on the Severn near Brosely, where this bituminous rock is found in the hills in the neighbourhood of coal seams. A similar treatment of oil sands near Wietze was described by TAUBE in 1769. A light and a heavy distillate were obtained during the distillation of this oil leaving about 25 % of bitumen in the stills. In the course of the eighteenth century quite an industry developed in Modena where the local oil from pools was recovered and distilled in copper stills. The distillate was used for lighting the streets of towns in the neighbourhood. The first oil of Poland was distilled in 1815.

But none of these local industries had any effect on the development of distillation methods, they used the contemporary stills as they found them. There was not yet any specialisation and no strict specifications of the marketed products existed which involved careful distillation. It should be noted that the young petroleum industry of 1860 was led to produce mainly kerosene stimulated by ARGAND's burner which formed the base of all later oil lamps. Though the industry was to profit from the distilling apparatus developed by the alcohol and tar industries, we still find such primitive stills as box-form "cheese-box" stills in use in Pennsylvania in the eighties and even later!

The production of "Dirschöl" or ichtyol from the Seefeld shales or Tyrol shows a very late survival of the old "destillatio per descensum". As early as 1350 a knight, BERTHOLD OF EBENHAUSEN (570) received the right to exploit these shales of the Oberinntal and thirty years later it was awarded to ULRICH VON MATZ. In 1576 ABRAHAM SCHNITZER obtained from the Archiduke FERDINAND these rights for twenty years because "he had by God's Mercy found through his diligence a certain Art to make good, true Oyle from certain specific Stones which are found in several mountains of the County of Tyrol and which are called "Tyrstenblut" ". This Seefeld shale was broken up into pieces of the size of a fist and put into crucibles, six or eight of which were placed upside down on an iron plate full of holes on a furnace. Under the plate, just under the mouth of each crucible, there was a funnel which ended in a collecting tube made of wood and leading to a receiver. Wood was then stacked between and over the crucibles and fired. About 13,3 % of oil was recovered. As late as 1803 MILLER devised separate receivers for each crucible but the ancient treatment persisted until late in last century to be displaced by proper distillation of the valuable shales in iron retorts (662).

Looking at the pictures of laboratories of the middle of the eighteenth century we see very little improvement of *distilling apparatus* whether one looks at a picture of the laboratory of the Capuchins in Paris (Fig. 128), where the grey friars made their concoctions, or

that of an English dispensary (Fig. 129) or of that of the court-
dispensary of Berlin (Fig. 130). Exception may be made for the small
adjustable tables (Fig. 128) on which receivers and other apparatus

Fig. 128. Laboratory of the Capucins at Paris
(XVIIIth cy.)

Fig. 129. Laboratory of an English chemist
(XVIIIth cy.)

rested, but they occur also in the works of KHUNRATH and CHARAS.
They were, however, generally introduced by the work of JOHANN
WOLFGANG WEDEL, a professor of medicine of Jena (1708-1757)
who wrote a booklet on the subject (652). Indeed, the apparatus had
changed little since BOYLE had his laboratory in Maiden Lane, working

with AMBROSE GODFREY (482) in something that was really a kind of semi-technical plant. Here we find the aludel, a kind of subliming pot also used as a section of condensing apparatus, and the athanor, a brick furnace, well distinguished from the Piger Henricus or Slow Harry, a slow heat furnace and the Furnus acediae or turris circulatoris, a tower furnace. Receivers of the round bottom flask type were called ampullae or bolt heads, the cucurbit distills into a matrass, a glass vessel with a long neck and oval form. Crucibles used for melting

Fig. 130. Court-dispensary at Berlin (End of the
. XVIIIth cy.)

are called crosslet. The Balneum Mariae and the "equi clibanum" (heated by putrifying manure) are often used and also the "egg", an egg-shaped distilling vessel.

This part-inventory of BOYLE's laboratory shows that no great changes had been made since the fifteenth century except for the better form and better material of these vessels, and the same is true for the centuries after BOYLE. We have already frequently mentioned that many scientists warned against the use of metal vessels for pharmaceutical products. After KUNCKEL had taught the chemists in his *Ars Vitraria* to blow glass, the possibility of producing more complicated glass apparatus was considerably increased and several well shaped eighteenth century pieces of apparatus have survived (Figs. 131 and 132) which show that the new art of blowing was speedily mastered. The earthenware vessel also received proper attention. POTT experimented on the correct clay mixture and DELIUS finally introduced porcelain into laboratory work in the final quarter of the eighteenth century. BERGMANN introduced the use of iron crucibles, KLAPROTH

that of silver ones and finally platinum found its way into the
laboratory though the propaganda of WOLLASTON many decades later.

Opinions were divided on the best way of *cooling distillates*. The
French distillers and scientists preferred air-cooling and DEMACHY
states that there existed stills in Paris with a hundred and twenty feet
air-cooled worm-condensers (165)! But the water-cooled Moor's head
was also very popular in laboratories and small distilleries in France. It

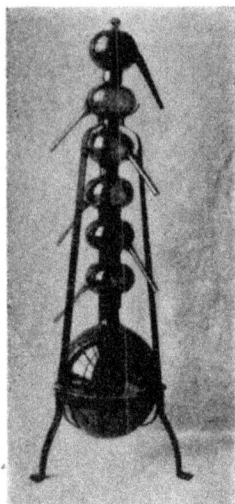

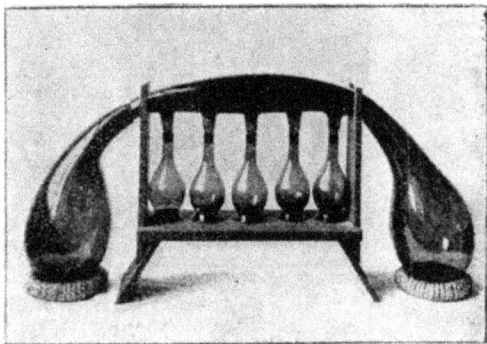

Figs. 131 & 132. Glass apparatus of the XVIIIth cy. (Münich Museum)

was propagated by BAUMÉ, CAMPY and DEMACHY and was used in
sizes of 6″ to 2′ 6″ diameter and 3′ to 4′ high. In England opinions
were also divided on this point. Y. WORTH (667) (668) mentions
both worm-cooler and Moor's head, though he says that the latter
is already somewhat obsolete. COOPER (139), however, mentions only
the worm-cooler both for strong waters like alcohol and for extracts
of flowers, etc. The old and false opinion that the cooling water should
be introduced at the top of the condenser where "the water became hot
first" remained very strong. This principle was still used by PAR-
MENTIER in 1805! On the other hand many saw through the fallacy
now and MUNIER (441) was the first to inject the cooling water at
the bottom of the tube, thus partly anticipating the counter-current
cooler.

Many distilleries still used intermittent replenishing of the cooling
water, though gradually the continuous cooling won away. The counter-
current principle is an achievement of the end of the eighteenth
century and not of LIEBIG as is often claimed (593) (594). Its in-

ventor is unknown. In a pamphlet entitled "Nouvelle construction d'alembic pour faire toute sorte de distillation en grand avec plus d'économie et plus d'avantage dans le résultat, en deux parties" an anonymous Frenchman mentions that it was discovered about 1770. A copy of the pamphlet of 1781 in the Casseler Landesbibliothek says in a note written on the margin that the inventor was one JOAO HYACINTHE DE MAGELHAENS or MAGELLAN, a Portuguese who got his inspiration from the Dariot stills in the cognac-distilleries of France. However, this last detail cannot be correct as no such type of cooling or anything like it is found in the books of DARIOT. This

Fig. 133. POISSONNIER's counter-current condenser

JEAN HYACINTHE DE MAGELHAENS, born at Lisbon in 1723, was an Austin friar who settled in London in 1764 where he died in 1790. In 1780 he published an *Essai sur la nouvelle théorie du feu élémentaire et de la chaleur des corps* (London, 1780).

The principle is certainly used by POISSONNIER in the condenser he designed for the still meant for the production of sweet water from sea water on board ships (Fig. 133).

Its adaptation to laboratory use is due to CHRISTIAN EHRENFRIED VON WEIGEL, the son of a doctor, born at Stralsund on May 24th 1748. He studied at Greifswald (1764) and Göttingen (1771) where he took his doctor's degree. After some years of practice at Stralsund he moves to Greifswald as teacher of botany and mineralogy, becoming director of the botanical gardens and the mineralogical collections in 1781. From 1775-1805 he was professor of chemistry and pharmacy and from 1794 to 1806 director of hygiene. In 1806 he was raised to the peerage dying on August 8th 1831. In his thesis of 1771 (650), under the heading *Observatio I, destillatio spiritus vini* he describes two pieces of tinplate bent round, one tube enclosing the other and joined together at the bottom by a ring of tin-plate soldered to both (Fig. 134). The water is introduced by means a high funnel and flows oft from the higher opening between the tubes. This condenser corroded quickly and the joints often leaked after some use. In the second part of his thesis (Pars Secunda, Gryphiae, 1773) he introduced a glass inner tube joined to the tinplate outer tube with gypsum. WEIGEL also indicated the clip in the form of a fork to hold the condenser. This clip was later improved by GAY-LUSSAC who made it mobile by introducing an axis round which it could be moved. JOHANN GADOLIN THE ELDER designed a condenser of the same type but with a rectangular section (232). He was of the opinion that a

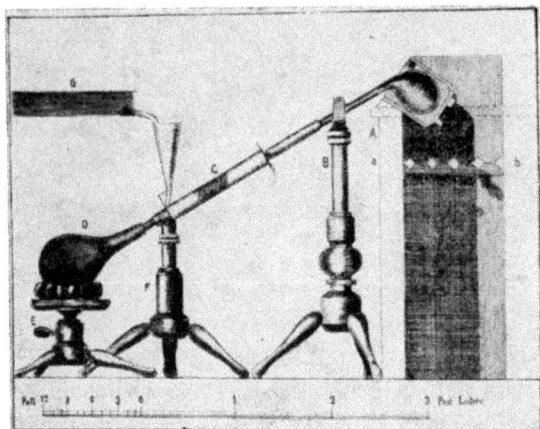

Fig. 134. WEIGEL's design

cooling tube should not be straight or spiral but should consist of straight pieces joined by knee-bends and easy to take apart for cleaning. Only in the case of this laboratory condenser he introduces a straight tube (Fig. 135). Incidentally he mentions that the Dutch distillers have used their slops to heat the intake of their stills and thus saved much fuel and obtained more alcohol from their product.

His son JOHANN GADOLIN THE YOUNGER (1760-1852), professor at Abö, whom we shall meet as an important figure in the development of the theory of heat, improved on the designs of his father. He wanted to introduce them in the distilleries and criticises the stills of his days. The tube conducting the uncondensed gases to the condenser should be at least as wide as the neck of the cucurbit. The rim of the alembic should be abolished and this part of the apparatus should never be used to condense distillate. This operation should be achieved by the counter-current condenser, which he finds WEIGEL had already invented before his father. The water from this condenser can be efficiently used for malting. The counter-current principle is not only necessary from the point of view of heat-efficiency, but also because hot water rises and cold water descends. This is the reason why the cold water should always be introduced at the bottom of the worm-tube or condenser. Further condensers of this type were designed by

Fig. 135. GADOLIN's condenser

him and published in the *Schriften der Kaiserl. freyen ökon. Gesell-schaft in St. Petersburg* for the years 1814 and 1818.

There were several attempts to improve the worm-condenser, for in practice it was difficult to clean it, which seeing the bad dephlegmation and the frequent puking of the stills, must have been necessary fairly often. NORBERG, inspector of the Swedish distilleries, experimented between 1780 and 1799 and finally designed a continuous condenser (458) with counter-current cooling and also another type of cooler, which he called "refrigerator". He admits that spiral coolers take up less room and cool efficiently, but they are difficult to clean. Therefore he joins to the still instead of alembic and cooler a large cylindrical or box-shaped vessel in which the vapours are condensed. He also designed "vapour-watchers" which were to prevent puking or foaming and which were attached to the top of the still.

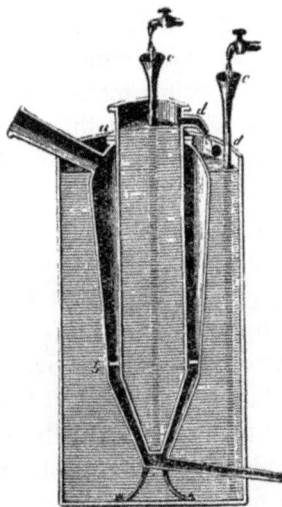

Fig. 136. Condenser designed by GEDDA

PETER NIKLAAS Baron VON GEDDA designed a conical condenser (Fig. 136) which was very popular for many decades. GEDDA, born at Stockholm in 1736, was member of the King's Council and of the Swedish Academy from 1786. He died at Stockholm on August 28th 1814. His condenser was easy to take apart and to clean (244). Still in the long run there were some difficulties and MITSCHERLICH improved the design. It had of course a good cooling capacity as the vapours passed between two water-cooled walls only one centimetre apart.

JOH. GOTTFRIED DINGLER, apothecary of Augsburg, designed a condensor with a tube running zig-zag, having screw-caps at the bends which permitted proper cleaning (1815). BEINDORFF from Frankfort has several spouts mounted on the alembic, each cooled separately and many other designs were tried out with more or less succes. The introduction of the counter-current principle in industry in the form of a "Liebig condenser" was due to HOLBERG. He describes it in his *Fabrikation des Grünspans.* (Leipzig, 1850). The Babo cooler (Fig. 194) is another early industrial form.

Some, like BRUGNATELLI, went back to old designs and worked with a second water-cooled Moor's head on top of the old one, thus obtaining two fractions (96) or with a long vertical cooling tube with several spouts, each giving a separate fraction (Marozio) (95), but these experiments did not meet with permanent succes.

The principle introduced by WEIGEL and falsely attributed to VON LIEBIG, who did nothing but design a counter-current condenser for the laboratory built up of glass and cork parts, proved its merits and stayed.

The results of the different experiments in designing condensers can be seen in a plate of BERZELIUS' *Handbook of chemistry* of 1830 (Fig. 137) (65). Here we see different forms of alembics, one of which indicated by fig. 4 was quite common in Sweden, but was superseded by that of fig. 6 until the Dutch idea of a semi-globular alembic with the vapour-line issuing with a semi-circular bend from the top of the alembic became general in the north. We also see in this plate NORBERG's condenser in fig. 8 and that of GEDDA in fig. 9, that of WEIGEL in fig. 10 and a combined Weichel-Gedda type in fig. 11. Fig. 12 shows a cooler designed by the Frenchman BÉRARD. BERZELIUS was the first to discuss intelligently what angle the neck of a retort should make with the body. In figs. 16, 17 and 18 of his plate he shows that this angle should be such that no condensate can flow back into the body of the retort.

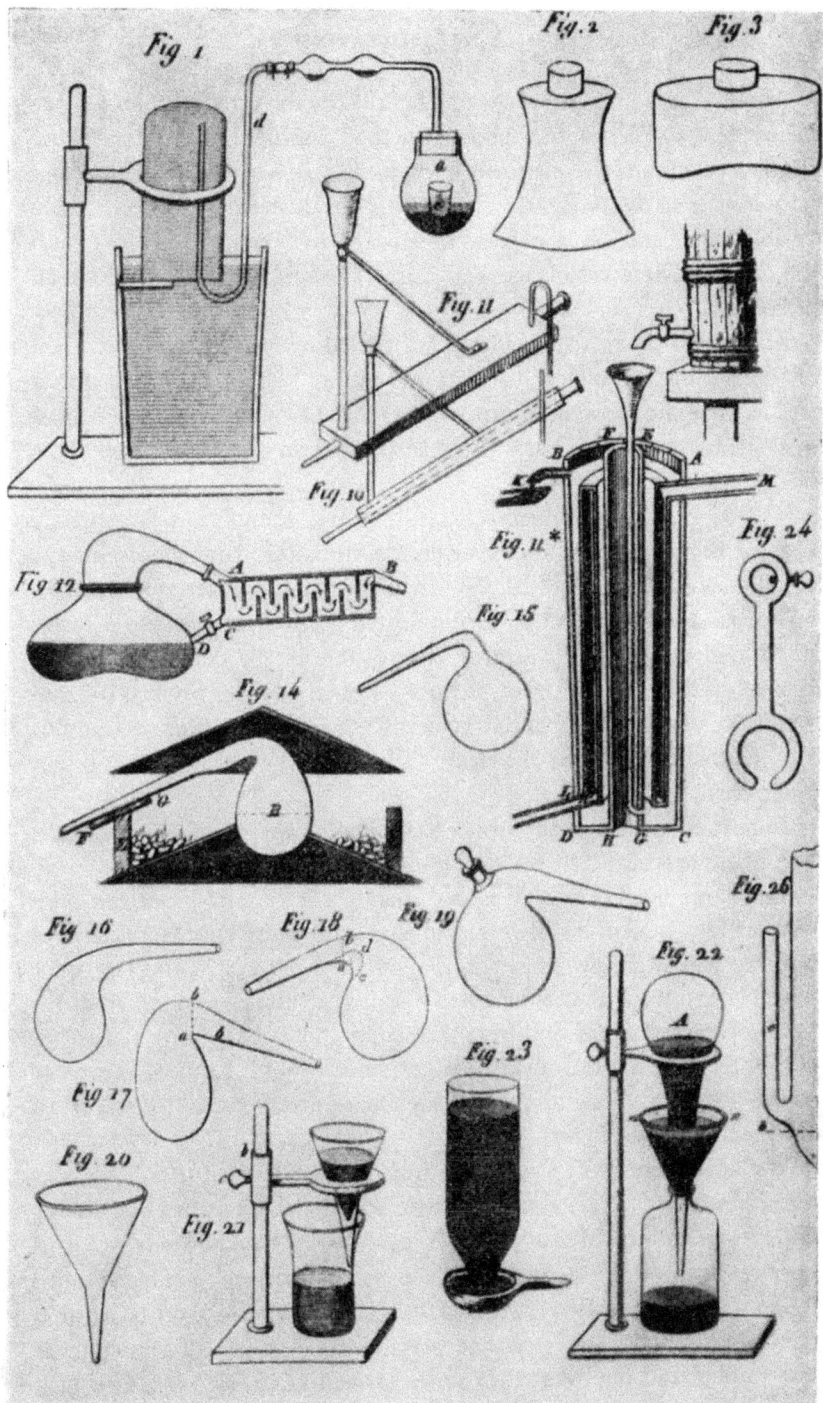

Fig. 137. A page from BERZELIUS' handbook illustrating laboratory apparatus such as alembics, retorts, condensers, etc.

There was some improvement in the methods of *separating the water from the distillate*. The first method was the use of a separating funnel which in the eighteenth century still had the appearance it had in the early sixteenth. ERNSTING, for instance, mentions a big vessel which ends in a narrow opening, which can be shut off with the thumb. These vessels are still called "pump" (Pumpe) in German literature.

NEUMANN seems to have been the first to mention the new method of separating the oil of distillate by adding salt to the water for it is said, that "he showed by experiments that he could make all essential oils float, or sink or keep them suspended in the water just as he liked".

The destillatio per filtrum was still in use, though it was generally applied immediately in the receiver. On the other hand the use of the so-called "Florentine bottle", which drew off the water continuously, was propagated by CHARAS who imported it from Italy, where it seems to have already been in use for a long time in the perfume industry. BAUMÉ and MACQUER give pictures of such vessels. From 1767 onwards we also hear of similar vessels for separating water and liquids heavier than water. But the name "Florentine bottle" is certainly not older than about 1850.

Important developments took place in the *testing of the distillate*. The many practical tests for the strength of the alcoholic distillate were summed up by SAVARY. We saw, that CHARAS introduced the gunpowder test, which is still found in the writings of DEMACHY. Several burning tests were developed in this period (GEOFFREY, PAULIN, etc.) which measured the residue after burning a certain amount of distillate. One of the possibilities of testing new base materials was fermenting them and distilling off the alcohol formed. Then of course in the eighteenth century the hydrometer practically swept all of these methods away.

We have mentioned that the hydrometer was known to SYNESIOS as the invention of HYPATIA and that PRISCIANUS mentions it in his *De Ponderibus et Mensuris* (sixth century). The knowledge of this physical method of testing distillates seems to have slumbered until it was revived by BOYLE, THOELDENS and other scientists of that generation. In the sixteenth and seventeenth centuries the instrument was studied and improved by a series of men such as BOYLE, HOMBER, VAN MUSSCHENBROEK, FAHRENHEIT, BAUMÉ, NOLLET, BRISSON, NICHOLSON, PONSELET and others. BAUMÉ propagated the use of the hydrometer, the fixed points chosen by him were those of a certain solution of sodium chloride and that of distilled water; for solutions with a specific gravity lower than 1 the scale was simply extended below zero. CARTIER more or less stole the idea of BAUMÉ and produced a scale, the points of which gave the strength of the alcohol in "degrees". This hydrometer was officially adopted in France by the decrees of April 3rd and September 4th 1771.

Apart from the scales of BAUMÉ and CARTIER many others were proposed, until GAY-LUSSAC produced his "alcoomètre" in 1824, on which instrument the strength of the alcohol could be read directly in percentages. This became the official French instrument by the law of June 26th 1824 and remained in use until 1880. GEOFFREY LE CADET had read a paper on his hydrometer or "spirit weigher" in the session of the Academy on April 27th 1718 and PAULIN sent in a paper on his in August 17th 1741, but neither method was officially adopted. In 1772 POUJET and BORIE OF CETTE conceived the idea of combining hydrometer and thermometer in one instrument. This was important as before that date little attention was given to the change of specific gravity with temperature, though it was well known to the scientists of the period. The subject was now, however, closely studied. DEMACHY wrote his handbook on the hydrometer and its correct use in 1774 (166).

The story of the introduction of the hydrometer in other countries, which followed France more or less closely, need not bother us here. With the advent of the hydrometer another control on the correct progress of distillation was introduced which in the long run had considerable effect on still design. Other methods mostly disappeared or remained only if they gave additional information on the quality of the distillate. The specific gravity read on the hydrometer still remains the most handy and quick way of controlling distillation on the spot.

Apart from the pre-heater, the counter-current condenser, the hydrometer and some experiments in still design this period therefore, brought no striking new ideas to distillation. But the eighteenth century had gathered many new scientific facts and experiments which waited for better times to be translated into practice. The French Revolution opened the eyes of many men to the profitable cooperation of science and industry. It had created the first rudiments of skilled labour and proper schooling of those employed in industry and the effects were wonderful. In next chapter we will discuss some of the prominent theories and facts which, once the tide turned, could be realised, but which at the end of the eighteenth century still remained "stopt in like a strong distillation with stinking clothes".

CHAPTER SEVEN

THE WEDDING OF SPIRIT AND SCIENCE

"Hast thou not learned me how to make
Perfumes? Distill? Preserve?"
(Cymbeline I, v. 13)

The tackling of three problems paved the way for new methods. The eighteenth century had certainly achieved some improvements by experimenting and reasoning, but it seemed to have spent all its forces at the outbreak of the French Revolution. Scientists discussed the old problems over and over again and no further development was found in the distilleries.

Still that same century tackled three new problems, which furnished a sound base for integral developments. It is true that the solution of none of them was reached in that century, but as their solution approached new ideas were evolved which permeated the minds of the practical distillers and showed them the way to improve their apparatus, even before the greatest of those problems, the theory of heat, had approached the stage that any mathematical solution of distillation problems could be attempted. In fact, many, not to say all, the principal elements of modern distilling apparatus were perfected several decades before the theory of heat had reached its final form to influence chemical technology.

These three problems were the manufacture of alcohol from potatoes, the theory of fermentation and the theory of heat. Problems, therefore, not only of a theoretical, but partly of a practical nature.

The demand for new base materials for the *alcohol industry and the use of potatoes* for this purpose was an old one. We have frequently mentioned experiments of this kind since the sixteenth century. Though the old prejudice against corn spirit had been overcome, there was still the practical need of finding a new base material for times when harvests were short and famine became a danger.

The possibility of using potatoes for the manufacture of alcohol was a fairly recent one. The potato was imported from the New World after its discovery, but not before the early seventeenth century does it figure amongst the plants generally found in the different botanical gardens. Its consumption on a larger scale did not begin until the end of the eighteenth century in England, France and Germany, much later in other countries. Its use as a staple food by the poorer classes dates from the middle of last century only. The cultivation of the potato in France gave rise to many juridical problems in the early eighteenth century as the peasants insisted that they did not have to pay the customary agricultural taxes on this new product.

The possibility of making alcohol from potatoes had been suggested by J. J. BECHER in his *Närrische Weisheit und Weise Narrheit* (1725) and some early experiments had been made. But the first serious attempts date from the second half of the eighteenth century. The *Transactions of the Royal Swedish Academy for the year 1747* contain several essays by DE LA GARDIE (242), SKYTTE (582) and others discussing this subject and stating that in general 1 Ton of potatoes will yield five pints of gin. In 1760 the first potato-distillery was opened at Monheim (Rheinpfalts) by DAVID MÖLLINGER and from that date onwards scientific essays and practical results abound. Between 1775 and 1785 we see the gradual transition from the burning of pure corn to that of pure potatoes, mixed batches being burnt in that decade. Then special literature follows, books and essays by FIEDLER in 1792 (220), in 1793 that of GABELANN (237) and those of MÜLLER (438) (439) in 1792 and 1797. The popular German word for gin "Schnaps" dates from the same period. We find it used for a "mouthful" since 1747 but have to wait until 1770 before it means "a pull of gin". The verb "schnapsen" does not occur before 1775.

The proper scientific study of the process started in Sweden and Germany but at the end of the century the Russian scientists cooperated too. One of the main difficulties was the proper and complete decomposition of the potato-flour. In 1811 KIRCHOFF (of St. Petersburg) discovered that this could be achieved by treating the flour with weak sulphuric acid. At the same time the famines of 1816 and 1817 further stimulated the treatment of potatoes. Still the growth of the industry was slow and it was not until 1830-1850 that the cheap potato-gin became the scourge of the labour-classes of middle and north-eastern Europe. The new handbooks on this art date from the same period. We have the handbook written by DOMBASLE in 1834 (187), that of GUTSMUTH of 1835 (276), the books of SIEMENS (567) (568) of the years 1829 and 1840 and many others. Potato-gin may be identified with the growing proletariat of the great manufacturing towns of last century.

After the decomposition under pressure was discovered (1871) and a special kind of yeast was found by HANSEN in 1881/1882 this type of gin was almost exclusively produced in north and north-eastern Europe. There was some opposition to this production from the side of the landowners who had profited from the spirit produced from the corn cultivated in their domains. In Germany we still find that no more than 5-10 % of the harvest of potatoes is converted into gin.

The manufacture of alcohol from potatoes, slow as its introduction may have been, created a new problem for the distillers. The earlier stills were designed for the production of alcohol from wine and they could be adapted with slight changes to the treatment of corn washes. But in the case of potatoes the washes grew so thick that special stills and other apparatus had to be built to deal with this new base material.

The second problem, the *theory of fermentation*, was considerably older. In the Middle Ages the words fermentatio and digestio were both used for all effervescent reactions. In the fifteenth century fermentation was still considered to be some refining away of substances which hindered or spoiled other products. At the same time the metals were often compared to seeds and thus the alchemists considered fermentation as one step towards the Stone.

LIBAVIUS was the first to distinguish clearly digestion and fermentation and considered the latter to be identical with putrefaction (1559). The ferment should be akin to the material to be fermented, so he thought, and the material should be finely powdered or liquid to give good effects.

VAN HELMONT a hundred years later found that the fermentation of grapes yielded a gas, which he called "gas vinorum", and which WREN later identified with the gas evolved by carbonates treated with acids (that is our carbon dioxide). But VAN HELMONT in his enthusiasm confused fermentation with the other reactions giving rise to the evolution of a gas. SYLVIUS DE LA BOË clearly distinguished principles of the former, but then LEMERY confuses the two again and calls them "a form of boiling" like CHARAS. BECHER on the other hand considered fermentation to be a kind of combustion.

BOYLE found that fermentation stopped in vacuum, so he concluded that air must contain some vital component as it did for combustion and respiration. STAHL used this idea to state that the spirit of wine consisted of phlogiston and water and CARTHEUSER says that alcohol did not contain any oil or acid.

KUNCKEL (340) mentions that he dissolved sugar in hot water and added beer-yeast and after fermentation had started, put in the cucurbit the flowers which he wanted to extract. After distilling off the contents he obtained an excellent "extract" of the contents of the cucurbit. He considers fermentation to be a kind of putrefication and he knew that acid or a low temperature prevented its occurence.

By primary fermentation he obtained spirit of wine from all sugar-containing liquids, but the formation of acetic acid was a secondary reaction by which part of the alcohol formed was again destroyed. On account of his experiments he considered many diseases to be fermentations.

MACQUER studied fermentation (397) from the practical side and gives several methods of extracting alcohol from different substances by "vinous fermentation". Also he studied the possibilities of improving the fermentation of some types of grapes, the wine of which remained sour (1776). Quantitative measurements of fermentation processes were started by MC. BRIDE (1764) and CAVENDISH (1766). Then LAVOISIER (352) discusses the material collected by VAN LEEUWENHOECK, BECHER, STAHL and the above-mentioned authors. He concludes that hundred parts of sugar with 9.7 parts of water give 59.1 parts of alcohol and 50.6 parts of carbon dioxide. Mistakes of minor importance in his reasoning were corrected by THÉNARD, GAY-LUSSAC and DE SAUSSURE who for instance found that 4 % of the sugar is consumed by secondary reactions. BERZELIUS was the first to discuss the functions of the yeast, which he considered to be a kind of catalyst.

The complete solution of the problem had to wait until the knowledge of the yeast-cell was complete. SCHWANN and PASTEUR then gave a definite outline of the fermentation process.

PAYEN and later DUCLAUX of the Institut Pasteur had already studied the diastase.

Towards the end of the nineteenth century the manufacture of alcohol from molasses gained more importance, it is rather dependent on the current prices of alcohol from grain and other farinaceous materials, which was the speciality of England, Holland, Germany, Russia and Italy, in this period at least. The manufacture of alcohol from cellulose is a late branch of the industry, it only became possible after the discovery of the conversion of hay into sugar by KRUGER (1875), SIMONSEN (1895) and BERKHAAN and GLASENAPP (1896). The fermentation of sulphite lye worked out by Swedish scientists is the most recent discovery in this field.

A great improvement of the malting phase was the decomposition of starch under pressure found by HOLLENFREUND in 1871 and introduced into the industry several years later (1875).

The better knowledge of the fermentation process meant not only the possibility of making alcohol from other base materials (310) but also better yields and more alcohol from the same quantity of wine and corn. With the growing demand for industrial alcohol it paved the way for larger productions. This meant larger stills or apparatus of greater capacity, which again led to new designs. We shall see that the cultivation of sugar-beets in the time of Napoleon, to compensate for the loss of cane-sugar from the colonies, was a strong stimulant for the design of new types of distilling apparatus, especially as the knowledge of fermenting the pulp of these beets made them an efficient base for the production of industrial alcohol.

The third and most important problem to be solved was the *theory of heat*. This was of course not a single problem, but quite a knot of smaller problems which had baffled many generations of scientists before the beginning of the nineteenth century.

From early times man had speculated on the *nature of heat,* but experimentally no efforts were made before the seventeenth century and therefore no serious attempt was made to find out what heat really was. The Ionian philosophers had a very vague conception of heat, though they conceived fire as the cause of heat and heat as its effect, really as a sort of condition of the bodies. PLATO says in his *Timaios* (58 d) that fire acting upon a body loosens the smaller parts of it and makes them mobile. If heat escapes during cooling, air takes its place and pushes the particles into their old places. Liquids that heat should contain fire, but hot and cold are simply relative judgements (*Philebos* 25). Both PLATO and ARISTOTLE agree to the fact that movement creates heat, but the latter does not consider movement and heat identical (*Meteor.* XII, 11) for it is the ether moved by sun or stars which creates heat (*Meteor.* I, 19).

These vague ideas continued to form the preponderant opinion about heat in the Middle Ages but gradually we see two different schools of thought arising. CARDANUS says that the essence of heat is movement and GILBERT claims that heat is the action of a very fine liquid, a kind of very subtile material ether. Cold is nothing but want of heat. Both TELESIUS and GALILEO were of the opinion that heat is only present in bodies as a movement of the parts, though they considered heat as a kind of liquid.

BACON writes in his *Novum Organum* (35) Book II, Aphorism xvii: "When I speak of Forms (laws) I mean nothing more than those laws and determinations of absolute actuality, which govern and constitute any simple nature such as light, heat, weight, in every kind of matter and subject that is susceptible of them. Thus the Form of Heat or the Form of Light is the same thing as the Law of Heat or the Law of Light. Heat is a sensitive quality, that is to say, it is a kind of sensation which certain phenomena produce in living things under certain conditions. Such heat is not in the so-called hot substance, rather one should say that the hot substance in question has the power of producing sensations of heat in living bodies, or it may be, increasing the volume of a gas, etc."

BACON also gives a Table of Degrees or of Comparison consisting of instances in which the nature under inquiry is found in different degrees more or less. In the case of heat the required instances would include the increase in animal heat by exercise, wine, fever, etc., the different degrees of heat in the various parts of a living body, variations in the intensity of solar heat with the different positions of the sun, with the amount of motion or blowing, with the distance of a burning glass from a burning body, etc. "The nature of which Heat is a particular case appears to be Motion. This is displayed most conspicuously in flame, which is always in motion and in boiling and simmering liquids which are also in perpetual motion. It is also shown in the excitement of increase of heat caused by motion as in bellows and blasts... in the extinction of fire and heat by any strong compression which checks or stops the motion... Heat itself, its essence and quiddity, is Motion and nothing else. It is a motion expansive, restrained and acting in its strong strife upon smaller particles of bodies" (Book II, aphorisms x-xx). In his *Interpretation of Nature* (37) he again stresses the idea that heat is not an expansive motion but a vibration of the particles. His *Sylvae* (36) mentions that "the power of heat is best perceived in distillation which is performed in close vessels or retorts" (§ 99) and "Distilled waters last longer than raw waters" (§ 347).

The philosopher HOBBES states that fire heats but is not hot in itself and BOYLE is most conclusive in his opinion that heat does not belong to the essence of things but that it is nothing but a motion of the molecules (86). In his essay on *the mechanicall origine of heat and colde"* (Section II, Exp. VI, pp. 59-62) he says: "In a various vehement and intestine commotion of the parts among themselves we formerly observed the nature of heat to consist." But BOYLE also repeatedly spoke of "atoms of fire" and showed the difference between combustion and other forms of heat by experiment. However, he was never inclined to regard cold as something positive and consisting of "peculiar frigorific agents". In a freezing mixture the cold was due to the salt dissolving in the snow-water.

ROBERT HOOKE (1635-1703) examined sparks under the microscope (306) and arrives at the conclusion that "heat is a property of a body arising from the motion or agitation of its parts". He ridicules with considerable gusto the idea of "fire-atoms" circulating and coursing through the pores of hot bodies. As "all parts of all bodies have some degree of heat in them", because "all parts of all bodies though never so solid do yet vibrate", nothing can be perfectly cold.

Most writers of earlier periods therefore believed that heat was due to motion, but the cause of the increase of vibration of parts of a body was unknown (60). Then the seventeenth century was strongly influenced by the opinions of DESCARTES who considered heat an accelerated motion of the air-particles which is mainly caused by light. The scientists were much concerned with the question whether heat was a substance or an accident, whether cold was a separate entity or a mere privation or negation of heat and in what substances they "principally resided" rather than trying to find out how the "matter of heat" entered into the composition of the bodies.

In the regions of the theory of "pure heat" decisive advance came quickly though it did not touch a large part of the experimental work. The experiments on radiation of heat conducted by BACON, MARIOTTE, HOOKE and NEWTON had considerable influence on theory and the comparison of heat and light. In his *Optice* (457) NEWTON asks: "Do not the vibrations of the ether cause the heat in hot bodies to be of greater intensity and more lasting? Do not bodies propagate their heat to cold ones that are near by the vibrations of the ether from the hot bodies into the cold ones?" NEWTON leaned towards a theory of heat radiation by means of this medium more subtile than air existing even when there is no air present. But his rejection of the imponderable ether in 1702 hindered a proper consideration of this suggestion relating to the manner of heat radiation. His school was thus thrown back to the tendency to regard heat as a material substance which we will discuss later on.

First, however, we should point out that other theorists did advance along the lines indicated in the earlier passages from NEWTON! Thus

LEIBNIZ maintains that light and heat are both vibrations of the ether differing only in fineness. But he considered cold as a separate motion whilst NEWTON said that as the particles of a body never rest in relation to each other there could be no absolutely cold body (349).

DAVID BERNOUILLI (1654-1705) gave an independent theory and proofs of a vibration theory of heat which is similar (65) to that for which his brother JOHANN obtained the Prix de l'Académie of 1746, and EULER wrote a prize-essay (213) in 1738, in which he vindicates the vibration theory and states that fire develops all the characteristics of heat in any body. But the mechanical theory of heat, here developed by eminent mathematicians, did not yet find acceptance until about 1840!

For in the eighteenth century it was for a period obscured by the material theory, which was caused by the rapid rise of the chemical knowledge, then dominated theoretically by the phlogiston theory. It is true, that "phlogiston" kept mostly the thermal aspects of chemistry in its wake, but as a large part of the experimental work was independent of the theory of the nature of heat we find that both light and heat are often considered to be elements of an etheric or fluid nature in this period.

Phlogiston was an idea related to the classical opinion that fire as an element possessed absolute levity (472). According to GREN (1787) "phlogiston is by no means an elementary body, but it is rather compounded of the bound matter of heat and light" and again in 1793 "heat as well as light is a materially and imponderable elastic fluid". Latent heat was considered to be only loosely attached to matter and the matter of heat exhibited chemical affinity and coherence with respect to bodies.

For LAVOISIER heat and light were constituents of oxygen, which in his theories takes the place of phlogiston and which is therefore not considered as an ordinary chemical element as Mlle METZGER showed, contrary to the popular belief about LAVOISIER. LAVOISIER coined the term "caloric" for the matter of heat in his *Méthode de Nomenclature Chimique* written in collaboration with DE MORVEAU, BERTHOLLET and FOURCROY (Paris, 1787, p. 30). This caloric was conceived as a kind of all-pervading, imponderable, highly elastic fluid, the particles of which repelled each other but were attracted by matter and flowed from the colder to the hotter body. HIGGINS (1775) says that fire is composed of phlogiston and light and according to ELLIOT (1780) phlogiston produces heat and light which were distinct fluids.

For HOPSON (1781) too phlogiston is the conjunction of heat and light. HUTTON (1792) claimed that phlogiston has more of the nature of light and heat than of gravitating matter.

The theory of CHR. WOLFF was of considerable influence. He took up the old ideas of GASSENDI and proclaimed that caloric was settled in the pores of the small particles composing the bodies. By its movements it became detectable. Thus he was able to explain the idea of latent heat which was by then measured by the experimentalists. BLACK and WILCKE had made the "specific heat" clear and obvious to everybody and they joined hands with WOLFF. LAVOISIER and LAPLACE in their *Mémoire sur la Chaleur* (351) did mention that there were still adherents of the mechanical theory of heat, but they could not themselves decide between it and the material theory of heat, they considered both practical solutions. For both explained the rules experimentally found on the mixing of different quantities of liquids of different temperatures and also the reversibility of the increase or decrease of the free heat of a system. The caloric theory was however heavily attacked by MACQUER, RUMFORD, DAVY, YOUNG and AMPÈRE. Still it survived the phlogiston theory and dominated the science of heat until the middle of the nineteenth century. This was largely due to BLACK; ingenious scientist as he was, he could not dissociate himself from the faulty theories of LEGHORN and IRVINE, which we will have occasion to discuss later.

The final blow to the caloric theory was delivered by RUMFORD and those after him. BENJAMIN THOMPSON was born in Massachusetts in 1753 and attended lectures at Harvard before settling at Rumford (now Concord) as a schoolmaster. His researches on the properties of gunpowder attracted attention and he was nominated to a post in the London Colonial Office. In 1779 he became F.R.S., but left London in 1784 to enter the service of the Elector of Bavaria. At München he became a Count of the Empire and left that town in 1795 for Ireland. He played a part in the establishment of the Royal Institution. A few years later he married LAVOISIER's widow and spent the rest of his life at Auteuil near Paris where he died in 1814. During trials on boring cannons at München he observed the large amounts of heat produced and experiments to measure it led to his essay (*Phil. Trans.* 1798, p. 80) (525) in which he came to the conclusion that heat was motion and nothing else.

At the same time the question whether heat was a material substance was decided. BOERHAAVE, BUFFON, ROEBUCK and others had tried to

weigh the substance of heat but failed. BLACK denied the question
and RUMFORD was now able to state (*Phil. Trans.* 1799, p. 179)
"I think that we may safely conclude that all attempts to discover any
effect of heat upon the apparent weight of bodies will be fruiteless".
But RUMFORD's experiments that friction could produce an indefinite
amount of heat were still inconclusve. However, he converted a youthful contemporary, DAVY, who carried out some friction experiments
with ice which convinced him and many others that heat could no
longer be considered as matter but was probably the vibration of the
corpuscles of bodies (*Works*, edit. 1839, vol. II, pp. 11-14).

By now an intimate connection between mechanical energy and heat
had been established. At this point SADI CARNOT joined the battle.
NICHOLAS LÉON SADI CARNOT was born at the Luxembourg at Paris
where his father lived as a member of the Directoire (1796). He
studied at the Ecole Polytechnique and joined the engineers corps at
Metz, but sent in his papers when he had reached the rank of captain,
to be able to spend his time on study and research. In 1832 he died
from cholera, but only long after his death was his genius properly
appreciated, first in England (115). CARNOT was able to understand
the nature of heat better because of the development of mechanics.
KEPLER had already identified work and energy, LEIBNIZ had coined
the expression 'living force" and EULER, LAGRANGE, LAPLACE and
GAUSS had developed the mathematical apparatus of the potential
theory. Now WATT failed to find a proper solution to the question
how much heat was required for a certain work.

GAY-LUSSAC had formulated the hypothesis that the internal energy
of a gas at constant temperature was independent of the volume and
CULLEN had given examples of the cooling of evaporating liquids
(1755). The researches of DARWIN, DALTON, CLÉMENT and DESORMES
were finished by CARNOT in his famous discussion of the ideal cycle
in which heat is transformed into energy. The pictorial expression of
his reasoning by drawing the cycle with adiabatics and isotherms, however, was the work of CLAPEYRON (1843). CARNOT in his earlier
writings still accepts the caloric but in the notes to his book (115) he
defines heat as the motion of the molecules. His most important
achievement was the law that energy is never lost but may become un-
available. The equivalence of work and heat as forms of energy had
already been suggested by RUMFORD, but now it was expressed by
ROBERT MAYER (1842) and calculated by JOULE. This was the beginning of the mechanical heat theory of CLAUSIUS (1822-1888) who

formulated the two laws of thermodynamics in POGGENDORFF's *Annalen* (vol. 79, 1850, p. 368). By that time the last followers of the caloric theory were convinced. CARNOT's researches definitely ended the confusion between the "level" and the "quantity" of heat.

For as the theories of heat and its nature were debated others were busy in finding out how to measure and express the degree of heat and the capacity of heat. At the time of BLACK the *temperature or degree of heat* was the only measurable quantity and entity. After the early water-thermometers of GALILEO and his time, REY was probably the first to make a thermometer with a closed tube (1632). The Grand duke FERDINAND II of Tuscany was the first to fill thermometers with alcohol (1641). Then HALLEY proposed the definite acceptance of mercury as a thermometer liquid (1693).

NEWTON conceived definite ideas of using an absolute scale of degrees of heat to mark the thermometer (*Phil. Trans.* XXII, 1701, p. 824). Finally FAHRENHEIT proposed his scale in 1714, RÉAUMUR added his in 1730, and the centigrade scale of CELSIUS was introduced in 1742. FAHRENHEIT was the first to use a boiling point as a definite point of the scale, but this did not solve the question of the *heat capacity* which the earlier authors often confused with the temperature idea.

In the seventeenth century the Accademia del Cimento had some notion of the heat capacity, or thermal capacity. The members made experiments by pouring equal amounts of liquids on to ice and they found that the quantity of ice melted in each case was different in spite of the fact that the liquids were all at one temperature. But most of the later writers like BOERHAAVE and RENALDINI believed that the thermometer indicates absolute quantities of heat (78) and therefore equilibrium of heat is reached in a system when there is an equal amount of heat in the bodies in equal volumes irrespective of the nature of that body.

Similarly CHR. WOLFF states that heating is easier as a body is more compact (1714). But KLINGENSTJERNA in his notes on MUSSCHEN-BROECK's *Elementa Physices* (1729, p. 604) concludes that thermal capacity is not identical with temperature. A better solution was reached from BLACK's experiments. BLACK was drawn to this study by his master WILLIAM CULLEN, a great friend of WILLIAM HUNTER. CULLEN was born at Hamilton on April 15th 1710 and after obtaining his M.D. at Glasgow (1740) became professor of medicine at his Alma Mater in 1751, to move to Edinburgh in 1755, where he taught chemistry. He was president of the College of Physicians from 1773 to 1775 and died on February 5th 1790. He studied the old Bengal method of "producing cold by evaporating fluids and some other means of producing cold" (148). He was however, like BOYLE, WATT and MARIAN, unable to give a proper explanation of the phenomenon (*Dissertation sur la glace,* Paris, 1794, p. 248).

JOSEPH BLACK was born of Scottish parents at Bordeaux in 1728, he studied chemistry at Glasgow under CULLEN and obtained his doctor's degree at Edinburgh in 1754. He succeeded CULLEN in the chair of chemistry at Glasgow (1756) and then at Edinburgh (1766) and died on November 26th 1799. BLACK's first researches on fluidity, his lectures of 1757, discussed experiments on the temperatures of mixtures of liquids or the melting of ice when mixed with hot water. BLACK not only formulated the "capacity of heat" (our specific heat) but also the heat "rendered latent" (our latent heat). For "calories, units of heat, he uses the word "degrees of heat". On these discoveries he read a paper to the Philosophical Club of the Glasgow University in 1762 and therefore his friend JOHN ROBISON (1739-1805) was quite right in defending his rights of originality against WILCKE, though SÉGUIN in his survey of the problem of heat (*Annales de Chimie* vol. III, 1789, pp. 148-242) can still afford to ignore BLACK.

But the mixing experiments conducted by BLACK had a long ancestry and they occupied the minds of many of his contemporaries.

They were started by JEAN MORIN (1583-1656), a professor of mathematics and astronomy at the Collège Royal, Paris. He attempted to discover the law of the resulting temperature when hot and cold samples of a liquid are mixed (437). Following DESCARTES and GASSENDI he believed in heat and cold as positive entities, but he thought that they were always conjoined and that neither existed entirely without the other but always in varying proportions. Each had a maximum and minimum degree. His reasoning in his *Astrologia Liber* VIII, cap. xv, p. 158 is entirely false, but he had the merit of attracting KRAFFT to the problem.

G. W. KRAFFT (1701-1754), a professor of mathematics and physics at St. Petersburg and afterwards at Tübingen, attempted to generalize MORIN's formula for the resultant temperature of water of different temperatures mixed, dropping all references to hot and cold and confining himself to degrees of heat as measured by FAHRENHEIT's thermometer. His formula was nearly exact but slightly low.

G. W. RICHMANN (1711-1753), a professor of experimental philosophy at St. Petersburg, killed in a thunderstorm by a discharge of his apparatus for measuring atmospheric electricity, improved on KRAFFT's formula. He had already experimented prior to KRAFFT but had dropped the question. His formula for the temperatures of mixtures was the result of abstract a priori considerations and he may have been the first to view temperature as a factor of intensity, perhaps even of level. However, he did not follow up the differences between his calculated results and those measured by KRAFFT.

His work was continued by J. C. WILCKE (1732-1796). WILCKE was professor of experimental philosophy and physics at the Stockholm Military Academy and later Secretary of the Swedish Academy. Early in 1772 he noticed (660) (661) that an amount of hot water would not melt an equal amount of snow, but that considerably more was wanted. This led him to his experiments in the course of which he finished RICHMANN's formula. He defined the specific heat as the amount of heat contained in one particle of water and determined the specific heats of many substances. BLACK's work anticipated his by some twenty years and WILCKE seems to have had some slight knowledge of it at second-hand.

Returning to BLACK it should be observed that his opinion on the nature of heat was still a material one, as he leaned towards CLEGHORN's views. CLEGHORN, who died a few years after obtaining his doctor's degree (1779) had discussed IRVINE's view. IRVINE had argued from BLACK's experiments that distinction between latent heat and sensible heat was unnecessary but also that specific heats of bodies changed and increased with temperature. CLEGHORN thought (134) that different substances attracted particles of heat with different intensities and having thus absorbed them attracted others until attraction was in equilibrium with their mutual repulsion. This theory could be extended to explain the latent heat absorption by the sudden increase in volume, when a body, say ice, was suddenly converted into water. BLACK though considering CLEGHORN's and IRVINE's heat-theories unsatisfactory had nothing better to present as he still believed in the caloric, though he had an aversion to theory and mainly conducted only experiments on heat.

BLACK's experiments were concluded by LAVOISIER in his icecalorimeter which he designed together with LAPLACE (351). In the paper on this subject written in 1780, read in 1783 and published in 1784 he discusses the different views on the nature of heat, without, as we saw, taking a special point of view. He adopts the term "chaleur spécifique" from MAGELLAN (404) who used it to denote the total heat in a unit mass of a substance at a given temperature. LAVOISIER and LAPLACE adopt the principle of the conservation of free heat as opposed to latent heat in their caloric experiments. Specific heats are not necessarily constant as they are only the ratios of the differentials of absolute heat and there is no warrant that these should be proportional to the absolute heats themselves. Like CAVENDISH, LAVOISIER continues BLACK's experiments on the phenomena of freezing and heating. BLACK's explanation of cooling by vaporization in vacuo and other similar experiments were of eminent importance to distillers.

Now the earlier views on *evaporation* were extremely confused (202). Early writers observed "vapour" rising from liquid surfaces and also evaporation at different temperatures, which became more rapid when air was moving. HALLEY (1690) imagined that "the Atoms of Water expanded into a Shell or Bubble, others add according to the larger or lesser Quantities of Heat blowing them up and carrying them off". KRATZENSTEIN (1743) overthrew this theory and cited as possible aids in the ascent of vapour the movement of heat particles and the power of air to dissolve water. NIEUWETYT (1719) suggested that fire is lighter than air and that it appears to cling to water. DESAGULIERS (1729) postulated repulsive and attractive forces between the liquid particles, the former augmenting with rising temperature. This would lead to the formation of an elastic fluid and "aggregates of particles" and vacuity would rise in the air. LAVOISIER especially introduces the repulsion of particles, ignoring ordinary evaporation and citing the action of the atmosphere. Then LE ROI (1751) definitely enunciates the solution theory, e.g. the similarity between the process of evaporation and the dissolving of a solid in a liquid. It appeared to relate to many observations.

Fervent discussions followed, for many forgotten experiments (BOYLE, PAPIN, HALLEY) justified the conclusion that evaporation takes place even without air. DE SAUSSURE (1790) distinguished evaporation (solution) at normal temperatures, but vaporisation (formation of an elastic pressure-exerting vapour) at boiling temperatures only. DE LUC in his *Idées sur la Météorologie* (Paris, 1786) was a bitter opponent. He pictured vapour being formed by the union of heat particles and water particles. Every vapour has a maximum density at a given temperature depending on the proximity of the particles composing the vapour, independent of the presence or absence of air and other fluids. DALTON then set out the same theory with more clearness and new experiments and completely over-shadowed DE LUC's work. CAVENDISH though toying with the idea of partial pressures was a thorough-going solutionist distinguishing (1778) evaporation at boiling temperature and that which is per-formed with less heat than that of boiling heat.

BLACK himself was not able to solve the question of the evaporation heat, this was the work of his pupil WATT. JAMES WATT, the famous engineer, was born at Greenock on January 19th 1736. He was of fickle health and went to London in 1754 as an apprentice to a maker of mathematical instruments. He afterwards tried to found his own business at Glasgow, but was prevented and got a similar job at the University helped by ANDERSON. There he got into contact with ADAM SMITH and JOSEPH BLACK. From 1768 onwards he worked in close contact with MATTHEW BOULTON and completed his trials with the Newcomen steam-engine sponsored by ROEBUCK. He retired from business in 1800 and died at Heathfield near Birmingham on August 15th 1819.

In a note attached to ROBISON's *System of mechanical philosophy* WATT tells us (518) that he did not profit so much from the lectures by BLACK as from the personal interviews, when he learnt to under-stand the idea of the "latent heat" and used it with much profit when constructing his steam-engine. CULLEN had made some experiments on evaporation in vacuo, IRVINE and others had followed and WATT himself to, but the difficulties were only overcome when discussing them with BLACK. It was only then that he could measure the heat content of steam and define the evaporation heat of water as a constant (649).

Now the right views on evaporation and the escape of steam grew quickly, the connection between boiling points and pressure becoming clear and the connection between vapour pressure and boiling point was investigated. Cooling by the expansion of gases and steam was already known. These phenomena are discussed with new figures on the vapour pressure by SOUTHERN in the same work (518). He con-siders this point "a very important one to the distiller and practical chemist, because it may make cheaper heating possible". Further work of RUMFORD, DESPRETZ, BIOT and BRIX and later REGNAULT led to the formula of CLAUSIUS for the heat of vaporization.

The fact, that JAMES WATT profited so much by BLACK's ideas shows the eminent importance of the latter's researches for the art of distilling. For we have already remarked that BLACK primarily was interested in practical experiments and not in theory. Not only did those who turned the steam-engine from a pumping machine for raising water from mines into a universal source of power look at BLACK as a master as WATT did. But all those industries using heat like the brewing, tanning and distilling factories could profit from the discussions of heat however incomplete they might still have been at the end of the eighteenth century. Especially the distilleries were booming at the time of BLACK as we have seen. In the early eighteenth century gin engulfed Britain in an overwhelming wave of drunkenness, as PLEDGE put it. And at the end of the century Scotland was engaged in its bitter resistance against the laws of excise. Therefore BLACK could remark to his students: "All of you know well enough how the operation of common distillation is conducted". The still propagated chemistry and heat. For not only is distillation one of the best methods of preparing pure substances but distillers, BLACK was aware, as users of fuel are conscious of a quantitative aspect of heat other than temperature. Hence he for instance measured the heating of the refrigeratory belonging to a still. Indeed PLEDGE is right in stating that BLACK ended another case of isolation of craft knowledge and theorist's knowledge. The union of these two traditions gave him the doctrine of latent heat of freezing and vaporization.

The consequences of the eighteenth century theories and experiments on heat were manifold, e.g.:
1. The destruction of the theory of the "weight of heat",
2. The rise of the mechanical theory of heat,
3. The attempts to determine zero temperature,
4. The investigation and application of the motive power of heat,
5. The rise and progress of calorimetry,
6. The investigation of the transference of heat by convection, conduction and radiation.

Though none of these problems was definitely solved in the early nineteenth century the practical distillers profited by the newly coined ideas and conceptions and they started to apply some of them.

Two of the most important applications to distillation were those of steam and vacuum. RUMFORD was the pioneer of the *application of steam to distillation*. It is true that GLAUBER proposed it much earlier (Fig. 108) and there is a patent of S. TH. WOOD on the distillation and preparation of spirits by steam (British patent No. 1492 of July 27th 1785), but RUMFORD could discuss the matter much more intelligently because of the new theory of heat (526)

(527). He says that the fault of earlier experiments was that the heat was not made to descend into the apparatus. For the steam should not be introduced at the lowest parts of the vessel but should come from above. Traps should prevent the liquid returning to the steam-boiler. He then describes the further lay-out of pipelines for steam heating and the arrangement of boilers, insulation of pipes being most important as is the joint between steampipe and vessel. Steamheating was already practically employed by the dyeing firm of Gott and Co., mercantile and manufacturing house at Leeds after the perusal of his *Seventh Essay*. The live steam used for heating their coppers had resulted in an appreciable economy of fuel. Other dyeing firms were adopting it. RUMFORD also mentions that he fitted the kitchen of the Royal Institution with steam for cooking purposes and adds several other possibilities such as drying houses, alumn works, central heating, etc. Both live steam and indirect heating are profitable.

He also says in this essay published in May 1802 and read to the Institut de France on June 9th 1806: "It may be employed in heating the fermented liquor from which ardent spirits are distilled. A proposal for introducing watery vapours (so he proposes live steam!) into a liquid from which pure ardent spirits are to be distilled or forced away by heat will, no doubt, be thought very extraordinary by those who have never meditated on the subject; but when they shall have considered it with attention, they will find reason to conclude that this saving of expense and this method of distilling bids to be very useful". Indeed he was right and the following decades show a growing application of live steam heating in England and Germany, though France always preferred indirect steam heating.

But then the difficulties of supplying open steam were not overcome before SAVALLE invented his steam-regulator (1857) and the use of live steam in distillation became hardly as common as it should before the latter half of the nineteenth century. But both live steam and indirect steam heating were most important means to avoid burning of the contents of the stills.

The earliest applications are found in England, though STONE used it in 1801 at MESLY near Charenton in a distillery. WYATT obtained a patent (670) in 1802 which provided for either direct or indirect steam heating. But the method suffered from considerable opposition from the Board of Excise. MOREWOOD could not obtain a patent including this principle on account of the opposition of this Board. In 1818 a Mr. BIRCH of Roscrea (Ireland) made trials with a still using steam jackets and solicited the Board's permission to work with steam, but his projects were ultimately rejected after the Board had tried out the apparatus on the principle that it was considered to afford facilities of smuggling under the existing "still-capacity" duties.

But the use of steam-jackets and steam-coils for indirect heating spread. TRITTON even used open steam against burning of the still bottom in his patent-still (629) and in 1822 Sir ANTHONY PERRIER used direct steam in his apparatus (477) blowing it in through small tubes which worked like bubble caps. A few years later a Swede, ROBERT LORENT, obtained a patent (1826) for "a method of applying steam without pressure to pans, boilers, coppers, stills, pipes and machinery in order to produce, transmit and regulate various temperatures of heat in the several processes of boiling, distilling, evaporating, inspissating, drying and warming and also to produce power". This was an example of indirect steam heating, in jackets. Later patents of COFFEY (1830), etc. propagated direct steam heating and especially the excellent distilling qualities of the Coffey stills, which many distilleries in England adopted, paved the way for the use of live steam in that country.

In France SOLIMANI was the first to take out a patent involving the use of a steam jacket and this method of heating attained a considerable vogue through the efforts of DUBRUNFAUT. French distillers were rather late in adopting the idea of using live steam, the effect of which was not so obvious in the case of distilling such thin liquids as wine, which contain no solids and which present hardly any difficulty in distilling.

The first to adopt RUMFORD's idea in Germany was a dyer APEL of BAUTZEN who boiled with steam in his factory. In 1810 TROMMSDORFF extracts with steam in his apothecary's shop and in 1812 STRUVE uses steam for inspissating at Dresden. But apothecaries were rather late in using steam for heating purposes for a boiler with steam-jacket was still shown as a novelty in the Apothecaries Hall at London in 1815. In 1818 BEINSDORFF introduced the steam-pot bearing his name. Then came the book by ZEISE (671) who since 1818 possessed the "Elefanten-Apotheke" at Altona. His propaganda and the fact that medicines burnt far less then by direct heating by a fire soon paved the way for the new method in the world of the apothecary. The German distillers PISTORIUS, GALL, KÖLLE and VON SIEMENS soon adopted it probably inspired by LORENT and SOLIMANI. GALL was the propagandist to whose efforts the general acceptance of the use of live steam by the German distilling trade is due.

Apart from TROMSDORFF, BEINSDORFF and ZEISE, DINGLER, who had invented several practical forms of apparatus such as a better separating funnel and a new Florentine bottle (187), was active for the introduction of steam in the apothecary's laboratory. BUCHNERS *Repertorium* (vol. XXIX, 1828, pp. 94-141) also mentions the contributions of a Dutch apothecary VAN DYCK (who also wrote his name VAN DYK), whose original essay appeared in the *Bijdragen tot de Natuur-*

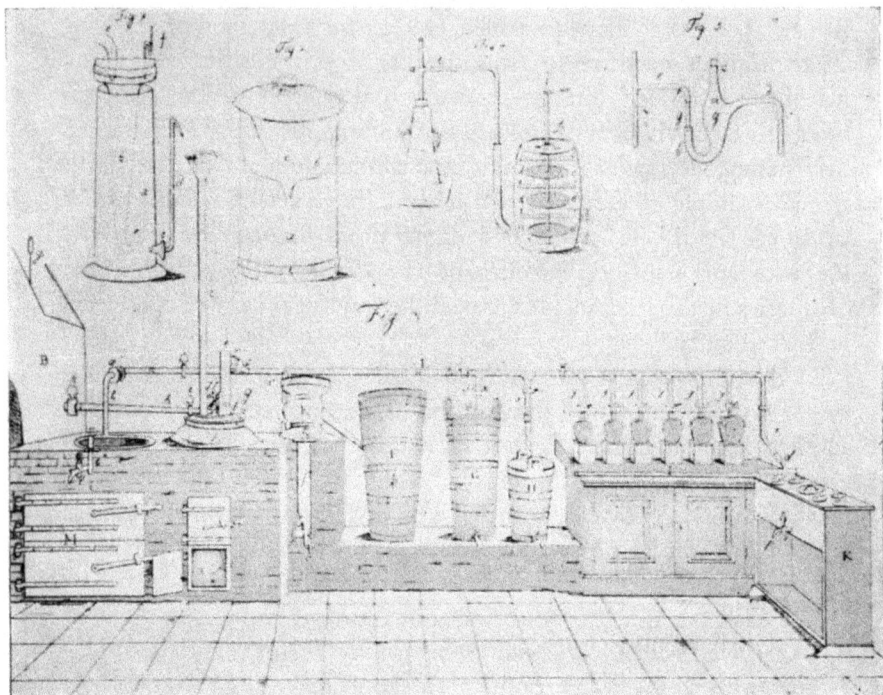

Fig. 138. VAN DYCK's steam apparatus for the apothecary

kundige Wetenschappen. He mentions the good results which steam heating gave in the paint industry and the dyer's trade of Holland. Therefore, he designs a steam-boiler with a constant water level (Fig. 138) which supplies steam to different apparatus in the laboratory such as a steam-bath, a still, a drying oven, etc. Both heating with a steam coil and live steam can be used with profit in a still, as shown by the still belonging to the Apothecaries Hall, London, which was described in the *Journal of the Royal Institution.* VAN DYCK criticises a fellow apothecary, DE HEMPTINNE, who in his

prize-essays of the Académie de Bruxelles (1817) on the question
"What are the applications in factories and home to which steam as
a means of heating can be applied with satisfaction" had denied that
it could be used in the chemists's laboratory with any profit.

The use of *vacuum in distillation* was first proposed by PHILIPPE
LEBON in 1796 (353). LEBON is said to have been inspired by the
essay on the composition of air presented to the Académie des Sciences by LAVOISIER in 1777. PHILIPPE LEBON was born at Bruchay in
1760, he studied at Paris and as early as 1785 we find him working
as "ingénieur des Ponts et Chaussées", afterwards as a teacher of
mechanics at the Ecole des Ponts et Chaussées. In 1797 he starts his
famous experiments at Bruchay and investigates the gases obtained by
distillation and combustion of wood, oil, tar and tannic acid. This led
to his famous patent of September 21st 1799 which meant the beginning of the lighting-gas industry. His "thermolamps" were a great
succes, though the smell and the candle power were still bad. But
LEBON died at Paris in 1804 while he was busy improving his patent
lamps. LAVOISIER had shown the spontaneous evaporation of ether in
vacuum and LEBON working on his ideas conceived the following
distilling apparatus. Two closed vessels, each in a separate water-bath,
are connected by a vapourline. This tube has a sidetube (which should
be at least 10 Meters long) ending in a water-seal. If the apparatus is
to distill water, one still plus the side-tube are filled with water and
the valve at the under end of the side-tube is opened. If the water has
run off from this tube and natural vacuum is obtained in this way, the
conduit leading to the second still is opened and distillation is achieved
by keeping the two water-baths at different temperatures, raising that
around the full still. LEBON also wanted to distil alcoholic liquors in
this way but it is not known whether his apparatus was ever tried out
in practice with these fluids.

TRITTON (629) proposed a patent still which is heated in a steam-
bath and uses vacuum in his apparatus and so did BARRY (48). TRIT-
TON's apparatus was first published in the *Annals of Philosophy* (vol.
XI, 1818, p. 445) and though he claims only a better taste of the
corn spirit thus produced, further letters in vol. XII of the *Annals*
show that the users of the still claimed certain savings in fuel consumption (Fig. 139). In BARRY's apparatus a retort is connected to
a receiver. The retort is either heated by a water-bath or by direct fire
(Fig. 140). The receiver is evacuated and the vacuum is kept up
afterwards by cooling and condensing the distillate on the same prin-

ciple which carried Howard's invention to a succes. Barry claims that the medicines inspissated or distilled in this way have a far better colour and taste than those distilled in the old way (1813).

An ingenious method was proposed by Smithson Tennant. He was born at Selby on November 30th 1761 and studied medicine at Edinburgh in 1781 where he came under the influence of Black. In 1782 we find him a fellow of Christ's College, Cambridge, where he studied chemistry and botany. In 1784 he travelled through Denmark and Sweden where he met Scheele. He became a F.R.S. in 1785 and obtained his M.B. in 1788 and his M.D. in 1796,

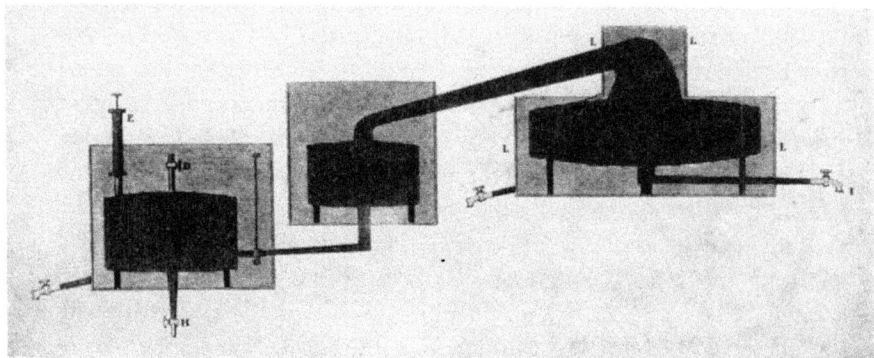

Fig. 139. Tritton's still

both at Cambridge. He lived in retirement but produced many important papers and was awarded the Copley medal of the Royal Society in 1804. In 1813 he became professor of chemistry at Cambridge and did important work in the school of Lavoisier. He was killed by a bridge falling on him at Boulogne on February 22nd 1815.

Tennant designed a double still for the distillation of seawater on board ship. The first still working under atmospheric pressure led into a worm-coil built into the second still, which with its cooler and receiver was evacuated before the distillation began. The condensation heat of the distillate of the first still was used mainly to heat the contents of the second still which by strong cooling of its distillate was supposed to maintain its own vacuum. Tennant thus hoped to produce double the amount of distilled water from sea-water by one heating only. Whatever the defects of his reasoning may be, the system contained the germs of the multiple-effect heating system which the later systems of Howard and others perfected.

In 1818 the apothecary ROMERSHAUSEN (of Achen on the Elbe) first distilled in an apparatus which was directly connected to a vacuum pump to maintain the vacuum during distillation. He also played with the idea, which TRITTON, BARRY and TENNANT had used and tried to apply the heat liberated by condensing the distillate to preheat the intake, in fact repeating the work of ARGAND, word of which did not yet seemed to have reached him.

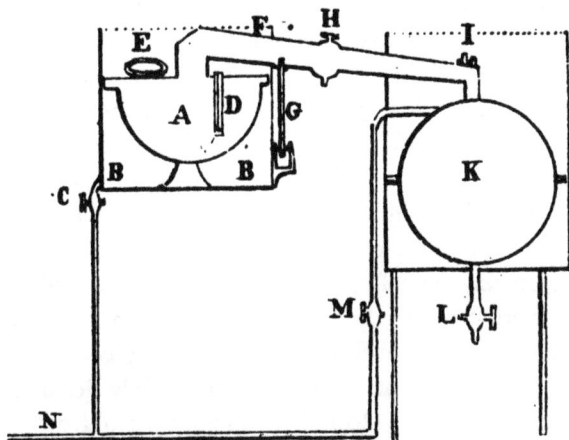

Fig. 140. JOHN BARRY's vacuum distilling apparatus

This new application of steam led to a revolution in the *evaporation technique.* In 1812 HOWARD invented (38) the evaporation vacuum pan using a steam jacket for heating and a device which was a combination of the surface condenser and a barometric jet condenser. This system was immediately adopted by all English sugar refineries, but not on the Continent because of the large size and unsatisfactory performance of the air pumps needed. The invention of the multiple-effect evaporation is a point over which there is considerable controversy. The French claim PECQUER's patent 1834 but his apparatus was never built. PÉCLET pointed out the principle in 1828 (probably inspired by TENNANT) but did not understand the use of vacuum. NORBERT RILLIEUX of New Orléans worked in Paris in 1830 on sugar problems, he conceived the idea and returned to the United States of America where he built the first apparatus in 1843 (U.S. patent No. 3237 of August 28th 1843). It was a horizontal evaporator having four bodies operating as a combined double and triple effect, the fourth body under the final vacuum was heated with steam from the first. In every case his machines number effects one less than the number of bodies. About 1850 Cail et Cie of Paris built vertical tube evaporators and according to Horsin-Deon the design was furnished by RILLIEUX. But it is more probable that the inventor was ROBERT, the director of the sugar-factory at Seelowitz, who invented the true multiple effect. His type enjoyed favour until 1879 when WELLNER-JELLINEK invented the horizontal tube evaporator, which has now disappeared in Europe but which is still much in favour in the U.S.A., or was so at any rate until a few years ago.

These applications of the power of heat show that the scientists had not struggled in vain. New possibilities were in the air and had the contact between theory and practice been as close as it is nowadays, the results would not have been long wanting. But as it was several branches of the distilling industry such as the manufacture of liqueurs were not particularly interested in better fractionation or the possibilities of manufacturing on a large scale, or even in heat economy. Here small batches were made to the taste of the public as the consumption was relatively small and the taste was the primary consideration.

Larger and better types of apparatus were mainly demanded by the growing trade in industrial alcohol. Gradually the old distilleries left off making their own alcohol, the malting houses disappeared and they grew accustomed to purchase the alcohol from the makers of industrial alcohol, who formed a new but very prosperous branch of the trade. The users of the industrial alcohol distilled their own liqueurs from it using their centuries-old recipes. This implied that they kept on using their old apparatus. If one should think that the old distilling methods were doomed, one is mistaken. The changes, even nowadays, are so slight that their "memory should be a fume and the receipt of reason a limbeck only".

CHAPTER EIGHT

NEW VESSELS FOR OLD

"One man with a dream, at pleasure
Shall go forth and conquer a crown;
And three with a new song's measure
Can trample an empire down."
(O'SHAUGENESSY, *Ode to the Dreamers of the World*)

In the first half of the nineteenth century Europe in its romantic mood still dreamt the dream of the ideals of the French Revolution which were translated into unexpected and stark realism about 1848. But steam and electricity rather than ideals changed the face of our continent and came possibly nearer to the ideals of fraternity and equality than did the governments of that period. Though the divorce of philosophy and science since the middle of the eighteenth century heralded the dawn of materialism, it meant freedom to science. From this period onwards science followed its own path in a youthful, creative and unhampered spirit which revealed itself in every field of science. Its enthusiasm both in the theoretical and practical fields made it blind to its limitations, for which it has sought help from philosophy and religion only recently.

The new science and the new technology changed the face of Europe. The Industrial Revolution was also an Economic Revolution. For a chain of shops and retail-merchants shoved itself between industry and the public, which hardly ever had any direct contact with the manufacturer of the goods. The class of the artisan, upheld by the traditions of the guild, disappeared before the new skilled labour. The new machinery asked for a different type of labour and for a special training which France first provided in the early days of the Revolution, but which England and other countries were not slow to imitate. Industry and technology were quickly becoming the main forces in the economics of Europe.

In the early nineteenth century France still remained the seat of pure science, for many English scientists, especially the physicists, were not aware of the faults of the Newtonian school to which they belonged and ran on in the old grooves. But on the other hand we saw that England was the mainstay of the Industrial Revolution and therefore long remained the country which supplied the rest of Europe with the coal, heavy machinery and metallurgical products necessary for the new industry. Still by the middle of the century Germany shook off its sleep and rapidly conquered the first place to the dismay of England and France. The struggle to recover the lost ground in pure and applied science in these countries fills the latter half of the century.

Now it may be generally true that England was supreme in the field of technology and France in the field of pure science. But such generalisations do not always hold when we go into detail and this dictum is particularly true if we look at the history of distillation. For here the French technologists were supreme in the early nineteenth century, they were the men "distilled out of our virtues" who led in the art. This new art of distilling was eminently practical for we have not yet reached the period of the mathematical and physical approach to the distillation apparatus. For science was not yet finished with putting its own house in order in those fields on which the proper handling of distillation problems depended. Still the distilling trade profited by the scientific facts as they were already known and taught in these days. The distillers, especially the manufacturers of industrial alcohol, were ready for the "distillment that was poured in the porches of their ears".

The production of alcohol in France advanced with enormous strides. CHAPTAL (124) mentions that in 1818 no less than 535,889,000 Liters of wine were distilled and in the same year the price of molasses rose to no less than 80 francs per quintal because of the high price of alcohol, whereas its intrinsic value may have been about 30 francs. In judging the references to the distilling trade in the French literature of this period it is important to know the meaning of such expressions as "trois-cinq", "trois-sept", etc. "Trois-cinq" means that three parts of the alcohol produced diluted with two parts of water give an alcohol of the strength of the standard Dutch gin ("preuve d'Hollande") that is about 29 % of alcohol. "Trois-sept" is therefore a stronger quality of alcohol. Other French writers often express the strength of alcohol in "degrees Cartier" or "degrees Baumé".

The French inventors were greatly privileged by the fact that wine remained practically their only base material upto 1850. This means that they had to deal with the distillation of a very thin liquid, from which the noxious substances like sugar, organic acids and colloids, yeast and fusel oil could be easily separated. The fluidity of wine made it possible to distill the liquid quickly and to adopt the pre-heater without any specific difficulty in the design of the apparatus once the primary ideas had been developed. Mashes prepared from potatoes or sugar-beets and pulp did not enter into competition seriously with wine until 1850 and even then they were at first distilled by most in the apparatus designed for wine, which was of course less suited to these thicker mashes. It was only in 1853 and the following years that the bad harvests and the high prices for industrial alcohol meant a premium on its production from sugar-beets, but even then the mashes were not as thick as those from potatoes or grain which the English and German distiller had to handle. The highest percentage of alcohol in thick mashes varies from 10 to 12 %, that in thin mashes from 8 to 10 %. The normal industrial alcohol of those early days had to have an alcohol content of at least 80 %. The early French distillers usually obtained this strength in two distillations, one directly from the wine brought the percentage of the raw alcohol to 50-52 %, and this distillate was rectified once to bring it up to the industrial standard of those days. Now the standard is of course 96 %. The English and German distillers, however, from the beginning tried to obtain the commercial quality in one distillation only, though they had perhaps to sacrifice part of the excellent heat-economy which the French always tried to obtain in their apparatus and to which they sacrificed the strength of the distillate.

Corn spirit was never of any importance in France. Even in the days of CHAPTAL people in France held that its taste was worse than that of spirit of wine and that the Dutch distilled their gin with juniper-berries to mask this bad taste.

Following up the new discoveries of the heat theorists, which we have discussed, the genius of the French distillers of the first two decades of the nineteenth century started an avalanche of patents of new distilling systems and apparatus. The following lines give a survey of the most important patents issued in France between 1801 and 1818:

1801 EDOUARD ADAM, LAURENT SOLIMANI
1802 BARRE (Nîmes)
1803 BARRE (Nîmes), J. B. FOURNIER (Nîmes)
1804 PIERRE GUY (Oléron)
1805 BRUGNIÈRE (Nîmes), CHASSARY (Montpellier), J. FLICKWIER (Cette), BÉRARD
1806 BASCOU (Montpellier), BERNARON (Carcassone), BONTOUX (Marseille), REBOUL (Calvisson), SIZAIRE (Carcassone)

1807 FILLY PÈRE (Calvisson), LELOUIS (La Rochelle)

1809 ZACHARIE ADAM (continuation of the patent of EDOUARD ADAM)

1810 RIVAZ (Sion)

1812 BAILLEUL (Auxerre)

1813 AGNIUS (Paris), DERIVES (Taillant), DUROSELLE (Paris), CELLIER BLUMENTHAL (Paris)

1814 BROUQUIÈRES (La Rochelle)

1815 PROSPER FABRE (Montauban), NICHOLAS NAZO (Marseille)

1818 CELLIER BLUMENTHAL (Paris).

Apart from the patents by CELLIER BLUMENTHAL and perhaps those of BAGLIONI all the others are more or less variants of the patents of ADAM, SOLIMANI and BÉRARD. Most of them were unpractical ideas which were never realised in practice or maybe only locally. We will therefore confine ourselves to the discussion of the basic patents of the more important inventors who actually saw their apparatus used on a large scale. The others may be of interest to the inventor or to the scientist, they are negligible as far as practical technology goes.

It will be seen from this list that most of these inventors hail from the wine districts of France and that Montpellier, Cette and Nîmes play a very prominent part in the development of the modern distilling column. Montpellier especially has cropped up throughout our story as a university town and a centre of distilleries fully interested in all aspects of the art.

There were, however, a few who continued the old traditions. There is for instance DUBUISSON (1763-1836) who wrote an excellent book (196) which treats the production of different beverages at great length but which contains no illustrations. Another writer of this kind was A. A. PARMENTIER, chief-pharmacist in the wars of 1757 to 1763 and then attached to the Hotel des Invalides. He became inspector of public hygiene and during the "Continental Blockade" he looked for a substitute for sugar and advised the use of glucose. His researches on this point stimulated CELLIER BLUMENTHAL and many others. PARMENTIER was a great philantropist, he died at Paris on December 7th 1813 at the ripe age of 76. His work on distillation gives little news, it contains the complete description of a distillery of the year 1802, the plate of which (Fig. 141) shows that there were few changes yet in the trade. PARMENTIER also contributed articles to the large works of ROZIER, and other similar compilers.

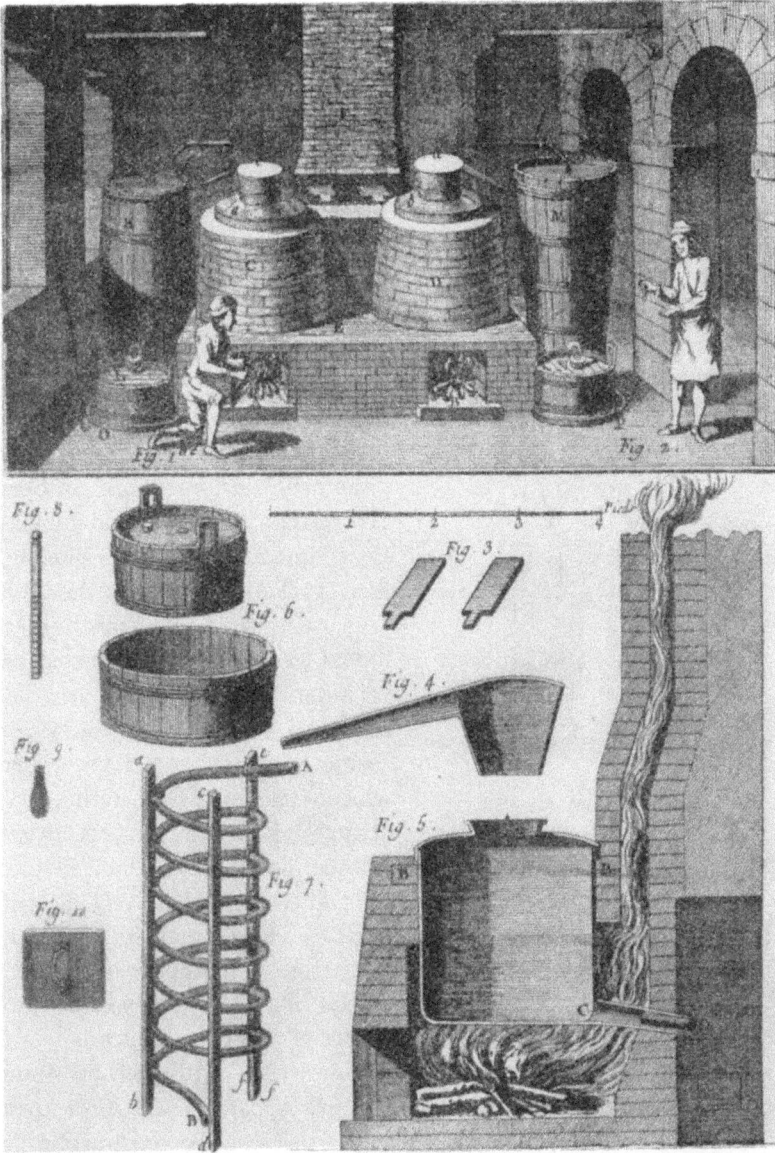

Fig. 141. A distillery of the year 1802 according to FARMENTIER

Of greater importance is the work (360) of L. S. LENORMAND
(1757-1839) who was born at Montpellier and who became chief
assistant ("préparateur") to LAVOISIER. During the Revolution he
was in charge of the production of nitre at Tarn, then he became
professor of chemistry at the Ecole Centrale and had many other posts
of importance as one of the most fruitful inventors in chemical technology in his days. LENORMAND's book is full of illustrations, many
of which we have used in this book. Through him we learn many
details of the early distilling apparatus of his century and his book is

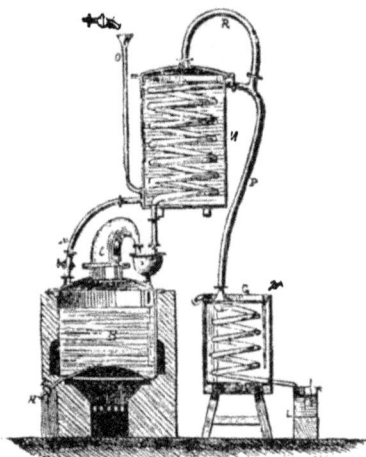

indispensable for anyone who
desires to study the art in France in
this period.

Among the theorists of the first
decade CURAUDAU was the most
prominent. He was born at Seez
(Normandy) and we find him as
an apothecary first at Vendôme,
then at Paris where he died on
January 25th 1813. According to
DUBRUNFAUT he constructed a
"distillation à chauffe-vin" or "distillation à double-effect", by which
he means a still with a pre-heater
which functions as a reflux-condenser (Fig. 142). CURAUDAU

Fig. 142. Curaudau still

claimed that his still could give all the grades of alcohol in one
operation only, that the pre-heater and cooler effectively heated
the intake of the still and that therefore little water was used in
the final cooler. This type, DUBRUNFAUT says, disappeared gradually in the twenties except in countries like Holland where corn
and potatoes were treated. The pre-heater of this type (which is really
that of ARGAND) is often called "cuve de vitesse" in French literature
because the velocity of distillation in this apparatus was more than
doubled. CURAUDAU also wrote several articles on the construction of
the brickwork of furnaces (149), on the fire-place (150) and on the
way in which the flue gases should be led round the still (152). In
the latter essay he claims that it is possible to evaporate twenty-five
units of water with one unit of coal, which of course is entirely
impossible even theoretically! He believed that a still should be deep
and that it should have a relatively small surface, contrary to the

beliefs of CHAPTAL and his school (151). In discussing the new systems of his age he says that ADAM pursued the idea of leading the vapours through alcoholic solutions to obtain the maximum amount of alcohol in the vapours, whilst SOLIMANI, FOURNIER, BÉRARD and others obtain the same effect by partial condensation of the water from the vapours. In both cases the heat efficiency is better than in the old stills. CURAUDAU's views are worth mentioning, because this "firework-maker" achieved something of international fame and was read even in Germany and Holland.

A typical still for the distillation of the lees of wine with a sieve built into it, which is also ascribed to CURAUDAU is shown in Fig. 143.

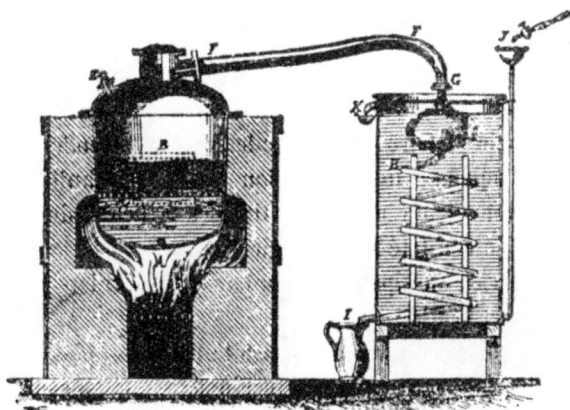

Fig. 143. Curaudau still for the lees of wine

The pioneer of the new still was EDOUARD ADAM (not EDMOND as some authors have it), a biography of whom was published by GIRARDIN (257) which unfortunately was not available to the author. ADAM was born at Rouen and URE calls him an "illiterate workman". He followed some courses in chemistry at Montpellier (possibly given by SOLIMANI) and there he heard the Woulfe bottle explained. The "Woulfe bottle" was not really designed by WOULFE, who was an eighteenth century chemist famous for his researches on the interaction of gaseous hydrochloric acid and alcohol (*Phil. Trans.* 1767) and the preparation of a new yellow dye by treating indigo with nitric acid (*Phil. Trans.* 1771). For WOULFE used a vessel with only one neck and a side-tube in absorbing gases (1767). The true inventor of what

we falsely call the Woulfe bottle was ANGELO SALUZZO, who as
early as 1759 used bottles with two or three necks to receive ammonia. LAVOISIER and BERGMANN propagated the use of this vessel for
other chemical reactions. ADAM used the principle of the Woulfe
bottle in his new distilling apparatus "to prepare alcohol in one
operation" for which he got a patent on May 29th 1801 (8) for a
period of 15 years, to which he obtained an additional patent on
June 2nd. 1801. The
main patent describes
his model refinery at
Cette consisting of 3
stills of 3 M³ of wine,
each with their own
rectifying apparatus.

Fig. 144. A simple Adam still (1801)

A simple form of
this apparatus is shown
in Fig. 144; a large
the further part of
the apparatus. ADAM
realised far better the
idea which ARGAND
and CURAUDAU may
have had, of making
the vapours rising
from the still meet
the returning phlegm
in counter-current to
dealcoholize this
phlegm as far as pos-

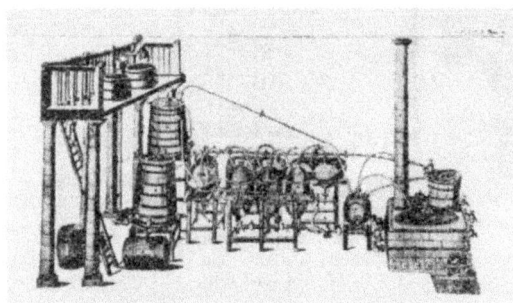

Fig. 145. A complete Adam distillery

sible. In Fig. 145 we recognise the same essential parts as in Fig. 144.
The vapours pass from the box-still into a vessel E, which ADAM calls
,.tambour" and which serves as a dephlegmator. Then they pass through
a series of "large eggs" (grands oeufs), which form the "distillatoire"
and which are filled with returning phlegm up to the lower rim. Afterwards the vapours are introduced into three pairs of globular vessels,
the "small eggs" (petits oeufs), each pair of vessels being immersed in
a tinned copper cooling bath filled with water. This part forms the "con-
densatoire", each of the eggs being again a Woulfe bottle in which the
vapour is led through the condensate before escaping to the next one.
The vapours then pass the wine pre-heater M and thence to the final

water-cooler N. The tubs on the platform are designed to fill the preheater. Each egg has a specific concentration of alcohol, which grows less as the phlegm, streaming in counter-current with the vapours towards the still, approaches the latter. The still is heated in a steam-bath thus saving money and fuel as some writers contend, though it would seem from the figures that direct heating was used. Slight individual changes in this system, however, seem to occur frequently. It is possible to draw samples from every egg and cooler to test the strength of the liquid in that particular vessel.

DUBRUNFAUT, without explaining this dictum, maintains that ADAM designed a cross between GLAUBER's retort and a Woulfe bottle! The heat economy of ADAM's apparatus was far better than that of any of the earlier stills, little of the heat of condensation being lost. The only drawbacks of the apparatus were that it was discontinuous and that it was very intricate, there were too many taps to handle safely.

ADAM was very enthusiastic about this invention and so was a commission appointed by the prefect of the dépt. de l'Herault to test it. Ure says that "he ran through the streets of Montpellier and told everyone of his invention. Therefore quickly many competitors arose, especially Solimani and Bérard". He succeeded in raising over a million francs and built no less than twenty new distilleries in the Midi. But his was the fate of many an inventor. His model-distillery at Montpellier itself and the freedom with which he talked of his invention were his bane. Soon many imitators and competitors arose and proposed new systems or simply stole his ideas.

One of these was LAURENT SOLIMANI, a professor of chemistry at the Ecole Centrale du Département du Gard at Montpellier and probably the man who demonstrated the Woulfe bottle at the courses during which ADAM conceived his ideas. A few days after ADAM SOLIMANI patented a new still (586) for a period of ten years (afterwards extended by decret of Oct. 17th 1814 until May 1st 1821) and two years later took out an additional patent (587). He built such a still in his own distillery at Calvisson and it was inspected by a member of the Académie du Gard who wrote a long and favourable report on it which is given in full by LENORMAND (Vol. II, p. 46). This still is remarkable for being the first authenticated still with indirect steam heating (Fig. 146). The vapours pass a "washing vessel" F and then an "alcogène" (see fig. 3 on the plate) whence they pass to a worm-cooler to be condensed and collected in vessel K. A hydrometer connected with a valve regulates the flow of the

water into the alcogène. The phlegm escaping from vessel F into
vessel M is pumped back into the still. BARRE later patented some
improvements on the Solimani still in his French patent No. 699
(without date!).

The distilleries built by ADAM at Cette (3 alembics, etc.), Toulon,
Perpignan and other places proved expensive and the competitors tried

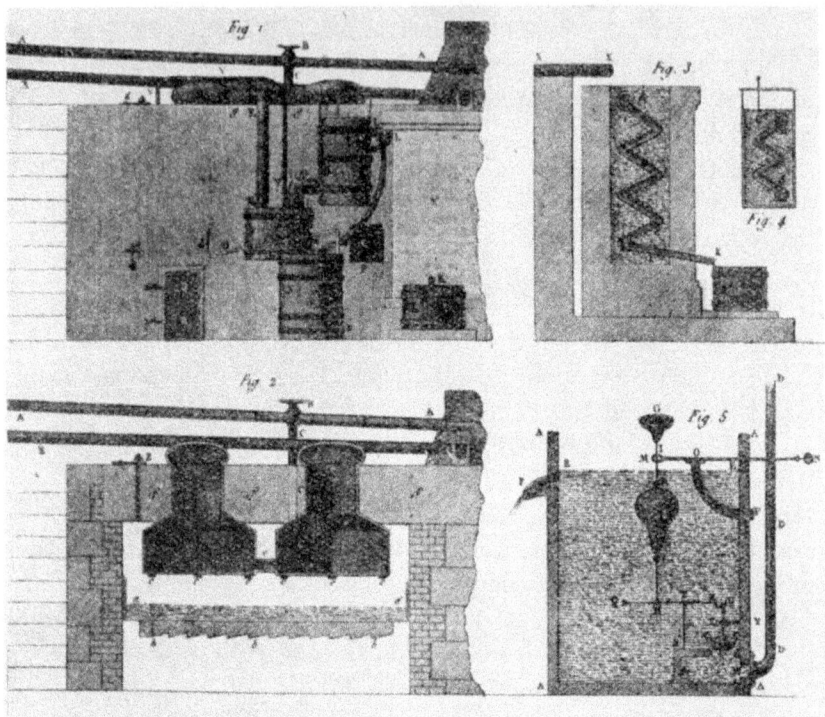

Fig. 146. SOLIMANI's still

to simplify his apparatus, sometimes giving only crude imitations like
SOLIMANI. Still his apparatus was used until far in the twenties and
BOUCHET VIOLS (French patents No. 2668 of July 15th 1824 and
additional patents of September 23rd 1824 and September 16th 1826)
and PASTRÉ (French patents of August 11th 1819, June 6th 1823
and No. 2706 of December 1st 1824) were among those who tried
to improve it. ADAM was soon entangled in law-suits which proved very
costly and to his own astonishment he lost most of them. Exhausted
and impoverished ADAM died in the winter of 1807 after having
given the Midi a new flourishing industry. He is certainly not iden-

tical with GASPARD ZACHARIE ADAM who took out an additional patent on ADAM's distilling apparatus on March 22nd 1809 and a

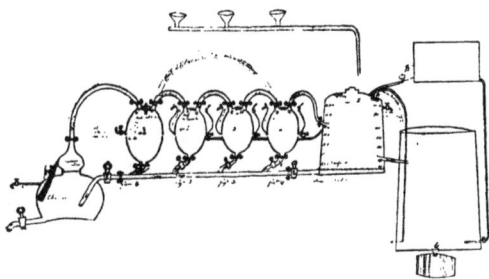

Fig. 147. An Adam apparatus of the year 1820

new one in 1817 (9) and who may be his son, but certainly was one of his heirs. The patent was extended by decret of Jan. 14th 1814, to the date of May 1st 1821.

But ADAMS ideas were not lost and many continued to use his still or slight modifications of it. Its fame spread and SCOTT in his *Visit to Paris* (London, 1816) mentions the "new distillery apparatus of M. Adam". The archives of the firm of Bols at Amsterdam contain a letter from their agent J. H. Beyerman at Bordeaux, dated August 1st 1820,

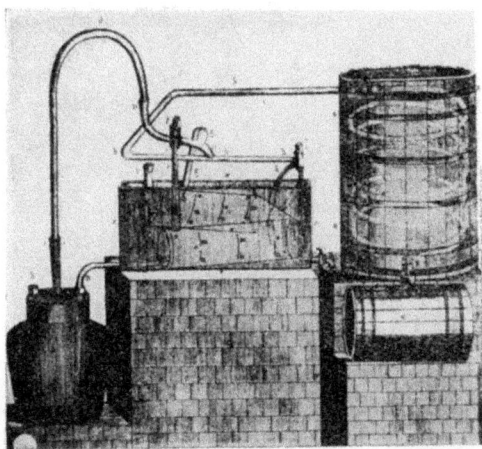

Fig. 148. The Bérard apparatus

which claims that the system of ADAM, MÉNARD and BAGLIONI are still favourites though that of SOLIMANI is now abandoned after some initial successes in Languedoc. This letter also contains a drawing of the Adam apparatus as it was then built (Fig. 147).

ISAAC BÉRARD was the inventor of a still of greater simplicity (Fig. 148). He was born at Lunel in 1770 and possessed distilleries at Grand Gallargues (Gard). He is not identical with JACQUES ETIENNE BÉRARD (born at Montpellier 1789 and dying there in 1896) who was professor at the University of Montpellier and a member of the commission that tested the Adam still, teaching chemistry and who may have been his brother. BÉRARD took out two patents (62), (63). The vapours from his still passed through cylindrical tubes which were divided into compartments by perforated plates. The condensing phlegm is conducted back to the still. This condenser is made of copper and cooled in a water-trough. Here we have the ancestor of the modern distilling column as BÉRARD's condenser can be regarded as a primitive horizontal fractionating column already used by FLICK-

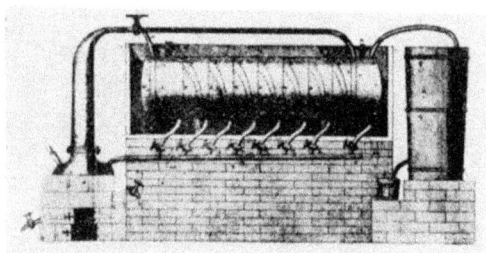

Fig. 149. The Ménard still (1804)

WIER (221). ADAM lost a very costly law-suit against BÉRARD, whose process continued to be very popular upto the thirties and even later. The system the development of which we can easily follow in the nine additional patents was highly praised by his contemporaries but there is much loss of heat because the still is often filled and emptied during the operations. But, as DUBRUNFAUT remarks, the apparatus may not be as economical as that of ADAM, it was much simpler to handle, hence its popularity. BÉRARD used the principle of partial condensation rather than that of the Woulfe bottle used by ADAM. BERTHOLLET and CHAPTAL who tried it out on August 30th 1809 were very enthusiastic about it (like the commission that tested the still at Grand Gallargues on June 9th 1805), but they do not formulate its defects.

A compact "Adam apparatus" was designed by a chemist from Lunel, AUGUSTIN MÉNARD, in 1804. He designed a horizontal column consisting of several compartments (Fig. 149), the first and the last of which were double the size of the others. Each compartment is a Woulfe bottle the condensate of which is not led into the former one, but straightaway back into the still. Thus he gave up the counter-

current idea which ADAM had introduced and partly adopted that
of BÉRARD for the sake of a simpler apparatus. MÉNARD never
patented his still but communicated its design and the method of
working it to anyone interested and even spent a lot of time in helping
the users over their first troubles with it.

PIERRE ALÈGRE, a distiller of Saint-Gelles (Gard) tried to combine the ideas of ADAM, BÉRARD and MÉNARD in his still of 1806,
later improved and patented in 1813. That is, in reality, he was not
himself the inventor, but he bought the patents of J. B. DUROSELLE

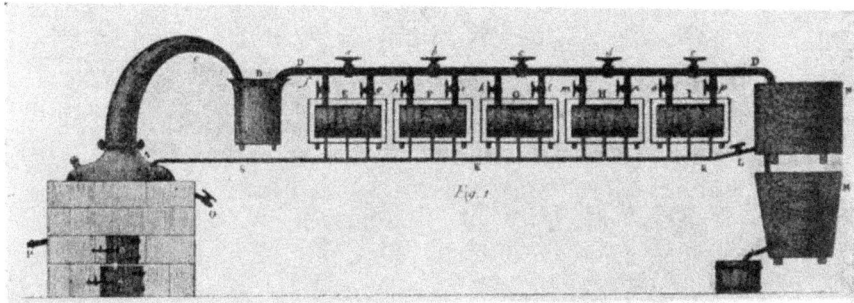

Fig. 150. The Alègre still (1806)

on March 4th 1816. In this apparatus (Fig. 150) the vapours from
the still pass through a hermetically sealed Woulfe bottle into a series
of condensers of the type Ménard designed and thence to the pre-
heater and the final worm-cooler. This apparatus was often rebuilt
and remodelled, another version being described by PAYEN in Vol.
III of the 1839 edition of La Maison Rustique (edited by Malapeyre).

In 1808 CARBONEL had already made his own version of the
Alègre still because of the frequent explosions caused by the wrong
handling of the many taps in the original apparatus. As regards this
point his apparatus (Fig. 151) may be easier to handle but it was
also too complicated to live long.

This was also the case with the still designed by A. S. DUPORTAL
described in his book (199) and criticised by LENORMAND (Vol. II,
p. 138). A small scale copy of this still is still in the possession of
the Conservatoire des Arts et Métiers of Paris.

According to DUJARDIN a similar still was also patented by LAU-
GIER in 1808, but this is not correct as the LAUGIERS (father and son)

took out several patents (April 22nd 1812; September 18th 1816; July 8th 1818) for the manufacture of essential oils and perfumes, but there is no still to their name in the patent register. A "Laugier" still of later date will be described later on.

CHAPTAL propagated for laboratory use a still of the old type but he states that the alembic should not be fitted to the neck of the cucurbit but to its widest diameter. The fire should enclose the still and prompt condensing of the vapours was the secret of obtaining a good distillate. But he also tried his hand at a still of the new type

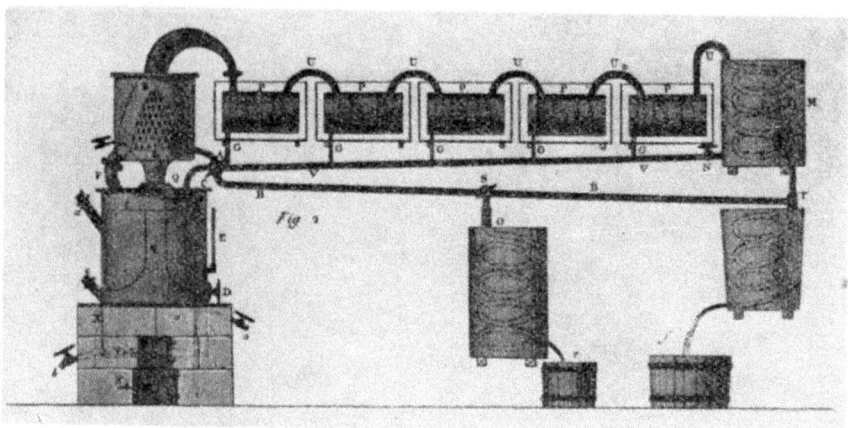

Fig. 151. The Carbonel still (1808)

and designed one (Fig. 152) in 1807 in which the principles introduced by ADAM and BÉRARD were combined, which did however not differ much from those discussed above. CHAPTAL (121) describes the patents of ADAM, BÉRARD and CELLIER in detail and states that it is now at last possible to regulate the heat far better than formerly, and also to build a battery of stills with a variable number of condensers of different forms as desired.

So ADAM, SOLIMANI, BARRE and BRUGNIÈRE had introduced the idea of running the first condensate counter-current to the vapours and thus enriching it to produce the desired strength of the alcohol in one run.

ADAM, BARRE, FLICKWIER, BÉRARD and SIZAIRE had used wine to absorb the heat of the vapours and to condense them, thus economizing considerably on the total amount of heat used.

A horizontal column was introduced by BÉRARD and the idea of continuous distillation was to be found in the apparatus of ADAM and FOURNIER, both using two stills which were heated alternately after draining the residue and refilling.

The final step of using the principles introduced by ADAM, BÉRARD etc. to build a distilling column was taken by JEAN BAPTISTE CELLIER BLUMENTHAL, whose fundamental inventions we must now discuss. But CELLIER was led to the design of a distilling column by his interest in manufacturing sugar from sugar-beets and therefore we will have to survey this subject first.

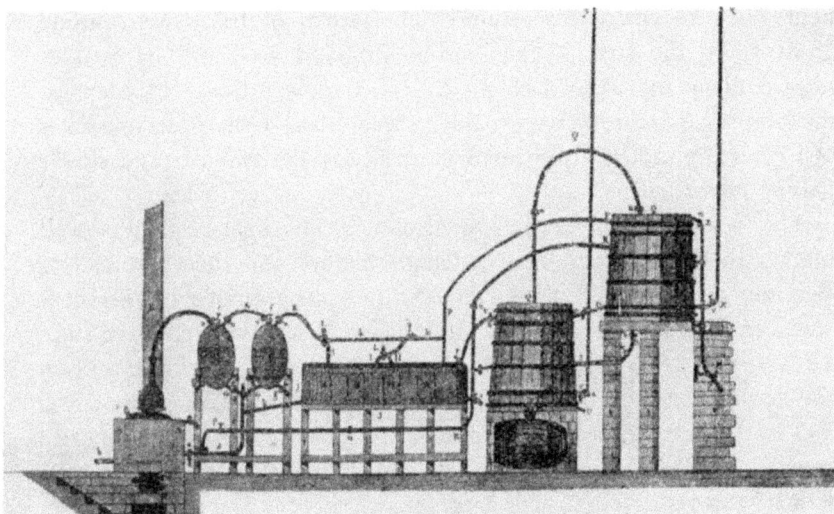

Fig. 152. CHAPTAL's combined Adam-Bérard system (1807)

The fact that sugar-beets contained sugar was observed by MARG-GRAF under the microscope in 1747 and reported to the Royal Academy of Berlin. This discovery forms the base of the beet-sugar factory of our days. The great promotor of this industry was ACHARD, who in 1786 collected his first small harvest of sugar-beets for production purposes and 1798 finished his researches. In 1802 the first beet-sugar factory was opened at Cunern (Silesia). During the Continental Blockade the attention of Napoleon was drawn to the production of sugar from new materials as the supplies of cane-sugar from the colonies were of course cut off by the English. Now Napoleon stimulated research and first he had been impressed by the possibilities of grape-sugar or glucose, the production of which seemed very simple in the laboratory. But it was exceedingly difficult, well-nigh impossible to realise these laboratory results in practice, though the best French scientists of that day were working on it. Just as Napoleon had given up his last hope of exploiting this new source of sugar his attention was drawn to the production from sugar-beets of which CHAPTAL was one of the promotors in France. After a famous visit to CHAPTAL's beet-sugar factory at Passy on March 25th 1811 Napoleon concentrated all his scientific batteries on the production of beet sugar in his empire. B. DELESSERT succeeded on Jan. 2nd. 1812 and the emperor decorated him with his own cross of the Légion d'Honneur when he visited his refinery. The factory of Passy was running from 1808, the first of the kind in Holland was built by Spakler near Arnhem in 1811. With the fall of Napoleon these first steps of the beet-sugar industry were held up, there was a sharp decline after 1815 but the industry recovered gradually in the twenties and slowly gained importance.

The possibility of producing alcohol from sugar-beets was well known in Germany in the eighteenth century and thence spread to Belgium and France. But the first man to make a serious and efficient study of the subject was A. P. DUBRUNFAUT, who worked on it since 1824 and in 1825 discovered the excellent effect of sulphuric acid on the decomposition of the pulp. Further improvements were introduced and by 1852 the problem was practically solved by DUBRUNFAUT and others. Among these workers we must enumerate DOMBASLE and CHAMPONNOIS.

MATTHIEU DE DOMBASLE (1774-1843) left military service for reasons of health and during the Continental Blockade tried to prepare sugar from beets. He lost much money in his projects and then left the industry. He then applied himself to the spreading of the new chemical theories and their practical applications by founding the Institut d'Agricole at Roville (1822), where he also conducted many successful practical trials, inventing a new plough and a new harrow. His results attracted a lot of attention especially in England. He died on December 27th 1843. He also worked on a better still, but the "combineur hydropneumatique" (French patent No. 1210 of November 6th 1816) was only a bad imitation of CELLIER's column.

HUGUES CHAMPONNOIS, born at Chaumont in 1803, studied chemistry and pharmacy. He came into contact with CHAPTAL who interested him in the manufacture of beet-sugar, in which he found his career. He invented the beet-washer and the distilling apparatus, which bears his name (1852). His many inventions and improvements found their reward in decorations and many important posts which he held. A considerable amount of his work was devoted to the proper conditions of the fermentation of pulp. His diffusion process was not used before 1880.

The first alcohol from beets was marketed at Paris in 1854 and by 1868 there were already 300 distilleries working on beets. DUBRUN-FAUT was a fanatical propagandist of the combination of sugar refinery and distillery as we shall see.

Many systems of distilling alcohol from fermented pulp were invented after 1850. LEPLAY (1854) cut the pulp into certain sizes before fermenting and afterwards distilled in a current of steam which left the fermenting properties of the bath intact. Other systems were devised by ROBERT (Vienna), VILLARD, SIEMENS, KAULER, RENARD, VAN VOLXEM (the collaborator of CELLIER), etc. But in Belgium and Germany the stills of CELLIER and SAVALLE remained favourites for a long time.

We also find a series of books on the subject in the same period. We draw attention to the *Traité de distillation* written by CHARLES BARRAL, that of J. A. BARRAL (the editor of the *Journal d'Agriculture* from 1849 to 1866) and that of A. CHEVALIER, the chemist (1828-1875).

Thus when CELLIER tried his hand at this problem in the first decade of the century, the industry of beet-sugar production was still in its infancy. J. B. CELLIER BLUMENTHAL was not a Belgian as many writers make him out, but a Frenchman, as he was born at Clermont-Ferrand in 1768. He was attracted to the problem of sugar refining with many compatriots by a prize of one million francs put up by Napoleon for a good method of obtaining a uniformly crystallised white sugar in large quantities. Now PARMENTIER published an essay on this subject (466) stating that PROUST had published in the *Journal de Physique* of 1807 a method of refining grape-sugar and other kinds of sugar with alcohol which PARMENTIER found indeed suitable for glucose and other sugars. He even applied it to honey and obtained an excellent white sugar. This idea of refining crude sugar with alcohol was in many minds. DEROSNE took out a patent on this idea (French patent No. 929 of May 13th 1808) and so did CELLIER (116) together with LAPORTE in 1811, designing a method of extracting and refining sugar with alcohol. Now this process involves the evaporation of large masses of alcohol from the sugar solutions and this must have led CELLIER to the search for a distilling apparatus capable of treating large masses of liquids at a greater speed than in the old Adam or Bérard apparatus. According to DUBRUNFAUT he had already invented his apparatus in 1808 but he certainly did not take a patent in that year but on Nov. 24th 1813 (117), with an

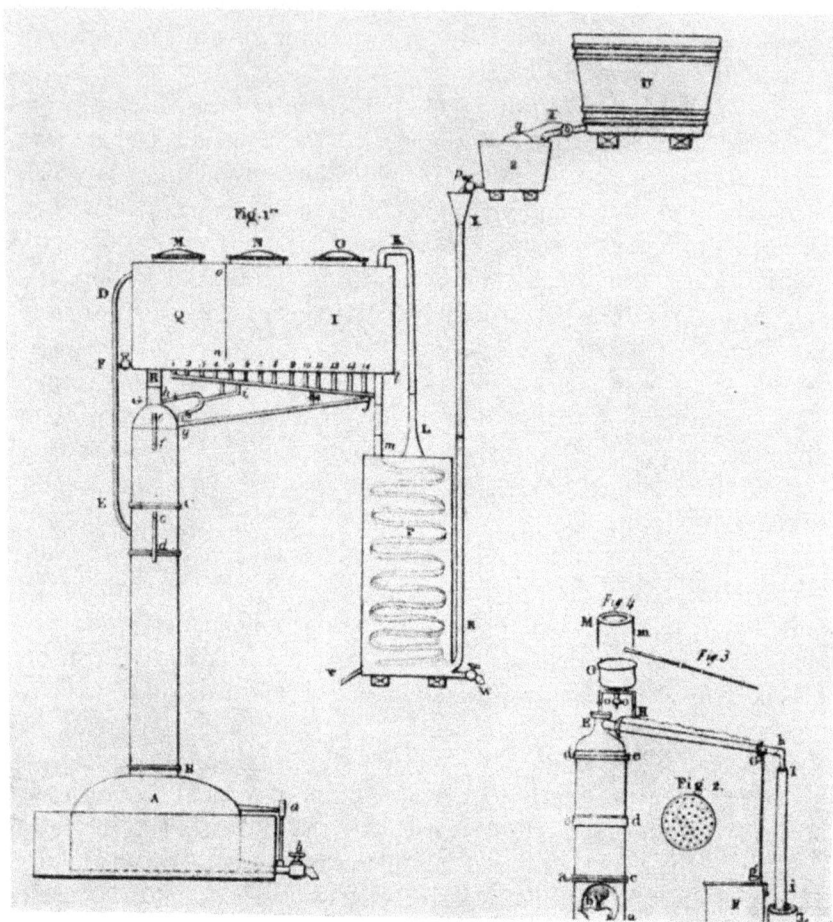

Fig. 153. The Cellier still of 1818

important additional patent in 1818. The apparatus which he designed was very simple, it consisted of a column, clearly derived from the Alègré and Ménard still, with bubble-plates, mounted *vertically* over the still and above it a condenser/pre-heater, which acted as a reflux-column too. The vapours escaping from this column were cooled in a large pre-heater, so that all the cooling was done by the intake wine (Fig. 153). Thus the ideas of ADAM and BÉRARD bore fruit, but the most important fact is that this apparatus was designed for continuous operation. Though the Baglioni apparatus was patented somewhat earlier (Aug. 24th 1813), it is certainly not of any use in practice having very bad fractionating properties. Both from the opinions of his contemporaries, who are definitely outspoken on the point (except BAGLIONI's pupils, PRIVAT, TAGLIONI, TACHOUZIN etc.)

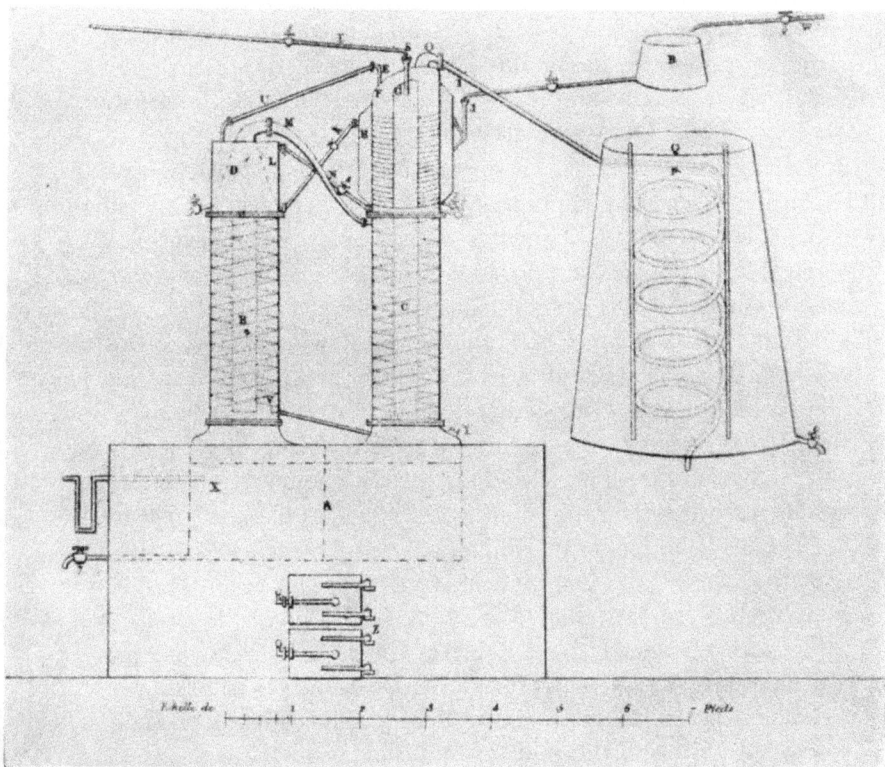

Fig. 154. The apparatus designed by BAGLIONI (1813)

and from the technical merits of this still, we judge that CELLIER may be truly regarded as the inventor of the modern fractionating column. For the earlier stills nearly always worked discontinuously or had to be emptied and refilled several times during the operation. CELLIER introduced the idea of a continuous stream of wine entering the pre-heater and a continuous stream of spent residue leaving the still. Great names like SAVALLE, DUBRUNFAUT and LACAMBRE state definitely that CELLIER's invention was earlier than that of BAGLIONI and that the French Government awarded him the basic 1818 patent after BAGLIONI lost his process against CELLIER.

CELLIER's patent of 1813 contains no sketch of the apparatus. In 1815 CELLIER applied for a Dutch patent and he tried to prove the merits of his apparatus by trials in 1816 but the jury, the "Eerste Klasse van het Koninklijk Nederlandsch Instituut" rejected his application on the grounds that "his apparatus had no particular merits for the distillation of malt-wine and gin over the stills in common use in this country. The apparatus is very ingenious in separating the spirituous part from the liquid flowing down in fine drops and even nearly achieves separating them all, instead of trusting in elevation." The file of this Dutch patent application is still to be found in the Rijksarchief at The Hague. It opens with a letter of March 7th 1815 announcing CELLIER's application for a Dutch patent for the invention for which he was granted a French patent in November 1812. Surely his French patent of 1813 is meant. CELLIER was invited to conduct some trials at Amsterdam, but he did not arrive before July 1816 and stayed at hotel "L'Etoile" ("De Zon"), Nieuwendijk, Amsterdam. The trial-runs of his apparatus were conducted at the distillery of J. A. Foppe on the Roeterseiland, Amsterdam. Here "sugarwater" was treated, that is the fermented residue of the stills and evaporators of the sugar-refineries, and this liquid was selected as it was the base-material for alcohol which was most like the wines treated in France but no longer in Holland.

The experiments were conducted in the presence of Messrs. van Marum and Vrolik of the "Koninklijk Nederlandsch Instituut" with the results mentioned above. Only in the case of malt-wine and gin was this type of still considered to have some advantages over the common type. A second trial with the common distillery mash was conducted in the distillery of Foppe, and again the conclusion was, that the apparatus "would only serve the purpose of redistilling stronger spirits". The patent application was now officially declined (September 19th 1817) in more general terms, as cited above.   .

In the same file we find a letter, dated August 4th 1817, which CELLIER had written in meantime to explain the bad results of the trial runs. This letter is most interesting because it contains sketches of two distilling apparatus (Figs. 155 A and B), the first of which seems to have been of the type used in the experiments and as we must surmise identical with that for which the French patent of 1813 was obtained.

CELLIER begins by blaming the bad results on his difficulty in making himself understood by the men working with the apparatus.

He mentions that the common types apparatus first give the strongest alcohol and then fractions of ever-decreasing alcohol content, while his apparatus will give a distillate of constant alcohol content as soon as the liquid has reached a temperature of 80° C and as long as there is any alcohol present in the liquid. Again the continuous distillation method will allow the distiller "to do away with the filling and heating periods as one could go on distilling all the year by constantly supplying base material to the still and fuel to the fire". Also no water is needed to condense the vapours and the only heat lost is that contained in the residue flowing from the still continuously.

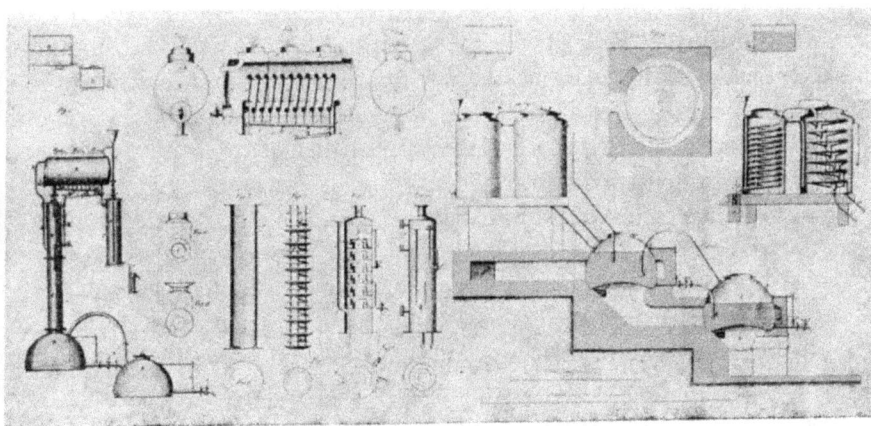

Fig. 155. The stills of CELLIER (figs 1-12) and LAUGIER (figs 13-15) according to LACAMBRE

By using not a simple condenser of the usual type, but that of his design it is possible to obtain every fraction (that is every degree of alcohol!) desired.

CELLIER works on the difference in boiling point between water and alcohol and believes that by maintaining the still at a temperature of 82-83° C all the alcohol will be distilled off. For this complete evaporation it is necessary to make the evaporation surface as large as possible. As "the liquid runs from plate to plate all the alcohol is evaporated and little remains in the liquid when it reaches the still". He states that he has observed this evaporation by introducing small glass windows in his experimental stills. He also mentions that his earliest form of bubble cap was simply a metal gauze stretched over the opening through which the vapours ascended.

The question of continuous distillation was certainly studied thoroughly by CELLIER, for in the same letter he states "he could easily submit 13 designs, which he executed to try out the principle, heated by direct fire, with a water-bath or with steam, either by introducing it into the liquid by means similar to the Woulfe bottle or by running the liquid over surfaces heated by steam (indirectly)". "In this way I have made beet-sugar and beef-tea cubes without any empyreumatic smell or taste". As to thicker mashes "I treated a kind of prunes that are used to make "quetsch-wasser" and also mashes of cherries which give the famous Kirsch-brandy". Therefore he does not doubt that by further experimenting he will succeed in treating the mashes as they are used in the Dutch distilleries. He ends his letter by stating that his apparatus have "yielded at least $\frac{1}{8}$ (that is 12.5 %) more alcohol than the older types, using nearly half the fuel and labour, practically no water, but still give a product of superior taste and smell".

Appended to the letter we find two sketches of apparatus which "have been made long ago by one not skilled in drawing". Fig. 155 A is not explained by CELLIER, but the file contains a "mémoire descriptif" of Fig. 155 B, from which we copy the following points:

A An ordinary still with a simple cover a instead of a capital.

B The first condensing vessel into which the vapours ascend (called alkoogène by CELLIER) and meet the condensed liquids. As the liquid stays at least 5-6 minutes in this vessel, all the alcohol vapours are driven from it into the worm condensers.

C and D are the worm condensers, the lowest points of the coils being connected by the lines rrrr and tapped by 5 spouts with the cocks 1-5 leading to the collecting line gg and thence to the condenser F to be withdrawn from the system.

E is the vessel that contains water, its contents are used to cool the worm condensers C and D. This cooling is regulated by a device iii which opens or shuts the cock h which feeds the cooling mantles of the worm-condensers.

The vapours have to pass 192 feet of coil before they reach the final cooling stage in F. This is still a water-cooled apparatus.

In the second drawing, Fig. 155 A, we find the germ of the later Cellier still. Though it still has the horizontal worm-condenser of his earlier designs, the idea of using the base material continuously as the cooling liquid is already fully developed. The worm-condenser C and the vessel B of Fig. 155 B have disappeared and the two parts of the later fractionating column begin to take shape, but they are as yet separate vessels, in which the proper contact and separation of liquid and vapour take place.

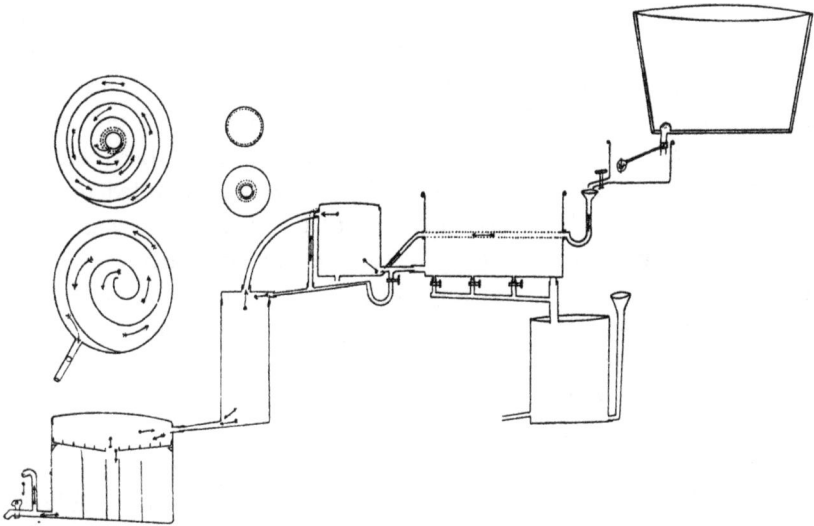

A

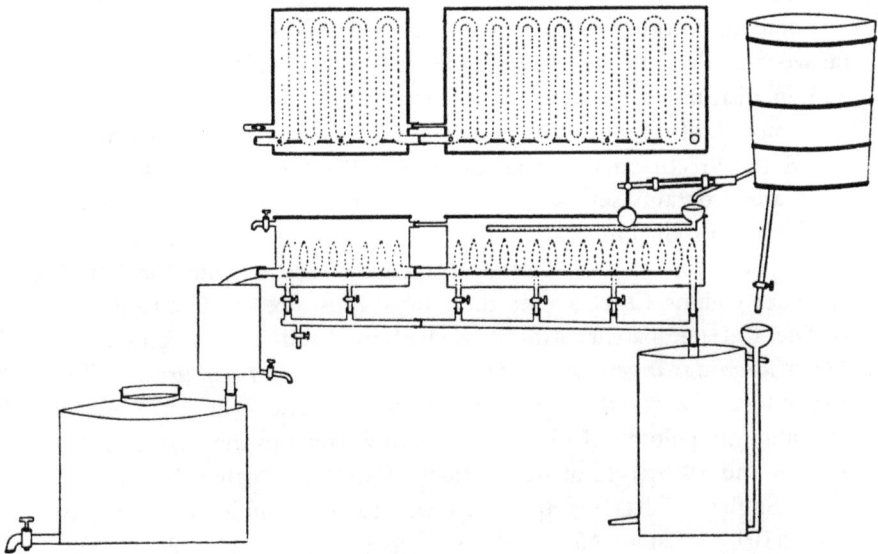

B

Figs. 155 A & 155 B. Designs from Cellier's Dutch patent application of 1817

As CELLIER himself states that these designs are old and as they served to cover an application of his French patent of 1813, we may be fairly sure, that the proper development of the apparatus patented in 1818 and bought by DEROSNE took place between 1815 and 1818. Indeed it must have taken shape when CELLIER and DEROSNE met at Amsterdam in 1816 and when the gold medal was awarded to him in the same year. It seems therefore plausible that the perfected Cellier still dates from 1816 or 1817, the proper understanding of the working of the fractionating column must have been conceived in 1816, the building of a successful experimental still may have cost some time and may have occupied the inventor while he was discussing the sale of the design to DEROSNE between 1816 and 1818, when he obtained his French patent.

Later CELLIER succeeded in obtaining a Dutch patent on his still (Octrooi No. 314 of June 5th 1828) and a second one (Octrooi No. 481 of March 16th 1830) both for a period of 10 years. He then lived at Kockelberg near Brussels and states that this apparatus, which differs only slightly from that illustrated in Fig. 158, was installed 15 days before the patent application in the distillery of Mr. CLAES at Lembeck working much better than those of DEROSNE. CLAES and he went to Paris to see the Derosne stills, both models at the Louvre and in distilleries. However, they ascertained that these used double the amount of fuel of the Cellier still. The top of this column is called the "rectificateur", the lower part the "colonne distillatoire". The latter contains several plates and the vapour is made to pass the liquid by introducing it through a series of small pipes from below which each end in two "horns" curved downwards into the liquid. In other colums CELLIER uses the bubble caps already described.

The CELLIER patents can be studied in detail in the volumes of *Description des inventions et procédés consignés dans les brevets d'invention et d'importation pris en France*, 1e série. The 1813 patent is found in volume **XXI**, Volume **XXV** contains no less than 55 figures and 100 pages of descriptions of CELLIER's patents!

In England CELLIER's apparatus was patented under the name of Dihl (British patent No. 3965 of December 5th 1815). As CELLIER shortly after his second application of 1818 (which is full of drawings) sold his rights to DEROSNE, we find that all further patents based on his invention refer to DEROSNE's patent, if they mention any original (for instance BRUGNIÈRES, French patent No. 1877 of June 23rd 1826).

Soon after 1813 CELLIER started to build three sets of apparatus with which he experimented helped by his friend SAVALLE, who possessed three sugar-factories and distilleries in the present Belgium (then still belonging to Holland). Some authors mention that they worked in The Hague, but neither there (where there never were factories of that kind) nor in any other city of Holland could their names be traced in the archives. Tradition has it that they tried out their apparatus in practice, and twice were nearly killed by explosions, but they persevered.

They were right, for the continuous distillation was much more economical than the old methods. Here the vapours rich in alcohol steadily meet mashes which are gradually transformed into spent residue because the alcohol is taken up by the vapours. Contrary to BAGLIONI, CELLIER took care to achieve close contact between these vapours and the liquid by introducing bubble-caps, whilst in the apparatus of BAGLIONI the liquid and the vapours run counter-current but have superficial contact only. In CELLIER's apparatus the distillate obtained is always the strongest possible concentration of alcohol and therefore it has a very small area per capacity unit. In the case of the discontinuous apparatus the first distillate is the strongest, but it becomes weaker as the operation proceeds. CELLIER's still lost very little heat like that of ADAM, for the only important amount of available heat lost is the difference in calories between the residue and the intake and even this was limited by later inventors using the spent residue as a pre-heating liquid. His apparatus was most suitable for the production of wine-spirit and it was soon adopted by the distillers, for even in southern France, the home of ADAM and BÉRARD, we find many specimens of Cellier apparatus by 1817. Because a rectification column was used the wine was not introduced into the still but into the column.

CELLIER obtained for his design the gold medal of the Société d'Encouragement pour l'Industrie Nationale of 1816, one out of four medals for the most prominent inventions awarded every ten years, and the French government granted him an additional patent of fifteen years free of rights in 1818 (118). Early that year CELLIER had grown weary of further experiments with SAVALLE, and as the column was already a succes without being perfect he sold the rights (on Jan. 20th) to make and sell his apparatus to CHARLES DEROSNE, an apothecary at Paris, for 1200 francs. On DEROSNE's further improvements of the still we will turn our attention further on. DEROSNE obtained additionel patents to that of CELLIER on August 28th 1818, June 19th 1821 and March 30th 1822.

After his return to Paris CELLIER was involved in many lawsuits for like ADAM he had many imitators and his patent rights do not seem to have been well defined. One of his most successful imitators, whose still for some time enjoyed considerable vogue in Languedoc was BAGLIONI (from Bordeaux). In his patent (29) a still is described with a column, which has an inner cylinder in which the wine is introduced at the bottom and which contains the reflux-condenser. The pre-heated wine flows over in the column (between the inner cylinder and the outer shell) where it flows down over a double Archimedean screw of thin copper-plate which is perforated. Here the vapours and the steam from the still rise and run in counter-current to the liquid flowing down in a thin layer, leaving the column at the top to be syphoned into the preheater and then into a final Gedda cooler. (Fig. 154). Ingenious though the apparatus was, it would not rectify well as the contact between vapours and liquid was not intimate enough, as modern trials with such columns have proved. Here the Cellier column was definitely better, because it forced the vapours through the liquid instead of over the liquid. DUBRUNFAUT already states that the apparatus was very difficult to handle and that it was often improved and modified. One of the followers of BAGLIONI, TACHOUZIN (French patent No. 1691 of December 17th 1824) gives an interesting survey of earlier methods in his patent and sees BAG-LIONI's column as the best of his period, he does not even mention that of CELLIER, but CELLIER's patent was in meantime bought by DEROSNE and is of course referred to as Derosne's still. BAGLIONI claimed priority over CELLIER, but in reality he took his idea from CELLIER, as the law-court decided. In fact BAGLIONI's column is not the logical development of the ideas of ADAM and BÉRARD and a perfected Alègre and Ménard column, but a version of that of CEL-LIER's and not even an intelligent one. The same can be said of the patent of DOMBASLE. BAGLIONI's still was for some time used in Languedoc, for instance at Privat (Mèze) and TACHOUZIN tried to perfect the apparatus for the distillation of lees of wine and corn. LAMOTHE (French patent No. 3771 of 1827) also tried to improve the Baglioni still, but it disappeared fairly quickly from general use.

It is interesting to read the opinions of contemporaries on CELLIER's column. CHAPTAL thought that CELLIER tried to extend the surface of the wine as far as possible and let the vapours pass thin layers of wine spread out over different plates. DUBRUNFAUT says that the process would not eliminate the taste from distillates obtained from thick mashes, which should always be redistilled. GIRARDIN (258) mentions that the most important point of the Cellier column was that only a small fraction of the wine or other base material was heated directly (or by steam) and that the alcohol vapours obtained thus did the rest of the fractionation work in the column.

In November 1816 CELLIER had already discussed the sale of his patent with DEROSNE, when they met in hotel "L'Etoile" at Amsterdam, probably when CELLIER was over in Holland to prove the merits

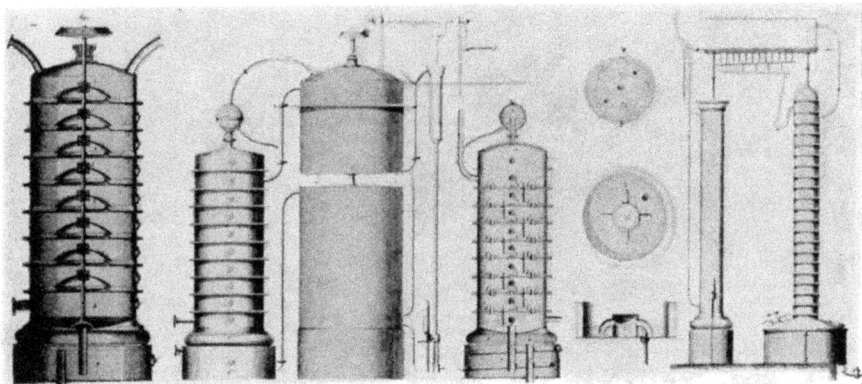

Fig. 156. Some later Cellier columns according to LACAMBRE

of his apparatus in connection with his application for a Dutch patent. He tells this in a letter to the Amsterdam firm of Van Zuylekom en Levert dated October 15th 1839. He seems to have used his stay at Paris, where he took the final measures for the transfer of his rights in November 1819, to supervise and inspect the improvement of his still by DEROSNE, but he must have left Paris early in 1820, for then we find him settled at Brussels, where he spent the rest of his life designing new stills and perfecting them for the treatment of thicker mashes obtained from grain and potatoes. The cause of the separation between CELLIER and DEROSNE is not clear, but it seems from contemporary literature that DEROSNE was a very clever business-man who had a gift for drawing to himself the glory of improvements and inventions made by his collaborators or subordinates. This may be the reason why the Cellier still is referred to as the Derosne still since the date that the rights were bought by DEROSNE. In the course of

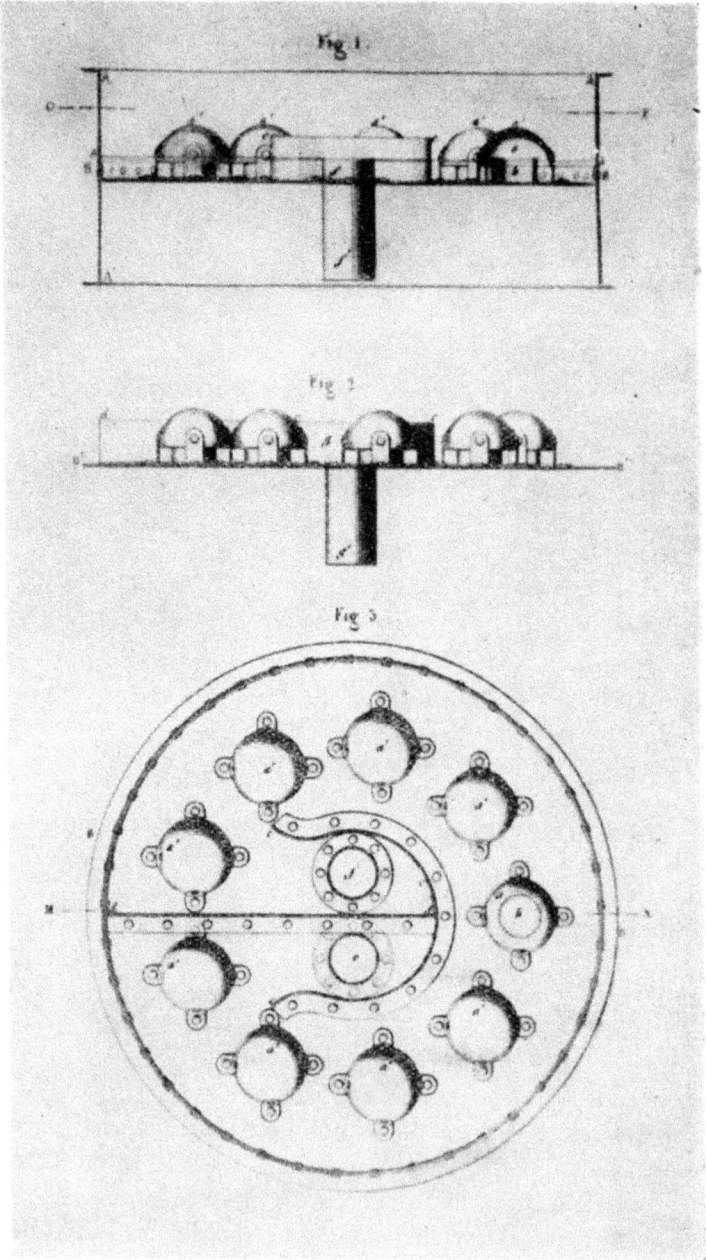

Fig. 157. Detail of the tray of CELLIER's column

these changes the column becomes rather more complicated. It is not certain whether CELLIER always mounted the column directly on the still or whether he preferred the more flexible construction of a separate column. The latter design is always found in his later stills, the former is usually said to have been introduced by DEROSNE (that is by his chief-engineer CAIL) who certainly always uses this construction to increase the stability of the whole (Fig. 155). We see in this Derosne design that the old Cellier reflux-condenser is separated from the column and is no longer of the old vertical worm-condenser type, but a worm-cooler, from which the condensate is drawn off at the lowest point of each bend.

The columns usually have 12 or 14 plates but this number sometimes went up to 20. CELLIER gradually changed the plates of his column which originally were simply perforated plates. In Fig. 155 we already find several other designs, such as square bubble-caps (one for each plate) (see also Fig. 156). Gradually he came to use nine bell-caps on each plate as shown in Fig. 157. DEROSNE increased the number of caps to 18 per plate. In Fig. 155 we also see the Laugier still as it was used in the days of LACAMBRE, that is about 1850. This apparatus also has the double still, which DEROSNE introduced and it was long in use in southwestern France.

CELLIER became weary of all these patent troubles and in 1820 he went to live in Brussels. In Belgium CELLIER worked on the adaption of his system to thicker mashes aided by the firm of Delattre and Dubois of Brussels. He first cooperated with the firm of Dooms at Lessines, then he experimented several years with VAN VOLXEM, the owner of sugar-factory and distillery in the neighbourhood of Brussels and inventor of a pre-heater shown in Fig. 156. This firm bought Cellier stills on May 29, 1829. The boilermaker and coppersmith Camal (of the rue St. Cristophe, Brussels) built most of his apparatus. Nor was his inventiveness confined to stills, for living in Brussels CELLIER took more than 15 new patents, e.g. a method of evaporation in vacuum (119) and in 1835 a beer-cooler, which was a prototype of the Beaudelot cooler used at present in the industry. His Belgian activities are mentioned in BRIAVOINE's De l'Industrie en Belgique, 1839.

CELLIER introduced live steam in some of his later apparatus. Fig. 158 shows a continuous still of 1828 which he claimed to give alcohol of any desired strength in one operation. It was used in the factories of Dooms at Lessines, Claes de Limbecq and many others. A recti-

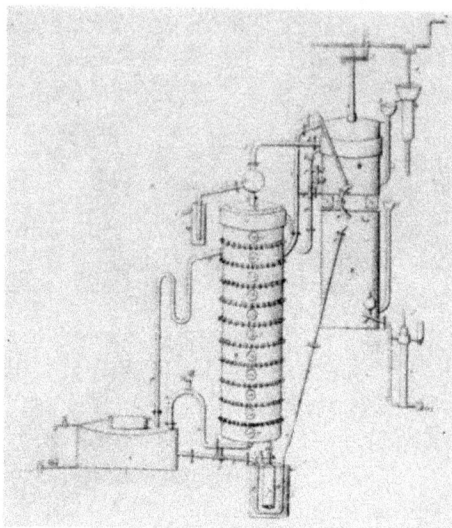

Fig. 158. A continuous Cellier column of 1828

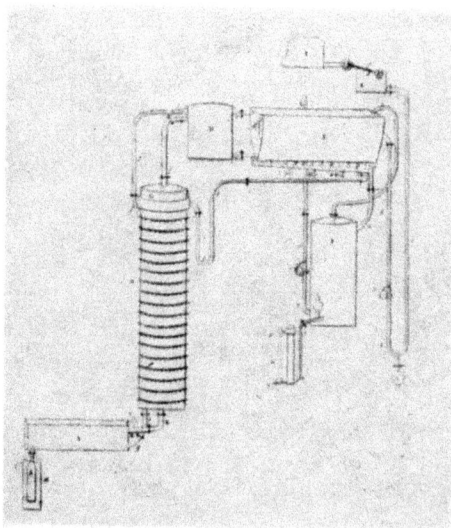

Fig. 159. A Cellier rectification column of 1828

fying column of the same year is shown in Fig. 159 which is said to produce at least "30 degrees Cartier" alcohol.

Shortly before his death at Brussels on August 30th 1840 CELLIER had a correspondence with the firm of Van Zuylekom and Levert, distillers at Amsterdam, who desired to buy a new distilling apparatus. It seems that a local copper-smith (from Cologne) had offered

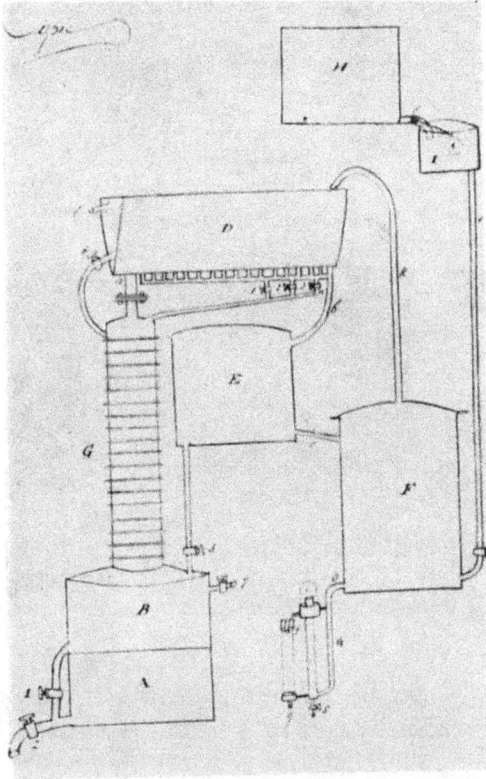

Fig. 160. A column designed for the firm of Van Zuylekom and Levert, Amsterdam, by CELLIER (1839)

to build a "Cellier apparatus", for CELLIER, in his letter of October 1st 1839, mentions that this coppersmith was an apprentice of Camal of Brussels who always executed his designs. CELLIER continues to say that he was the true inventor of the distilling column and that DEROSNE, who sold his apparatus at the price of about six francs a kilogram, had made no essential improvements in his designs since CELLIER left for Brussels. He denies the claims of DEROSNE that his

"Derosne apparatus" would be able to make alcohol of 38° or 40° Cartier (94 % vol.), as during the time of their cooperation they never produced anything stronger than 33° (84 % vol.) and DEROSNE did not improve his still after. If the Dutch firm wishes an apparatus distilling 6000 to 7000 litres of grain-mashes per day he submits a new design (Fig. 160) expressily made for that purpose. Using one still only it can be operated discontinuously in runs of 90 minutes, but

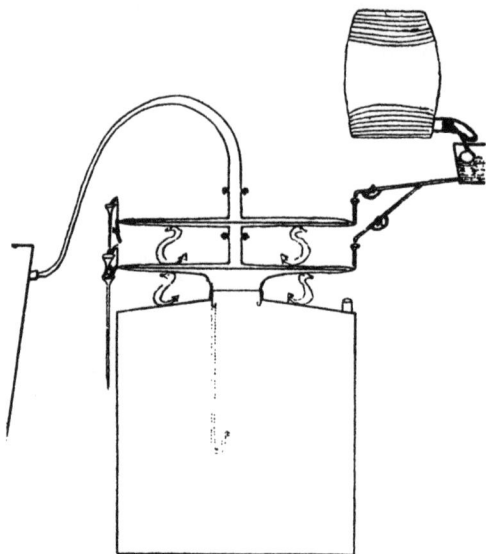

Fig. 161. Design of the still built by van Zuylekom and Levert, 1840

with two stills it can be run continuously. On obtaining further details of their wishes he notifies Van Zuylekom that the apparatus will weigh about 1200 kilograms and that the height will be about 6 meters.

Van Zuylekom replied that they did not think that the cooling capacity was sufficient and they thought that the apparatus was too large and unwieldy so that he proposed to discuss it at Brussels. However, the same archives tell us that nothing came of this interview and that the Dutch firm built locally an imitation of the Pistorius still, designed and executed by the local coppersmith Wellinghuyzen, after a design already installed for the Amsterdam distiller Hoppe. This apparatus was finished on January 24th 1840 (Fig. 161).

CELLIER died in his house (117 rue de la Régence, St. Josse-ten-Noode, Brussels) on August 30, 1840 leaving behind a widow with three children.

The *Manuel du Distillateur* (408) mentions a long list of patents issued between 1820 and 1850 on new distillation systems, mostly derived from CELLIER, though this is hardly ever stated in the patent. We cite the patents of JULIEN (French Patent No. 1623 of July 28th 1819), GIRARD et TANSIZIER (French patent No. 1297 of September 17th 1821); LELOUIS (French patent No. 1389 of January 6th 1821), LANTHELME (French patent No. 1413 of March 27th 1823); MAILLARD and DUMESTE (French patent No. 2036 of June 2nd 1826); JACQUES TULLIÈRE (French patent No. 2367 of August 4th 1818); BOUCHET VIOLS (French patent No. 2668 of July 15th 1824 with additional patents of September 23rd 1824 and September 16th 1826), DEBEZIS, MAGNAN, HUORT, PRAJET, S. ALLEAU (French patent No. 742 of October 24th 1817), REBOUL, CASTEL, GUGNON, SERTON, PELLETAN, BRUGNIÈRES (French patent No. 1877 of June 23rd 1826); MONMORY and DE SMETZ; LAUGIER, SABOUREAUD, VILLARD, MARESTÉ and EGROT, of which only the last two had some success in practice. The Maresté still was used in Cognac upto 1870, the Egrot still will be discussed later on as a modification of the Derosne still.

A curious effect of the Cellier still was the enthusiasm with which it was greeted by all circles of practical chemists. Thus CLÉMENT and DESORMES, impressed by DEROSNE's trials with the Cellier still hastened to spread the good news in their courses at the Conservatoire des Arts et Métiers at Paris. They themselves designed a similar column, which they called "colonne absorbante" or "cascade chimique" consisting of a column filled with glass balls of 2-3 cms. diameter, with which they proposed to produce bleaching-liquor continuously. They insisted that such a column would also be quite useful for alcohol, seeing the close contact of vapour and liquid and the thin film of the returning liquid and they took a patent for a similar distilling column.

It seems that CELLIER toyed with the idea of heating his still in a water- or steam-bath (See Fig. 153), but his later stills are usually made for direct heating with fire or steam. The earliest news on CELLIER's still was published by the French periodical, the *Bulletin de la société d'encouragement pour l'industrie nationale* of 1817 (p. 256) and these lines were copied by the *Jahrbuch des polytechnischen Instituts* of Vienna in 1824 (p. 486). The improvements by DEROSNE were first described in the *Mercure technologique* of December 1823. In this way the details about the new still came to Germany and other countries.

The man who bought CELLIER's rights and improved the still, was

CHARLES LOUIS DEROSNE, born at Paris in 1780. He studied chemistry and early showed a particular aptitude for translating laboratory processes into practical installations for factories. At the death of his father he took over the famous family dispensary, which he managed with his brother (1806). He was then already working on industrial processes, patenting a new process of refining sugar (French patent No. 929 of May 13th 1808) and improving the Achard process (1811). Shortly afterwards he translated ACHARD's famous work on the manufacture of beet-sugar (Paris, 1812). This interest probably brought him into contact with CELLIER BLUMENTHAL, with whom he later cooperated. As early as 1817 he suggested some improvements in the Cellier still, after opening discussions about taking over the rights in November 1816 at Amsterdam. These improvements got him the silver medal at the industrial exhibition of 1819. He then seems to have left the old dispensary of the Rue Faubourg St. Honoré to his brother and founded a workshop at Chaillot. Here his pupil and collaborator, the engineer CAIL, suggested further improvements and in 1825 the firm of Derosne and Cail was founded and the shop at Chaillot was extended. DEROSNE died, a famous man, at Paris in 1846.

JEAN FRANÇOIS CAIL, born at Chef Boutonne on February 2nd 1804, helped to build up the fame of the Chaillot workshop, which for twenty-five years was the most important maker of complete installations for sugar-refineries for the colonies (Dutch East Indies, etc.) Later he took to the building of locomotives with great success. After the twenties it seems that DEROSNE and CAIL did little work on their distilling apparatus, at any rate no further improvements were introduced, hence the remarks which CELLIER made to van Zuylekom. CAIL and DEROSNE wrote a handbook on the refining of cane-sugar in the tropics (1844).

On buying CELLIER's rights DEROSNE immediately started trial runs to compare its effect with that of the old familiar types. Thus he built a still at the factory of DUNAL (professor of chemistry at Montpellier) at Massilargues (Lunel, dept. de l'Herault) where it was compared with a Bérard still and showed the enormous economy in time, money and labour. A second experiment at the Planche and Gaudion factory at Pezenas compared the Cellier apparatus with an improved Adam still. PAYEN give the following figures from the report of the committee which conducted the trial:

	Adam apparatus	Cellier/Derosne apparatus
Amount of wine distilled	12,768 L.	12,678 L.
Labour	two workmen	two workmen
Operation finished in	75 hours	53 hours
Coal used	990 kgrs.	420 kgrs.
Cooling water used	17,000	nil
Alcohol 3/6 obtained	1,622 l.	1,640 l.

It will be clear that the committee consisting of practical distillers waxed very enthusiastic over the new still. These were the trials which inspired CLEMENT and DESORMES to construct their "cascade chimique".

We have already seen that DEROSNE placed the Cellier column directly on the still, a feature which seems to be his own (or CAIL's?) as the Cellier designs usually have column and still apart, while they are always combined in the Derosne apparatus. (Figs. 162, 163 & 164). In the early Derosne still we see a rather heavy pre-heater balanced on the column, the reflux of which is not yet returned to the top of the column as in later models (Fig. 164). DEROSNE always uses bubble-cap trays in his column, but the form of the bubble cap varied from the rectangular in the older ones to the round cap or the double-tube form (which Perrier had introduced) in the later models. The circulation of the liquid on the

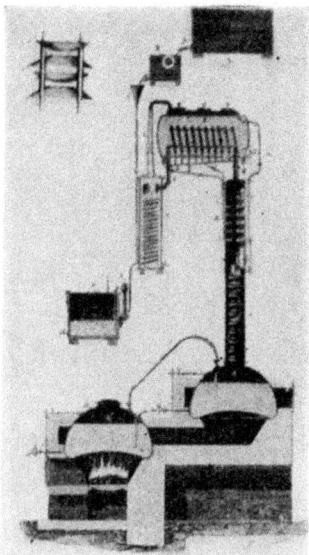

Fig. 162. DEROSNE's new Cellier still (about 1820)

tray is already quite good as the number of caps is increased, first by CELLIER from one to nine, than by DEROSNE to 18. The intake flows from a vessel placed above the still and passes a simple regulator or valve which can be adjusted to insure a proper and constant supply of base material. The entire Derosne apparatus was built to obtain the highest heat-economy possible, hence all the coolers are pre-heaters and no water is used. As usual with such French distilling apparatus this means that first alcohol of 52 % is produced and hence a rectification is necessary, whereas the German and English distillers aim at obtaining the quality of alcohol desired in one operation, even if they have to loose heat by this method.

All Derosne stills for wine, mashes of pulp, etc. use indirect steam in the lower still if possible. One of the features of the Derosne apparatus is the two coupled stills, which he introduced instead of the one still of CELLIER, an unnecessary feature which we find in DEROSNE's designs and all those of his imitators.

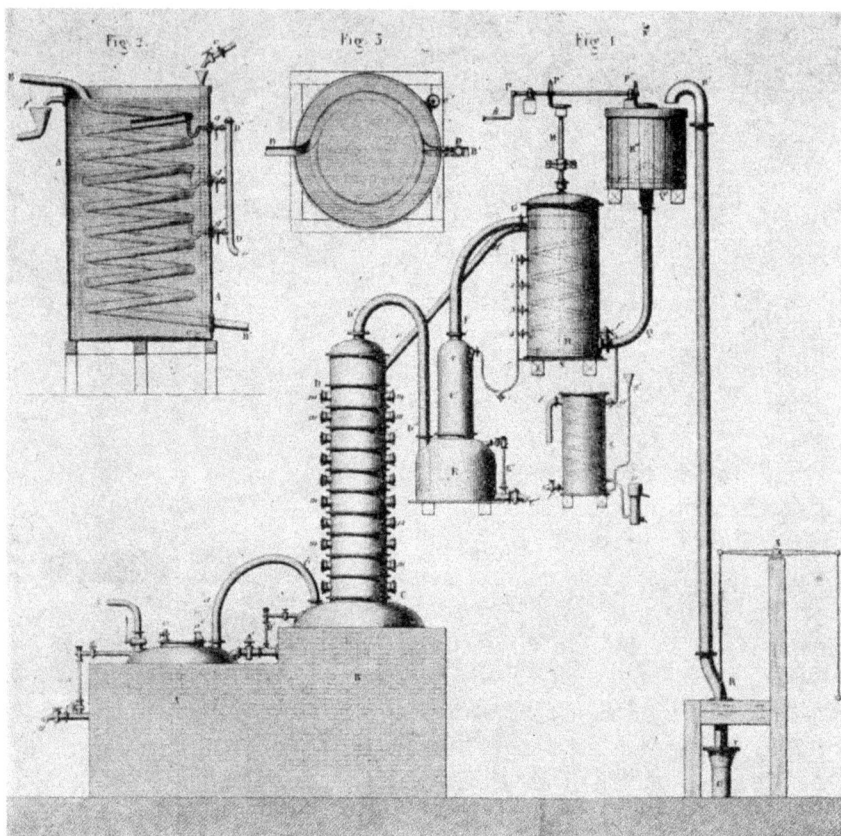

Fig. 163. A Derosne/Cail still for thick mashes

The French stills long remain rather small and therefore they work more expensively than the larger English stills of the Coffey type. The still diameter is usually 85 centimeters and that of the column 25,30 or 35 cm. in which case the daily intake is 4000-5000, 6000-8000 or 10,000-12,000 l of liquid. In 1825 such an apparatus cost about 1000, 2600 or 5000 francs and PAYEN states that the price of a Derosne still of 4000 to 100,000 l capacity varied

from 2500 to 30,000 francs in 1858, which shows that by that time
larger capacities were already in demand. Still DUBRUNFAUT states
that the Derosne still did not satisfy the distillers entirely in the
case of strong wines. Apart from the unnecessary complications of
the two stills, the original design was a rather unstable construction
with the heavy pre-heater placed above the column. Hence we find

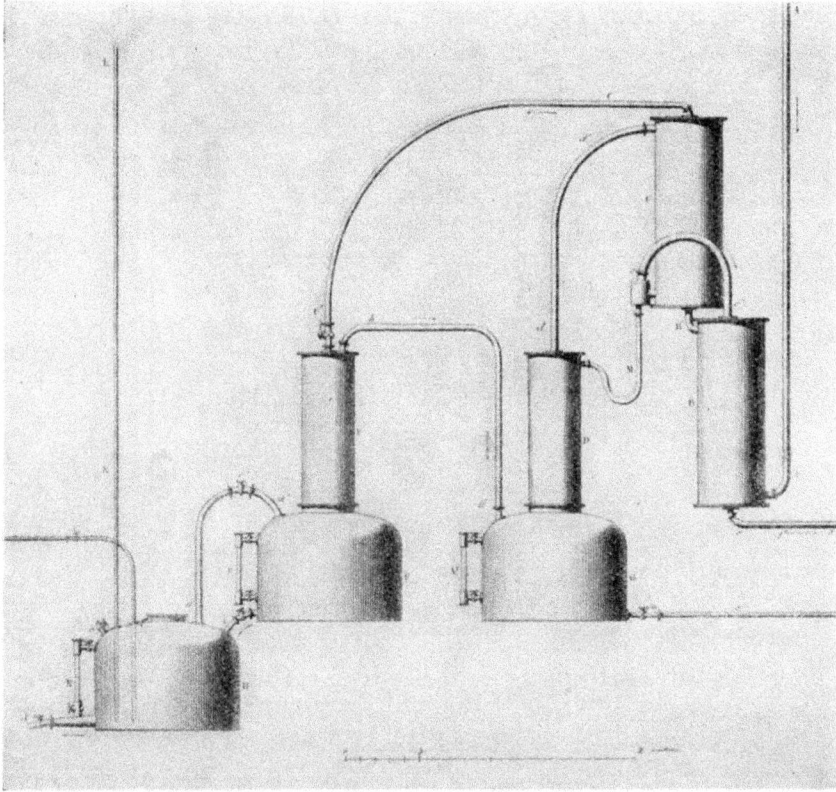

Fig. 164. A Derosne still for molasses

the apparatus split up into smaller units in the later designs such
as the Derosne-Cail apparatus for thick mashes (Fig. 163) and the
apparatus for molasses. The first one was built for the Belgian
distillers, constant supply of liquid to the column being ensured by
a stirrer in the supply-tank. The German distillers who used this
apparatus introduced CHRISTOPHE's regulator instead of the stirrer.

The Laugier still was designed as an improvement of the Derosne
column (Fig. 155) but it contains no original idea and worked badly.

Another still derived from DEROSNE's construction was that of
EGROT which DUPLAIS descusses in detail (198). He mentions that
the construction is more elegant and more stable than that of DEROSNE,
which often leaked and bent when the parts had not been erected
with the greatest care. The Egrot apparatus was better balanced and
easy to take to pieces. The inventor was confident that he had found
the ideal apparatus for the smaller farmer. But the distilling me-
chanism seems to have been too complicated for the small user and

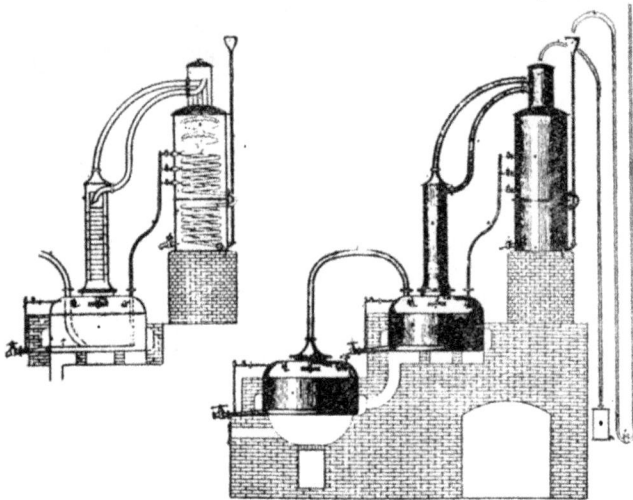

Fig. 165. The Egrot still (1854)

the fact that it was heated directly caused it to produce mostly second
quality alcohol. Hence the Egrot still disappeared from use in the
sixties, after having been propagated in a new form on the 1867
Exhibition at Paris.

CELLIER's earliest cooperator, PIERRE ARMAND DÉSIRÉ SAVALLE,.
was born on March 3rd 1791. Part of his life was spent in Holland
(that is in the present Belgium) where he did his first experiments
with CELLIER in his sugar-refineries. After they had several times
risked their lives CELLIER went back to Paris and sold his rights to
DEROSNE but SAVALLE worked on. SAVALLE was a brilliant man,
gifted in mathematics, physics and engineering, and his successes
came soon. He was assisted by his son FRANÇOIS DÉSIRÉ SAVALLE
(born at Louvain on March 17th 1838 and when the prices of
alcohol went up in 1852 he went to Saint Denis to found a distillery
(1855) in which the experiments were carried on for the firm of

Savalle and Cie. SAVALLE died at Lille on April 17th 1864 after having reached fame for his distilling apparatus and especially for his steam regulator. The date of this invention is usually given as 1857, but we know that the germs of this invention came to him after an explosion of his experimental still in 1846 and that its final form was reached after 1850. PAYEN does not yet mention it in 1859, though it was certainly completed and sold by then. This means of regulating the steam supply ensured regularity and stability to the whole operation, though of course it was only applicable to live steam as there did not yet exist any regulator for exhaust steam(Fig. 166).

After his father's death in 1864, DÉSIRÉ succeeded as the head of the firm and published several books on the apparatus made by him (551) (552). SAVALLE understood that there should be some connection between the cooling surfaces of pre-heaters and condensers and the amount of distillate obtained, but he could not calculate this. He also knew that an intimate contact between liquid and vapour in the column was essential and hence he changed the ancient CELLIER perforated plate for a tray of bubble-caps, generally of the bell-type, using rectangular columns. But in the case of thin liquids such as wine he retained the perforated tray as did COFFEY. As condenser and pre-heater he introduced the tube-condenser, invented by GRIMBLE. In general his apparatus is more sensible and more stabily constructed than that of DEROSNE though he too insisted on a perfect heat economy, cooling as little as possible with water. So in this case too the alcohol produced in the primary operation was too weak and had to be rectified. The Grimble tube-condensers were either used as pre-heaters or as water-coolers, not with air-cooling as GRIMBLE originally intended. Fig. 167 shows an installation for the treatment of mashes of cane-sugar and Fig. 168 one for the distillation of thick beet-sugar mashes. In the first case indirect heating by live steam is applied in the still, in the second case the contents of the still are circulated through a tubular heater standing apart from the column. By this well-conceived construction SAVALLE soon obtained a throughput seven to eight times as high as CELLIER.

The younger SAVALLE mentions in his book (551) that in 1873 there were already 16,000 distilleries working in Germany, whereas this number reached only 700 in France because of the heavy duties and taxes in the latter country. He then describes apparatus for the distillation of alcohol from the mashes of molasses, sugar-beets,

malted grain and potatoes, cane-sugar and wine. In the latter case
100 l of alcohol are produced at the cost of 40 kgrs of coal. The
column designed for this purpose by the firm of SAVALLE was a

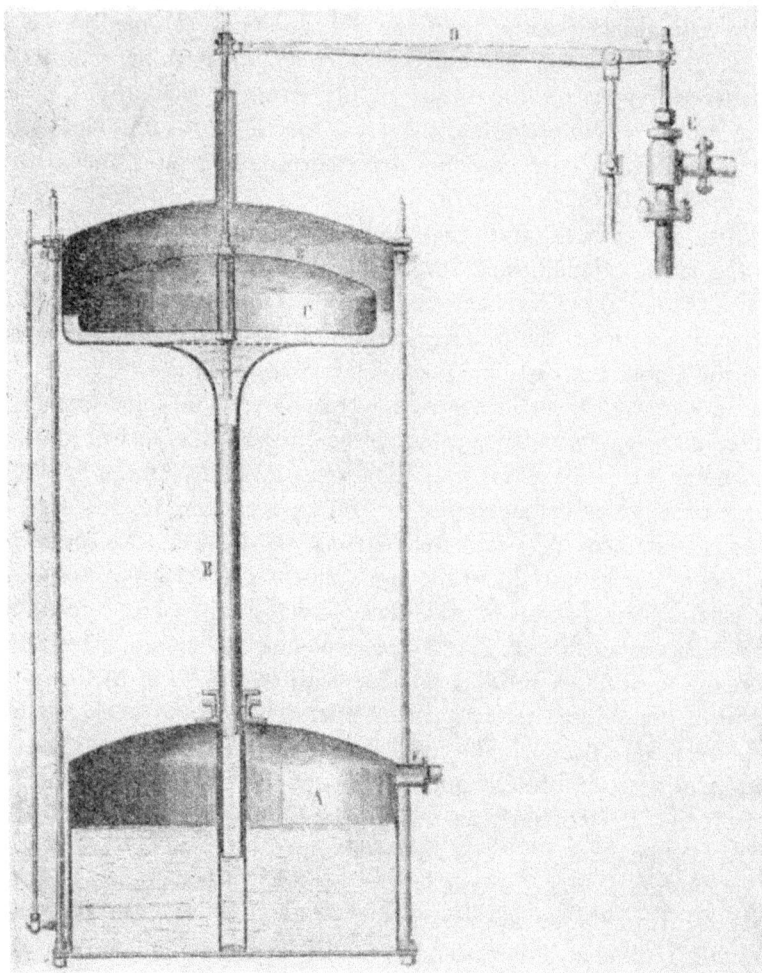

Fig. 166. ARMAND SAVALLE's steam regulator

favorite in Spain and Portugal. For the production of alcohol from
fermented fruit juices the same still is used.

The rectification columns of SAVALLE usually contained 30 trays
and condensed and refluxed $^2/_3$ of the total distillate. These columns
producing 96 % alcohol had a daily capacity of 500 to 20.000 l

SAVALLE also describes columns for the rectification of methyl alcohol and for the fractionation of crude benzene. Two columns of the latter type were built at Ludwigshafen for the Badische Aniline

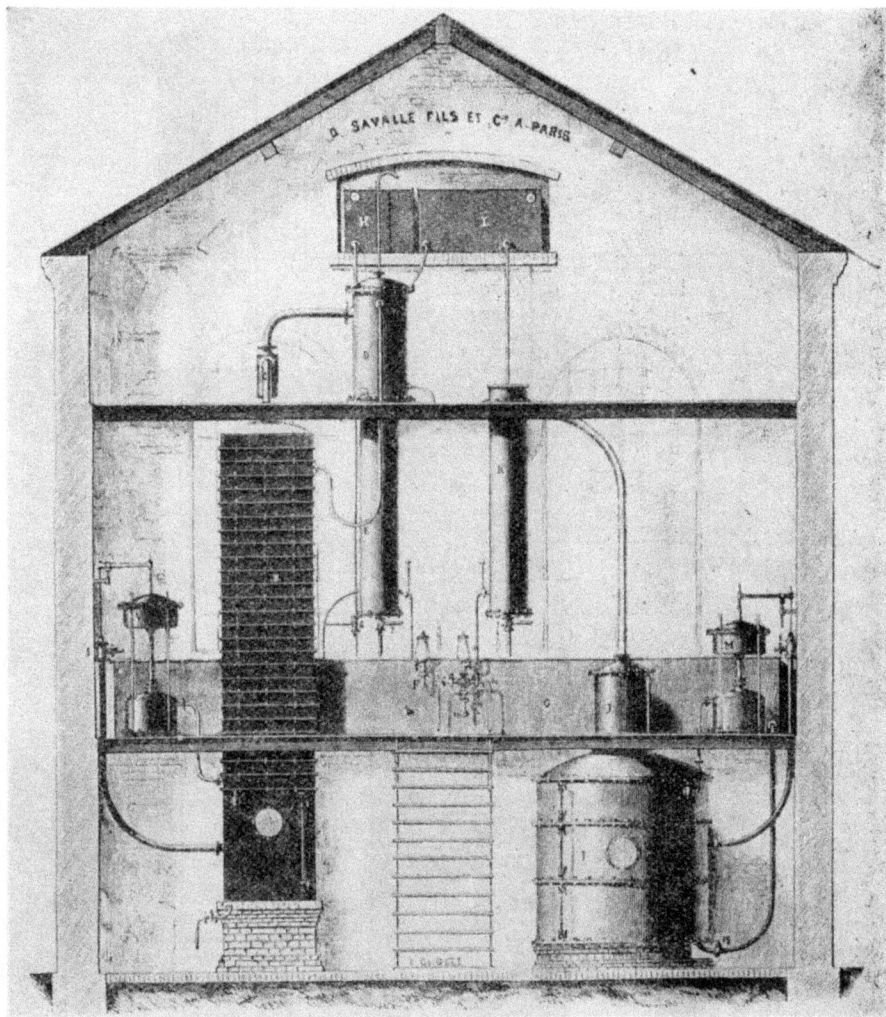

Fig. 167. A Savalle installation for alcohol from cane-sugar

und Soda Fabriken, which easily delivered pure benzene, a fraction 85-105° C and toluene. The residue is distilled in old-fashioned still to yield the heavy benzole fraction boiling upto 168° C. The third disciple of CELLIER was A. P. DUBRUNFAUT, born at

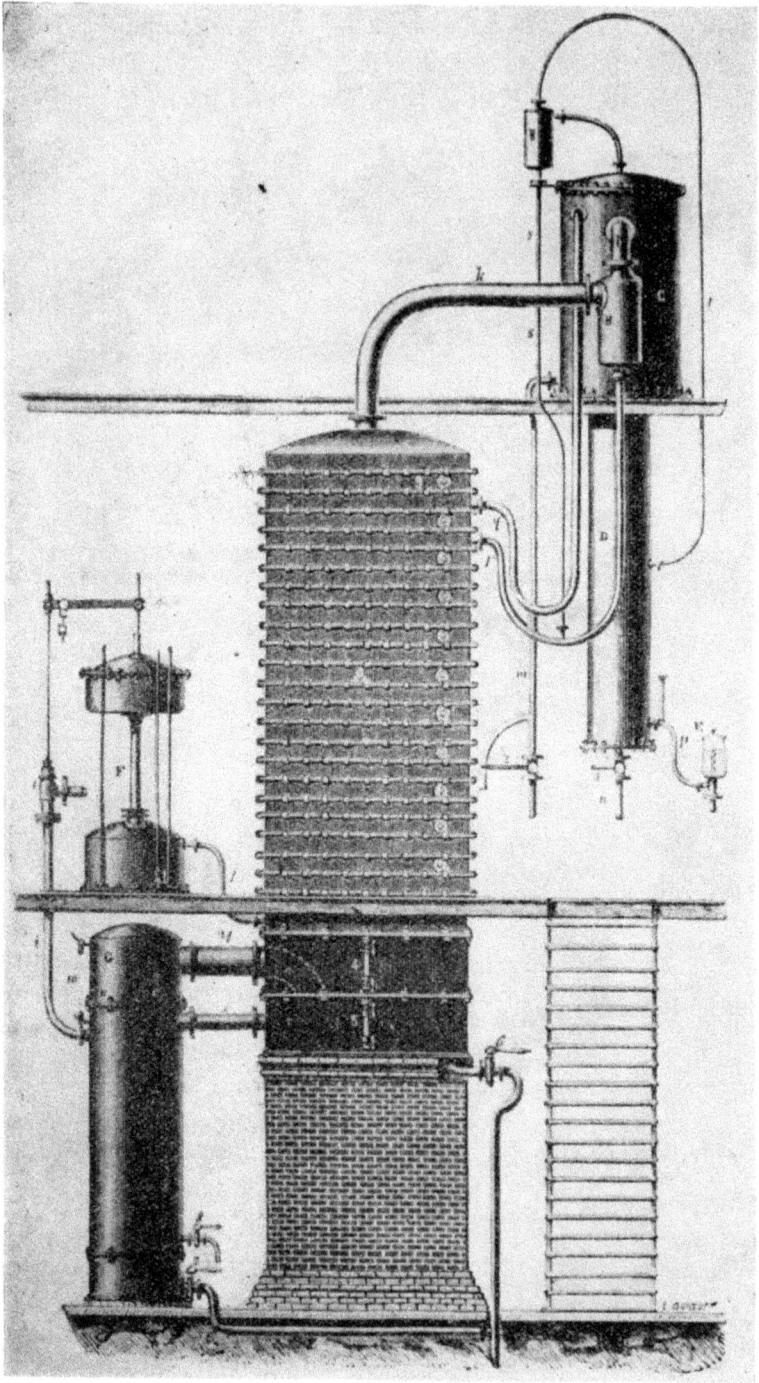

Fig. 168. A Savalle still for beet-sugar alcohol

Lille in 1797. He taught chemical technology at the school of commerce at Paris and became interested in the production of beet-sugar. He designed a still based on the types of ADAM and BÉRARD (Fig. 169) which has historical importance only, as his later stills are based on the Cellier type (Fig. 170). In 1833 DUBRUNFAUT left his post to devote himself to his new hobby. Since 1824 he had studied the possibilities of fermenting pulp and in 1852 he discovered the beneficial of acid on fermentation. In 1832 he designed his new still based

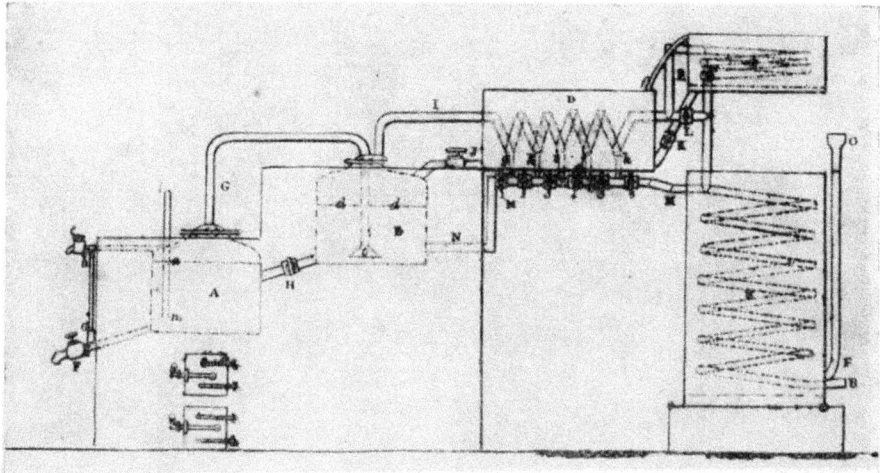

Fig. 169. DUBRUNFAUT's early "Adam-Bérard" still

on the Cellier-Derosne model (Fig. 170) which was able to distill the mashes of grain, molasses and lees of wine. A few years later (1837) he built a distillery at Valenciennes according to these plans and devoted many years to perfecting this system, publishing his results in 1845. Between 1845 and 1852 his distillery produced no less than 12,500,000 l of 92-94 % alcohol from molasses and he stressed the possibilities of combining the manufacture of sugar and that of alcohol from molasses and pulp as an economic proposition. This idea was particularly fruitful when the harvests of corn, wine and potatoes were short in the fifties. The first trial run at the Pétiot factory near Châlons showed that it was very economical when using the periods between the working-seasons of the sugar-factories for distilling (1852-1853) and further trials in 1853 and 1854 proved this again. DUBRUNFAUT also improved DEROSNE's still for the treatment of pulp. His efforts found their reward in the medal of honour awarded to him during the World Exhibition of 1855. He died at Paris on October 8th 1881, after having discovered with DUPLAY a method of direct fermentation of pulp in 1854, which they reported to the Société d'Agriculture.

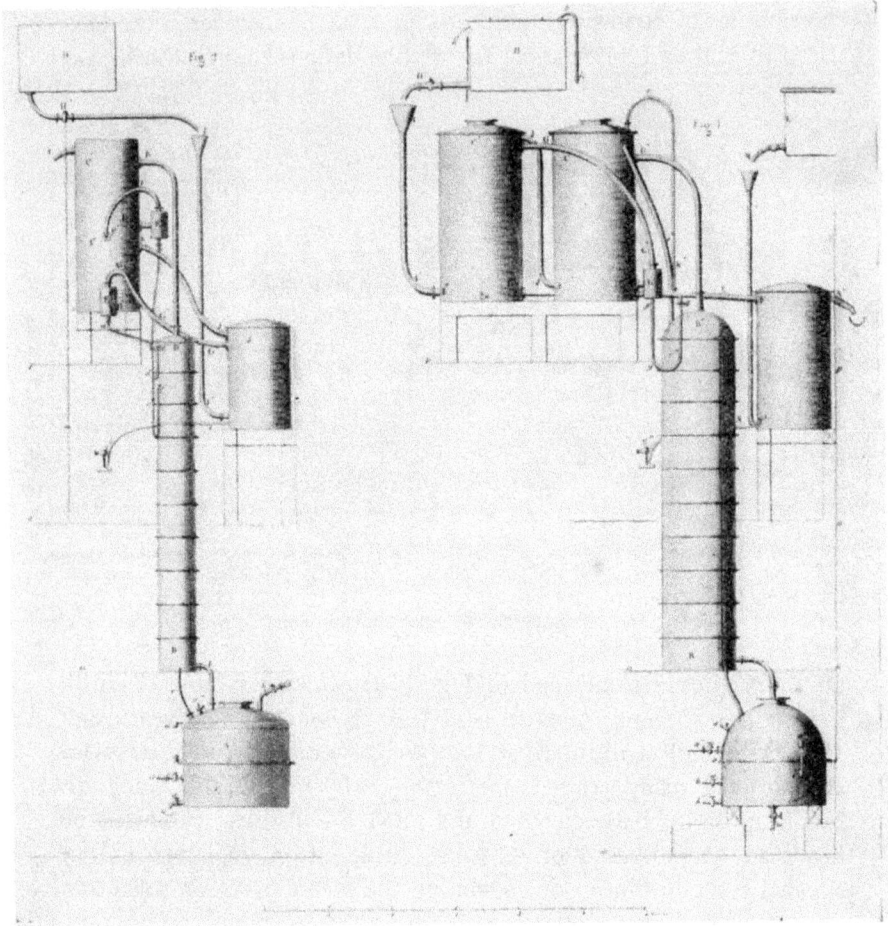

Fig. 170. Different later Dubrunfaut stills

DUBRUNFAUT introduced a large still, the larger the better the taste of the alcohol, he said. His column usually has 18 trays of a diameter of 80-100 cm. The two coolers-pre-heaters condense a reflux which is reintroduced in the column on the third plate from the top. The vapours that pass the pre-heaters are condensed in a water-cooler. In some of his designs the column is split up into two smaller ones. The

Dubrunfaut apparatus was designed for a daily intake of 50,000 to 120,000 l. and a yield of 200 to 4800 l of alcohol (92-94 %). The fusel-oil is separated in a rectification column. DUBRUNFAUT adds that a trace of hard sodium soap will keep the fusel oil back in the residue.

CHAMPONNOIS simplified DUBRUNFAUT's still by introducing valves and by giving a greater cooling surface combined with a smaller total weight of the apparatus. These columns (Fig. 171), sometimes designed with a special cooler (Fig. 172) were built for small distilleries and farms. There have good rectifying properties and were used as models for the later German rectifying columns of HECKMANN, etc.

DUBRUNFAUT's book on distillation (193) is our most important source of the development of the art between 1800 and 1815 as LENORMAND discusses only the apparatus, but DUBRUNFAUT discusses the entire art. He distinguishes two classes of base materials. first those that can be easily fermented like grapes, beets and fruit and secondly those that want saccharification (or malting) before fermentation like corn, maize, rice and potatoes and describes the fermentation processes fully. After the historical chapters he discusses the physical entities like specific heat and the laws of heat transmission and conduction. He distinguishes four kinds of distillation, e.g. the "distillation simple" in the old cucurbits, the "distillation à chauffe-vin" of ARGAND and CURAUDAU, the "distillation à vapeurs et à rectificateurs" of ADAM, BÉRARD and their contemporaries and finally the "distillation continue" of CELLIER. His book closes with an attempt to calculate columns. Starting from the vapour pressure of two components he calculates the specific heat, density and boiling point of the mixtures and uses the formula given by BIOT for the calculation of vapour pressures at higher temperatures. He discusses three cases: 1) when only one component is condensed from the vapours, 2) if both are partially condensed and remain in contact with the resultant vapour and 3) if vapours produced at certain temperatures come into contact with the liquid at a lower temperature. Though these calculations are full of mistakes, this early attempt is well worth mentioning.

DUBRUNFAUT's book was also translated into English (194) and published with a valuable preface by P. JONAS. It exterted considerable influence on the designs of English distillers.

Another source of the development of the art around the forties is

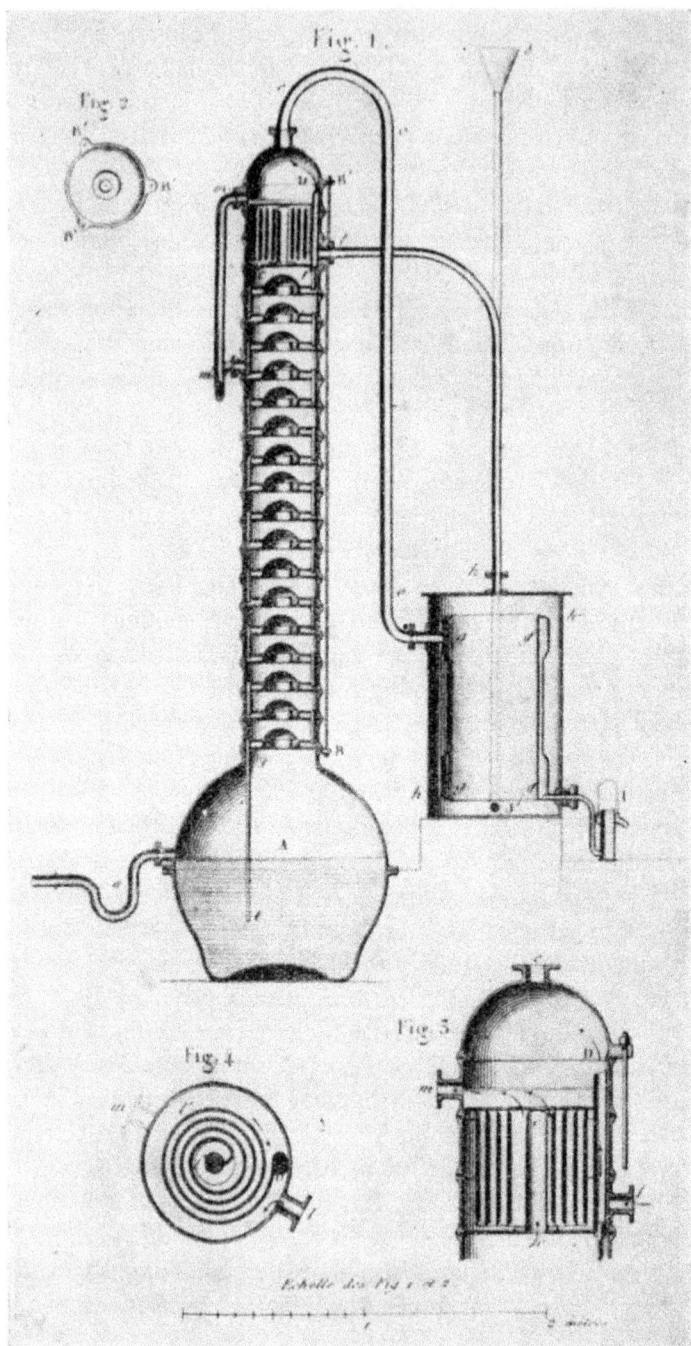

Fig. 171. A Champonnois still (1854)

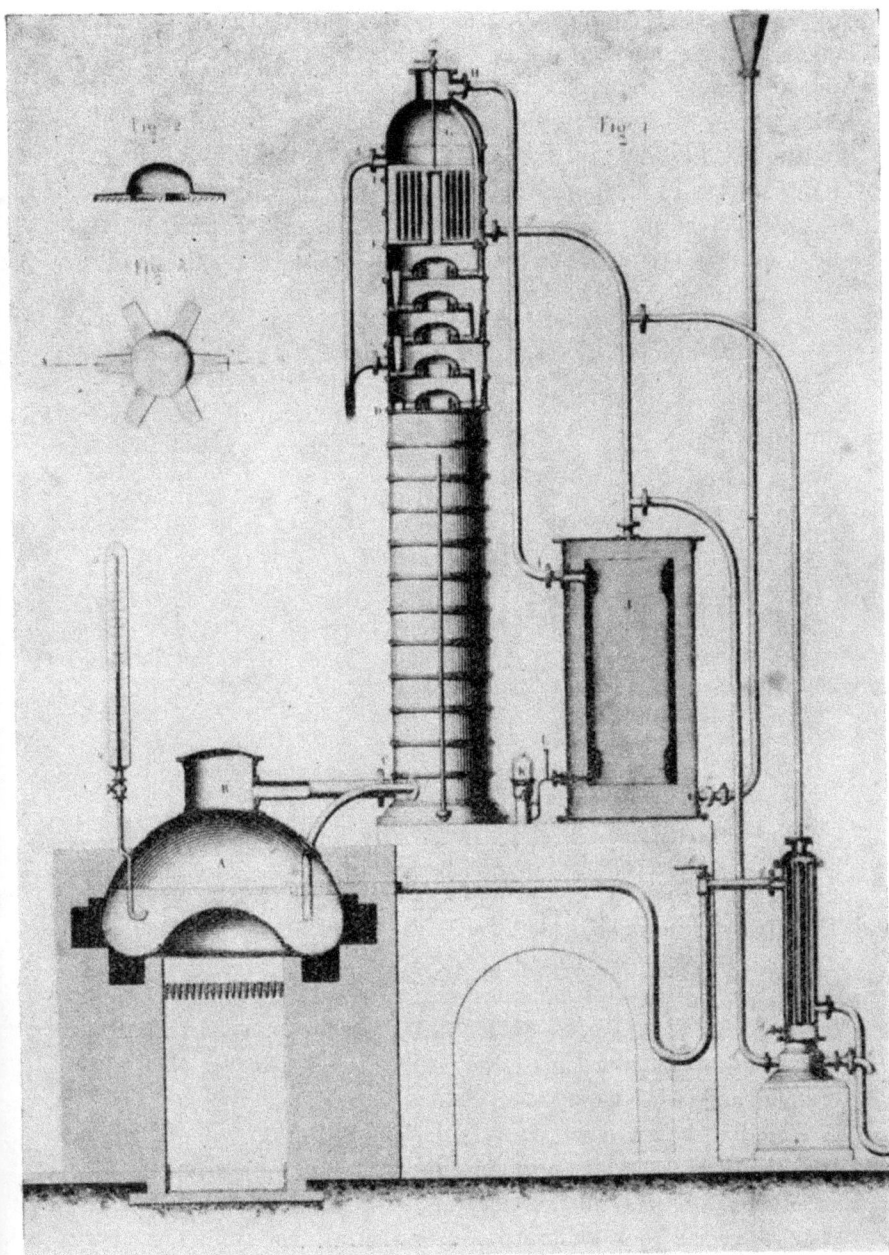

Fig. 172. A Champonnois still with special cooler

the book by JEAN GIRARDIN (258), which was translated into many languages, also into Dutch (259). He mentions that the small still used to try out different base materials for the possibility of manufacturing alcohol, which is usually ascribed to GAY-LUSSAC and which was modified by DUVAL and SALLERON, was originally designed by DESCROIZILLES of Dieppe, who died in 1825 after having promoted the adoption of the new chemistry of LAVOISIER in France. He also mentions that the "gunpowder test" for alcohol was still used in France in 1846!

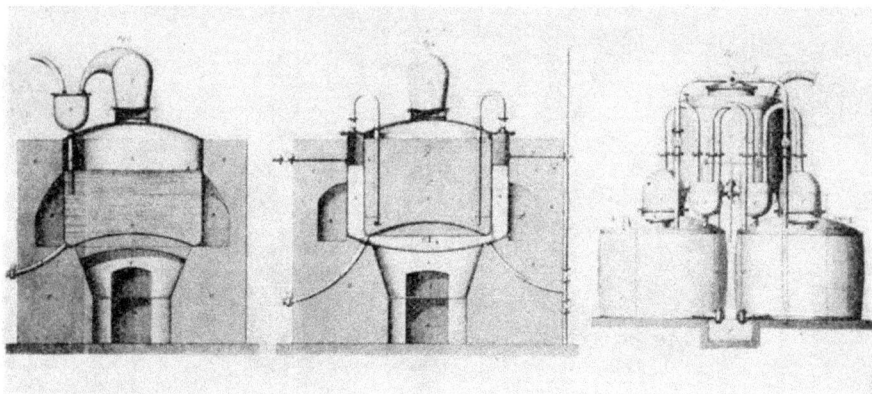

Fig. 173. LACAMBRE's designs for stills with different methods of heating .

The earliest of the last two handbooks, dating from this period, was that of LACAMBRE (342). He designed a simple still which could be used with direct fire, with a water-bath or heated by live steam. (Fig. 173). His book gives a detailed description of the stills of CELLIER, DEROSNE and LAUGIER.

The latter of these handbooks, that of PAYEN (474) is far more complete. ANSELME PAYEN (1785-1871) was a professor at the Ecole Centrale, who succeeded his father in 1825 as a manager of a sugar-refinery, and who wrote many books on chemical technology and dietetics. He was a great friend of BOUSSINGAULT with whom he unravelled the composition of cellulose, discovered by them. In 1835 he succeeded DUMAS in the chair at the Ecole Centrale. PAYEN also wrote a special book on the production of alcohol from sugar-beets (473), and his book on chemical technology contains many details on the art (475). He discusses the stills of ARGAND, ADAM, CELLIER, DEROSNE, DUBRUNFAUT, CHAMPONNOIS and others in detail and many data in these pages were taken from his works.

It will be clear that the combined efforts of ADAM, BÉRARD, CELLIER and DEROSNE completely changed the art of distilling within twenty years and that the period from 1820 to 1840 was devoted to the perfectioning of the continuous distilling column in France. Therefore the first two decades of the nineteenth century must be reckoned as the most important period in the history of distillation and the advance achieved then could only be outstripped by the intro- duction of mathematical physics in the design of distilling apparatus in the last decade of the nineteenth century under the influence of the well-known works of HAUSBRAND (288) and SOREL (589) (590) (591).

CHAPTER NINE

THE AFTERMATH

"Discipulus est prioris posterior dies"
SYRUS.

The tempestuous development of the new still in France long left the distilleries of Germany and England unaffected. At the same time it was beyond question that the experiments on development of the old still of the cucurbit type were not continued. It may be true that these old stills only gave a weak distillate, that had to be redistilled several times and the taste was often spoiled by empyreumatic oils. It is true that by 1850 they were only used by small distillers of industrial alcohol in Holland and Germany, whereas the distillation of wine, malt-wine and mashes of potatoes was generally carried out in the new large continuous apparatus. But the old cucurbit, with a simple worm-condenser and a dephlegmator (oeuf de retour) for returning puked residual oil to the still remained in favour with the distillers of brandies and liqueurs who feared that the taste and colour of their product would be spoiled by the larger apparatus and whose normal output was too small to run the larger stills economically. The low cost of the old still (the bane of every inventor) is the reason why these have survived upto the present day for distilleries where liqueurs are made (Fig. 174).

We could cite many patents and designs of these old stills published between 1800 and 1850, but we shall confine ourselves of citing a few examples of stills of the first decade of the nineteenth century. In Germany FISCHER, JOH. CHRIST. HOFFMANN and STÖCKEL designed new stills, in Sweden NORBERG (1799) and Ritter VON EDELCRANTZ (1805); in England ACTON, BURKITT and SHANNON should be mentioned. In France we find the still of LELOUIS (distiller of La Rochelle, 1803), FIRMIN BARUE (of Nîmes, 1802) and others. BORDIER-MARCET in vented a portable still for the production of cherry-brandy for his estates of Versoix (1804) which GAY-LUSSAC demonstrated before the Société d'Encouragement pour l'industrie nationale (see their *Bulletin* of May 1807). HERPIN tried to speed up distillation by blowing air into the still, but the only effect was a greater loss of alcohol. In Italy BRUGNATELLI designed a new still for brandy. As late as June 26th 1819 WILLIAMS took out a patent on a wider alembic which was to yield a stronger distillate, but only with the loss of much heat and alcohol.

Holland was one of the countries which was very late in adopting the new technique. One can easily prove this by citing a few prize-essays from the period which we are discussing, and which show how little the better technique had penetrated to Holland.

Fig. 174. The apparatus of the Van Zuylekom en Levert distillery, Amsterdam

In 1818 the Nederlandsche Maatschappij voor Nijverheid en Handel, a scientific society for the promotion of industry and commerce offered a prize for the best essay on the question whether a new base material could be found instead of grain which would yield gin of the same price as that from or preferably lower (449). On September 26th 1818 an essay was sent in proposing the sugar-beet "already well known from the period of our oppression" (the Napoleonic period), but no prize was awarded as the matter was considered out of date, because the first trials in France had proved the alcohol thus produced to be too expensive.

In 1821 the Hollandsche Maatschappij der Wetenschappen at Haarlem had invited designs for a better still, heated with steam, like the new French stills, which was to yield a corn-spirit of better taste. An essay was sent in before the appointed date of January 1st 1822 but as the writer seemed to have misunderstood the question the term was postponed until January 1st 1824. The supplement which the writer sent in to his first essay was, however, considered to be unsatisfactory. No prize was awarded and another essay was invited on the question of the separation of the fusel-oil spirit and a description of its properties (1832). After delaying the term within which the essay had to be sent in for another two years until January 1836 a paper was written at last by A. H. VAN DER BOON MESCH, professor of chemical technology at Leyden, which was awarded the gold medal of the society. In this essay the properties of fusel-oil are well discussed and the author proposed either to refine the spirit with charcoal or to redistil it. The still which he proposes to use is taken from HERMBSTÄDT, which author was very much in favour with the Dutch distillers. The ignorance of the professor from Leyden in the field of practical distilling can not be better illustrated than by quoting his dictum that "the stills of DORN, DEROSNE and CELLIER are just modifications of the Pistorius still" (sic!) (424).

In the same year the Nederlandsche Maatschappij voor Nijverheid en Handel asked the distillers to try their hand at the design of a better still which should have a large distilling velocity and yield a purer spirit (448). On September 21st 1837 the only essay was sent in, and it proved to be entirely unsatisfactory. The Committee which discussed it remarked, that the author did not seem to know the stills of ADAM and PISTORIUS, that he had no inkling of the recent discoveries in France and England and that he did not even seem to have heard of the practical results of these new stills in the distilleries of Germany, France, England and the colonies. So they remarked: "This proves in what deplorable conditions our distilleries in the province of Holland are, they do not even seem to know of the trials with the Pistorius stills for potato-mashes in the province of Groningen."

Nor do the archives of private firms such as the firm of Erven Lucas Bols of Amsterdam show anything different. This firm possesses a manuscript written by their former manager G. Th. VAN 'T WOUT between the years 1830 and 1847. It contains not only a description of the Bols distillery in those days but more particularly directions for distillers written by an insider. It is typical of the general knowledge of the Dutch distiller of the period. VAN 'T WOUT bases nearly all his directions on the old cucurbit or the Adam and

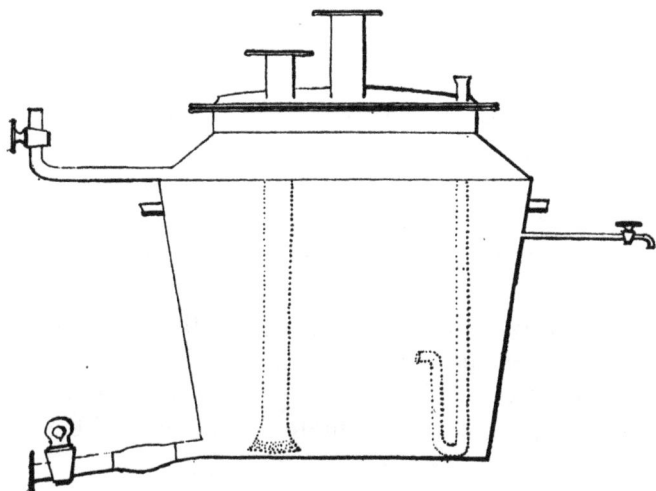

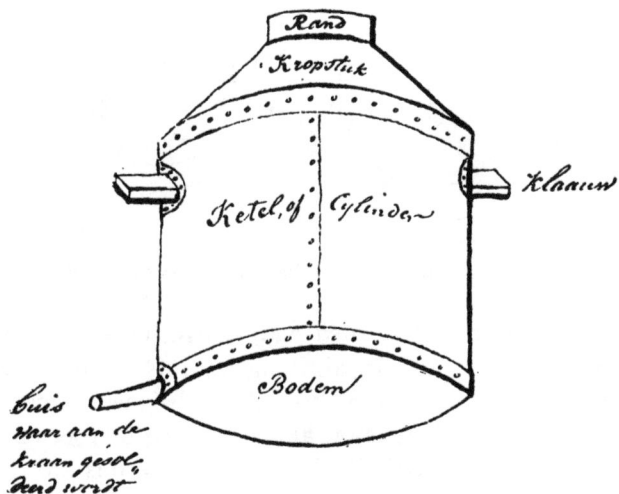

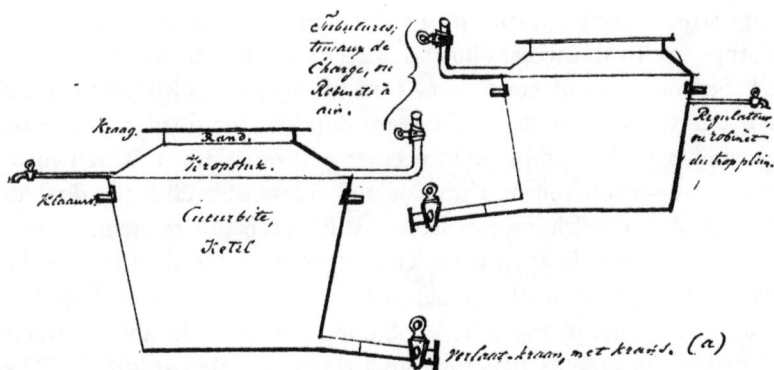

Fig. 175, 176 & 177. Still designs by VAN 'T WOUT (1830-1847)

Bérard apparatus though he knows the Cellier and the Derosne stills quite well and states that "they are equally useful for all kinds of mashes". Most of his knowledge is taken from DUBRUNFAUT's book. He states that the continuous stills are not so important when one produce liqueurs as here neither time nor heat but taste and reputation are the main factors.

The stills were made of 3 mm copper plate, they still have the old cucurbit shape except that the bottom is concave (Figs. 175, 176 & 177). The fire-grate is only small (Fig. 178) and just behind the firedoor. The copper alembic is adjusted in the neck of the cucurbit

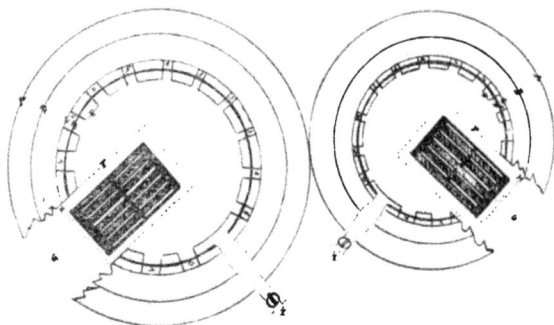

Fig. 178. Fire-grate for a distillery-furnace (VAN 'T WOUT)

with about ½" overlap and luted with dough, usually an iron band keeps it down on the still. In distilling liqueurs the herbs and fruit are put in a special tray or basket called "baigneur à claire voie" (Dutch = beun) which rests on a wooden ring or foot on the still bottom. A waterbath is only used when clear liquids are to be distilled. Sometimes the inner parts of the stills are tinned. A water-bath is preferred if small quantities of pure distillate are to be obtained. In other cases careful handling of direct fire will do the job just as well. Several types of condensers ("rectificateurs", „alcoogènes") are used, but generally none of these are required for distilling liqueurs and cooling with a simple worm-cooler will do. VAN 'T WOUT mentions steam distillation and says of the Adam apparatus "it doubles the quantity of spirit in the second still" probably referring to the later designs of this apparatus. The strength of the distillate in the Adam still depends on the amount of "eggs" used in series. BÉRARD, he says, makes use of the different boiling points of alcohol and water and makes the vapours pass through a through cooled at 80° C. "The famous CHAPTAL seems to have been the first to combine the advan-

tages of both systems", probably referring to the still in Fig. 152.
He was so impressed by it that in the Bols distillery, "het Lootsje"
the common name for "still" is "Chaptal". In VAN 'T WOUT's eyes
the "Cellier apparatus is nothing but a worm bathed in hot water,
and therefore a Bérard condenser and a pillar of basins piled up on
the still so that the vapour passes from one to the other just as through
Adam's eggs".

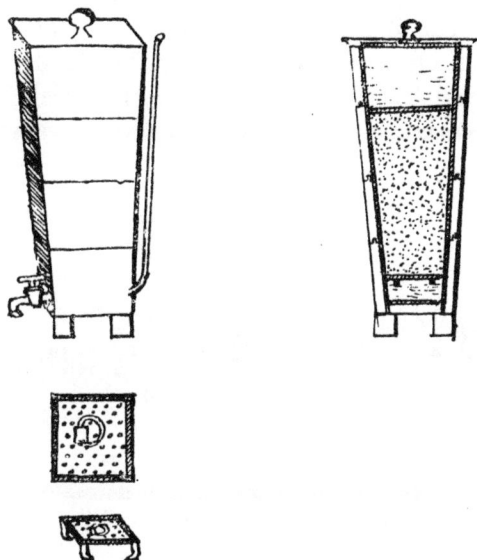

Fig. 179. Dumont filter used for refining alcohol

Though VAN 'T WOUT considers GEDDA's condenser the best he
does not deny that worm-condensers have some advantages, but
he prefers them rectangular, because he has read somewhere, that
"the force of the steam is broken by sharp corners". He tries to calculate the surface of the condenser and reminds the reader of the
noxious properties of copper, which BERZELIUS proved, tin is better
than lead for condensers.

The construction of the furnace is discussed along the lines of
thought of the "fire-work-philosopher Curaudau". Charcoal is the
best fuel, as coal sometimes gives off bad fumes which spoil the
distillate. If the distillate should need to be refined the easiest way
is filtration over charcoal, for which operation he recommends the
Dumont filter (Fig. 179). VAN 'T WOUT also describes at length the
other apparatus necessary in a distillery such as hydrometers, thermometers, balances, etc.

The archives of the firm of Bols also contain a few letters from their Bordeaux agent, a Dutchman J. H. BEYERMAN. In a letter which we have already quoted (dated August 12th 1820) he advises them on the type of new still which they intend to build at Amsterdam. The Ménard still is generally praised. The Solimani system is used at Languedoc but the Adam, Ménard and Baglioni stills are preferable. There is also a good Alègre still but the writer could not get any description of it and the patent of the still of BAGLIONI is nearly at an end. He sends a drawing of a double Ménard unit (Fig. 180) but it seems that the Amsterdam experts decided to try out an

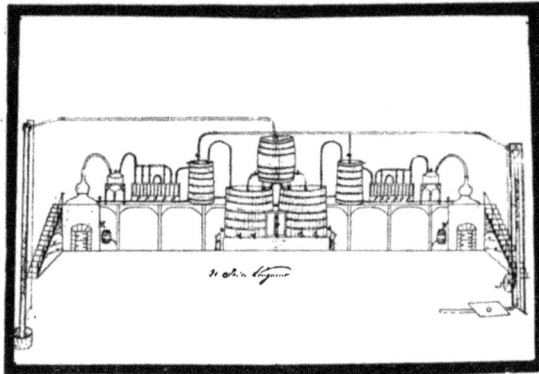

Fig. 180. A double Ménard unit from Bordeaux (1820)

Adam unit and this is referred to in a letter dated December 5th 1820 which announces its expedition. A letter of April 10th 1821 contains many complaints about this still, it does not work well and leak so much that there is question of sueing the Bordeaux coppersmith who made it. But the agent replies to state that this coppersmith has an excellent local reputation and that the leakage is due to transport and should be repaired at Amsterdam. In the same letter he adds a few details on the operation of the Adam apparatus, but it is doubtful whether this ever came out of the experimental stage.

When ten years later the firm of van Zuylekom and Levert correspond with CELLIER on their plans of installing a new apparatus they finally decide to build a Pistorius still, such as was recently set up built by their competitor HOPPE (Fig. 181) who bought it in Cologne. They have one made by the local copper-smith Welling-Huyzen (Fig. 161) which differs hardly from the design made twenty years earlier by VAN 'T WOUT in the Bols manuscript (Fig. 182).

These lines will be sufficient to show how slowly the new French

Fig. 181. Design of a Pistorius still for the firm of Van Zuylekom en Levert (1840)

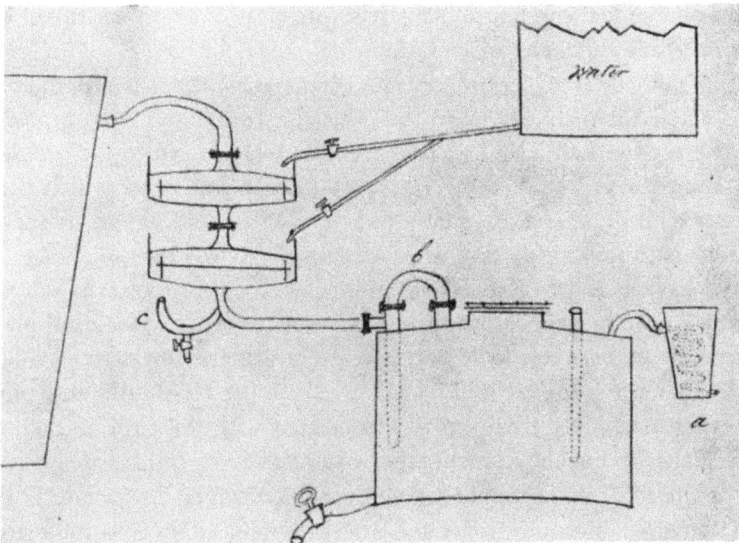

Fig. 182. Design of a Pistorius still by VAN 'T WOUT (1820)

stills penetrated to the north and the same can be said for *England*. The stagnation of the English distillery in the first decade can be measured from a book by MOREWOOD (436) who himself seems to have dabbled in distilling apparatus about 1804. But even upto the present day the Scottish distillers use pot-stills of the ancient type in sizes from 3000 to 8000 gallons though 5000-6000 is the usual size. In Irish distilleries the pot-still is still a favourite though the size is usually much larger and goes up to 20,000 gallons. A special feature of the Irish pot-still is the "Lynne-arm", the tube connecting the still with the worm-cooler. It usually goes up vertically 10′ to to 20′ from the still and then runs horizontally for some 30′ to 40′. This horizontal branch has a slight depression which is sometimes cooled in a trough. It forms a kind of crude reflux-condenser, the condensate of which is led back to the still, while the rest of the vapours pass on to the worm-condensor.

In the spirit industry the new French stills could not be adopted as such, for though the English distilleries worked with rather thin mashes of grain, these still contained too many solids to be handled in the new apparatus without difficulties. The French pre-heater was discarded altogether, as the hot cooling-water could be used with profit in the malting-house, and therefore heat-economy on this point was only a secondary consideration.

The new laws of excise drew a sharp line between the malting houses and the distilleries and then the new continuous stills began to become economical. The English inventors were drawn to the problem and the crown of their efforts was the Coffey still, after which there were several decades of stagnation of the English still design. A survey of the most important English patents will show this.

We have already discussed the patent taken out by CHARLES WYATT in 1802 (670). Apart from the application of steam he took out many other patents between 1790 and 1817. He was the son of the famous JOHN WYATT (1700-1766), who invented the fly-shuttle and other spinning machinery when he was associated with the firm of Boulton and Watt. In his above mentioned patent WYATT states "Steam is to be produced from water or other liquids contained in a close boiler and conveyed through proper tubes or channels, either into the matter intended for distillation or through it, or wholly or practically round and beneath, and in contact with the external surface of the still or vessel in which the matter is contained, or in any of these methods combined, that of introducing the steam wholly and totally into the body of the liquor to be distilled so as to be mingled therewith, being the most advantageous and effectual. The liquid being thus heated by the action or contact of steam, the subsequent parts of the process proceed in the same way, or nearly in the same way as in common distillation over fire". He also designed a still the cover of which was used as a condensing element, but it was rather inefficient and unimportant.

JAMES MILLER's patent of 1815 (428) describes an old-fashioned still in which the wine, pre-heated in the worm-condenser, is intro-

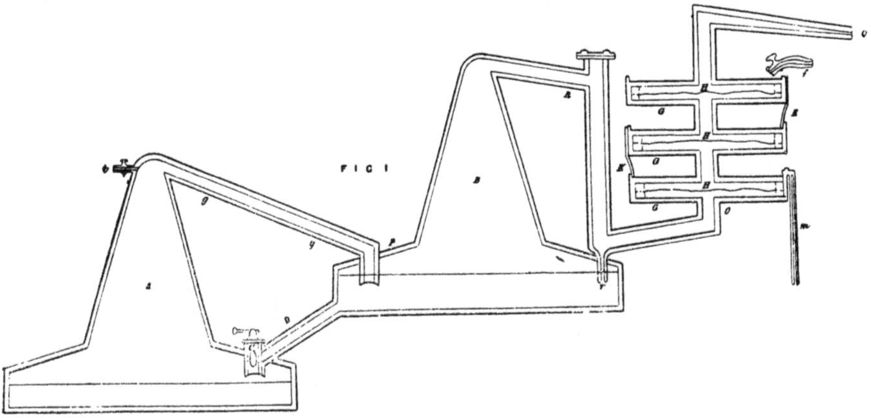

Fig. 183. The Corty still (1818)

duced in a kind of worm, which rather complicates the whole apparatus and does not look very efficient.

The patent of JOSEPH CORTY (144) of 1818 is more important (Fig. 183). It consists of two stills with flat bottoms and conical alembics, one of which is placed slightly higher than the other. The preheated liquid flows back from it into the first still, whence the vapours pass through the liquid of the second one, enrich themselves and are stripped of heavier components in a peculiar cooling system before passing to the final worm condenser. The reflux is passed back into the second still. The reflux-condenser with its three water-cooled boxes ressembles the Pistorius condenser in many ways and the point whether there is any connection between the two apparatus should be investigated as the Corty still was described in detail in a German paper of 1819 and the Pistorius still is said to have been designed in 1818 or 1820.

Sir ANTHONY PERRIER's specification of 1822 (477) is interesting

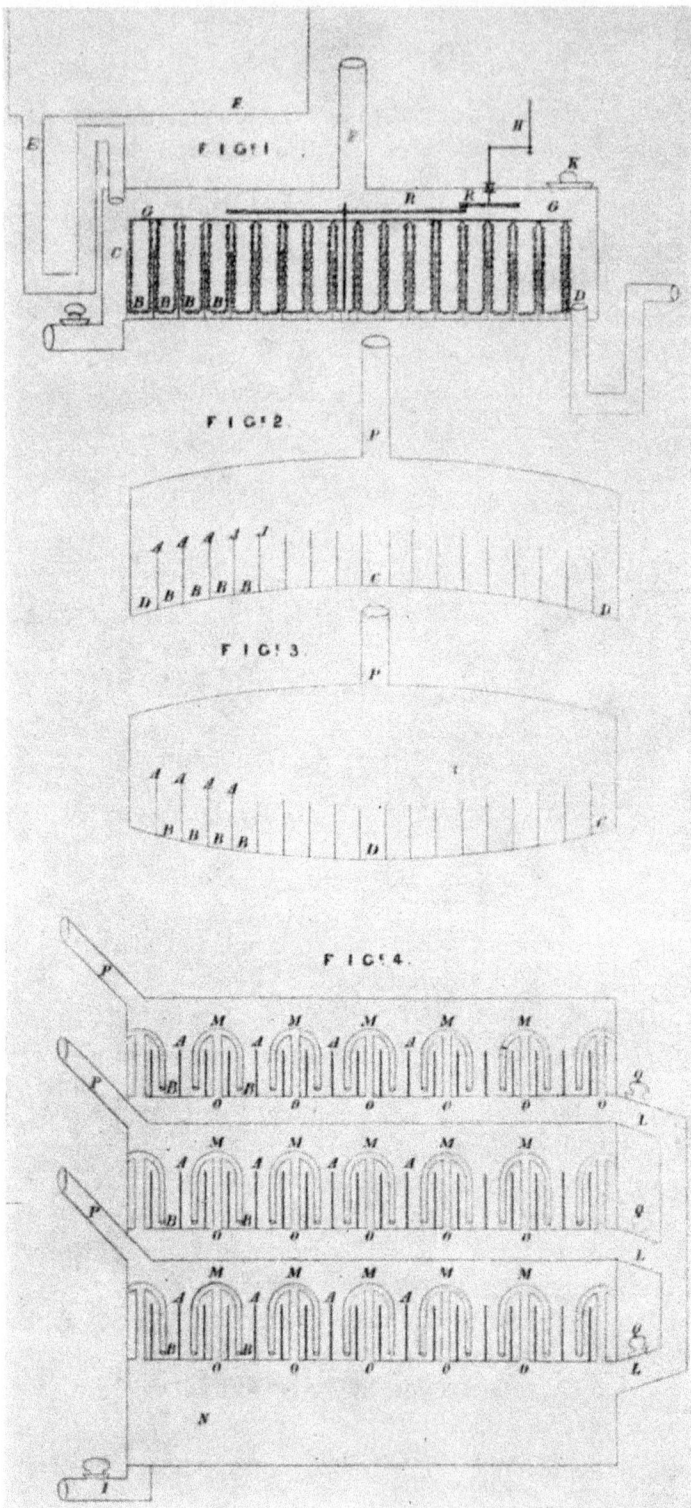

Fig. 184. Sir ANTHONY PERRIER's dephlegmator (1822)

in many ways. It was also patented in France (French patent No. 2157 of June 21st 1822). His still is divided into compartiments by ledges forcing the liquid to run a long way before leaving the still. Stirrers with chains attached improve the circulation. If one wants to blow in steam a device shown in fig. 4 of his application (Fig. 184) should be used. At the same time a pile of these flat vessels can be formed if one wants to use this invention as a rectifying column, and the recurved tubes, which we have just mentioned play a part in the columns designed by SAVALLE and others. It seems that PERRIER got his idea from the Scottish still of the end of the preceeding century and thence insisted on a larger heating surface to evaporate as quickly as possible.

In 1825 WILLIAM GRIMBLE claimed "certain improvements in the construction of apparatus for distilling spirituous liquours" which consists of a dephlegmator to be mounted on the still. It is composed of parallel series of vertical tubes to be cooled by air, cold or tepid water, etc. In the upper space where the vapours are assembled there are further air tubes for cooling. SAVALLE used this idea of the tube condenser with some change in his apparatus and it has become a standard apparatus since (Fig. 185) (269).

JEAN JACQUES SAINT MARC, a veterinary surgeon of Napoleon's army left his post after the battle of Waterloo to live in France but settle in England in 1823 where he worked on his patent still of 1825 (410). This still was the first one of continental design in England (Fig. 186) and it was used as a very effective apparatus for thin washes and wines. The figure shows the column with outer water condenser to be mounted on a rectifying still, mostly a double still according to the inventor. From the form of the caps and other details MAERCKER concludes that it shows influences of the Pistorius still, but this seems very doubtful, as these were also in use in several French stills of the period. The apparatus worked discontinuously at first, but later inlet and outlet were opened periodically and therefore a continuous stream of distillate was obtained. It was long in use in the British colonies of the West Indies for the production of rhum.

Wo should also quote for curiosity's sake the patent of ROBERT LORENT, a Swede (British patent No. 5323 of January 19th 1826) who invented a "Method of applying steam without pressure to pans, boilers, coppers, stills, pipes, and machinery in order to produce, transmit or regulate various temperatures of heat in the several processes of boiling, distilling, evaporating, inspissating, drying and also to produce power". (Fig. 187).

In 1828 EVANS designed a still that rotated and distilled through a hollow axis (214). To refill it the fire, carried by a grate mounted on wheels was drawn away. The still had of course no practical merits.

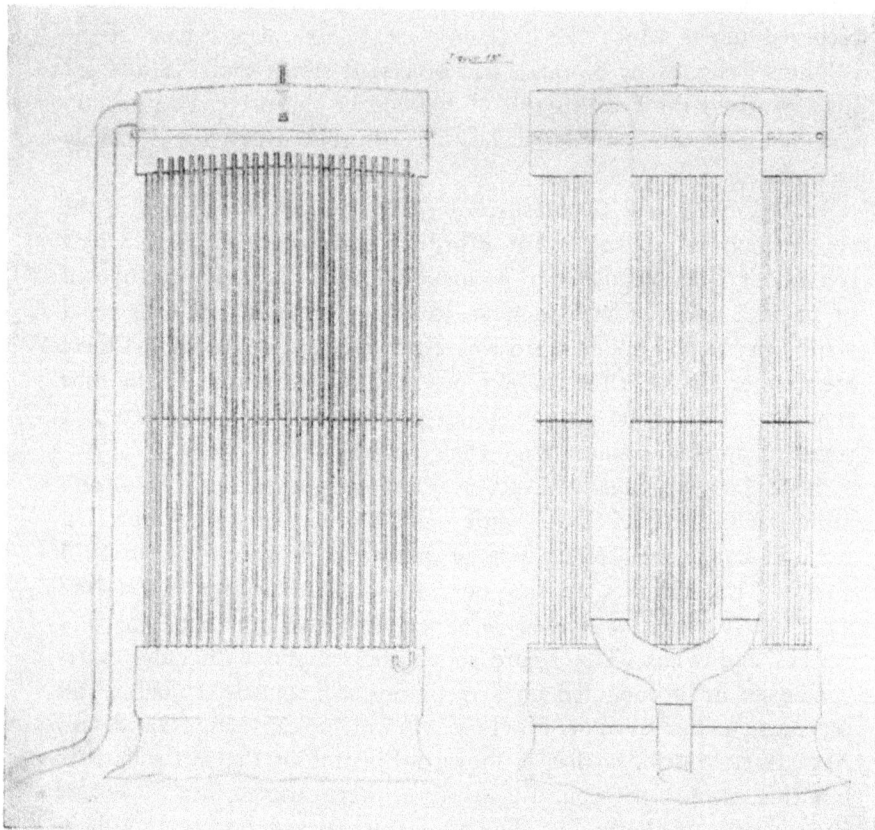

Fig. 185. The Grimble tube-condenser (1825)

A very ingenious apparatus was designed by ROBERT STEIN (602) who "exposed a continuous supply of wash or liquid to be distilled to a current of hot steam, while such wash or liquid was in a minutely subdivided state like a mist, shower or spray". This apparatus worked on the multiple effect principle and was very clever mechanically, but the distillate was too weak and it was never effective in practice.

WILLIAM SHAND inserted between the still and the worm-condenser three more small wooden stills as shown in Fig. 188, with water-cooled

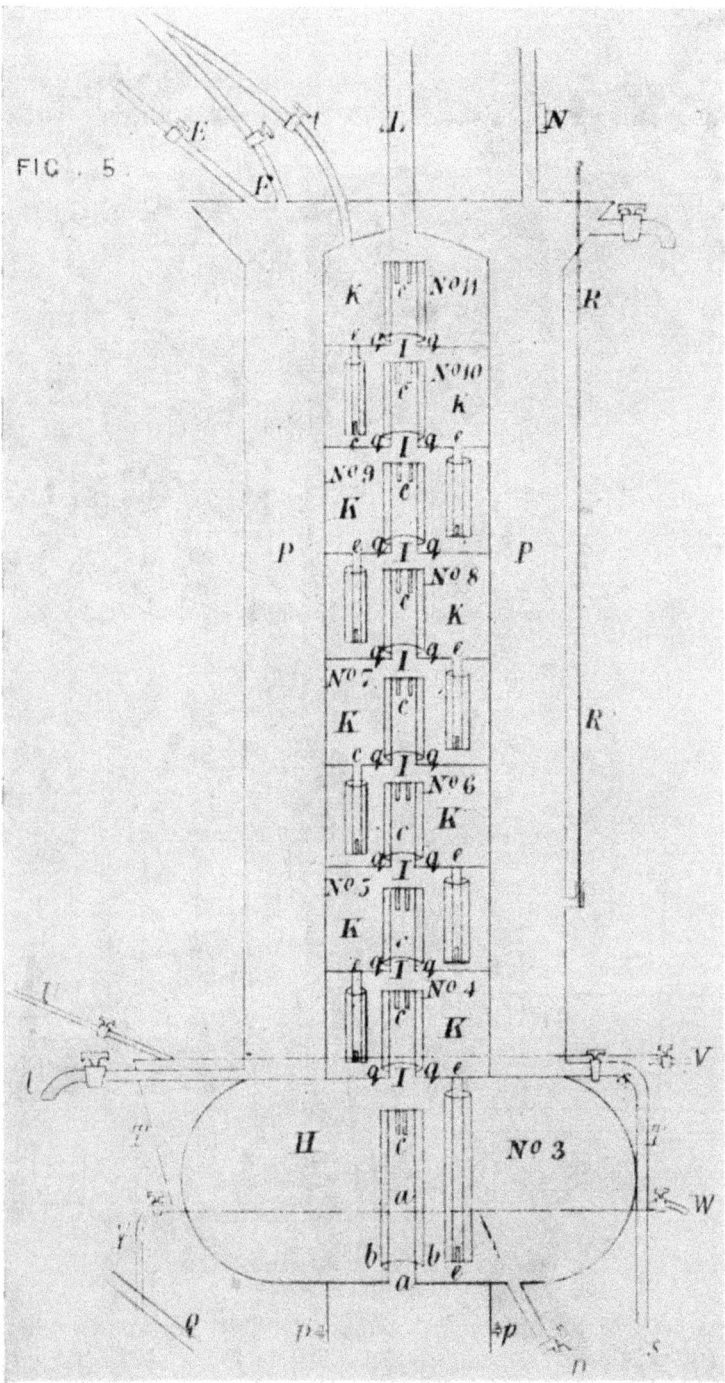

Fig. 186. SAINT MARC's still (1825)

copper covers (572). The vapours passing through these series of "dephlegmating rectifiers" loose their heavier components which are

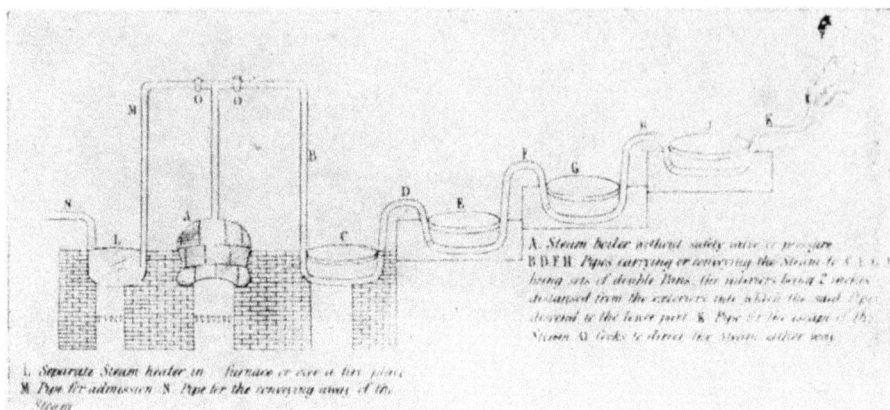

Fig. 187. LORENT'S method of heating with indirect steam (1826)

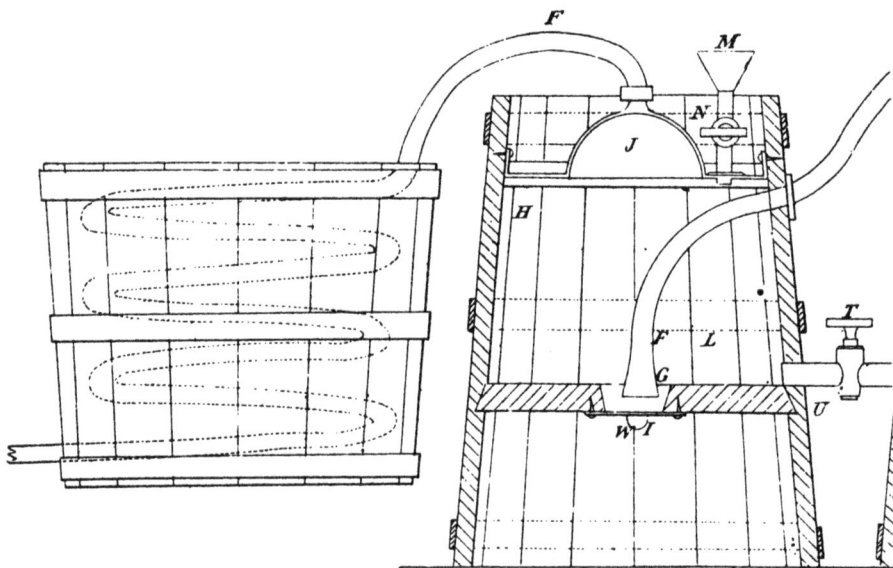

Fig. 188. WILLIAM SHAND'S still (1828)

passed back to the still in counter-current. It is not known whether this still was ever used on a larger scale.

The most influential design of the period was that of AENEAS COFFEY patented in 1830 (135). His first patent had not yet the two-column design of the later types, but it is already a very practical still (Fig. 189). COFFEY took out an earlier patent on February 5th 1830 which was decidedly improved by his patent of August 5th of the same year. His column has plates with valves through which the return liquid cannot flow but through which the vapours ascend. The liquid has to drip down from plate to plate and thus cover a long way as the plates have an overflow with a gaslock, preventing the vapours from ascending that way. The wash is forced rapidly through a pipe or series of pipes of small diameter during which time it is acquiring heat and before it reaches its boiling point. The most important point is "the practice of causing the wash after it has come into contact with the vapours to flow into a continued and

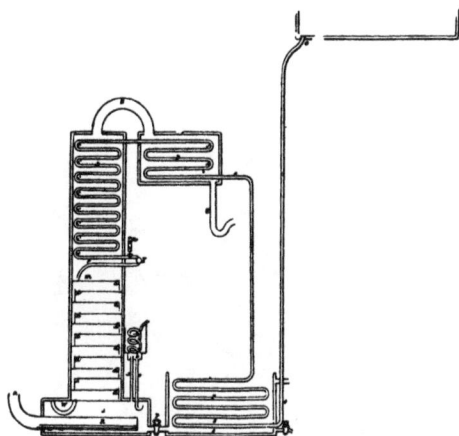

Fig. 189. The Coffey still (1830)

uninterrupted stream over numerous metallic plates furnished with valves opening from below for the ascending vapours. The velocity of the wash in the pre-heating pipes should prevent any deposit".

The Coffey still was an immediate succes. DINGLER reports that in 1836 there were already stills treating no less than 13,600 L. of wash per hour with 6.5-7 % of alcohol and in the days of PAYEN hourly capacities of 20,000 L. were no exceptions. They easily gave the 80 % alcohol in one operation and therefore worked very cheaply as compared with the continental stills. Soon they were used all over England. In the later designs the column is split up into two smaller ones, the first ("analyser") delivers the spent wash, the second ("rectifier") delivers a condensate to the feint cooler, and the spirit is cooled separately in a spirit cooler.

In the same year DANIEL TOWERS-SHEARS, copper-smith of Bank-
side, invented a bad imitation of the Pistorius still (573), with three

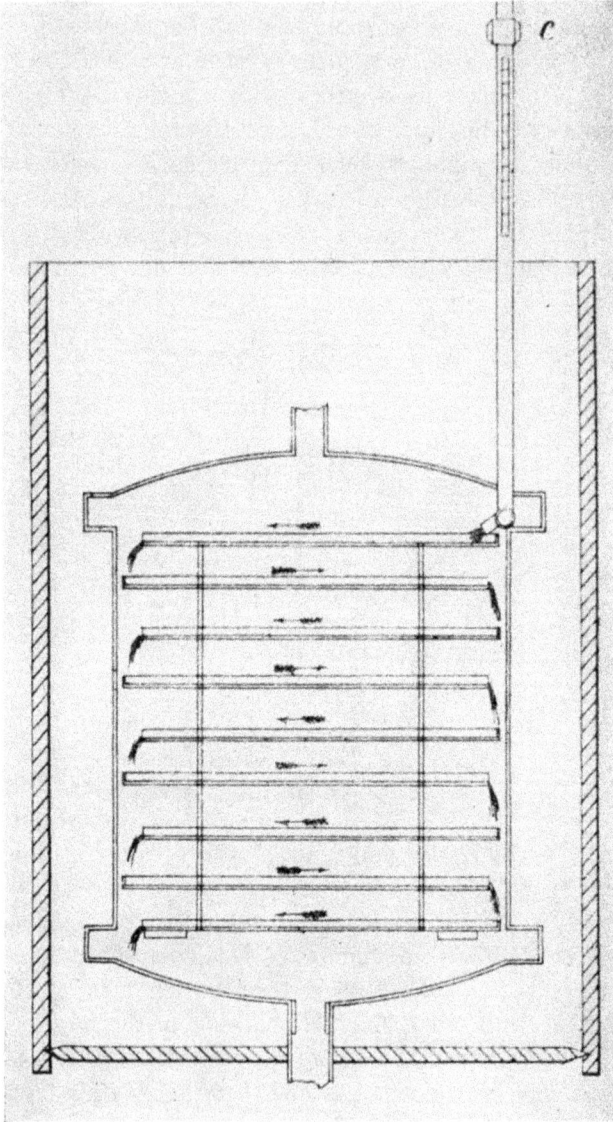

Fig. 190. ANDREW URE's still

Pistorius dephlegmators and a wine pre-heater. Though URE mentions
it as a combination of the Coffey and Corty stills and thinks it rather
important, it was soon forgotten for it gave no proper rectification.

ANDREW URE himself produced a new distilling apparatus in 1831 (637) in which, in his own words, "the resources of the most refined French stills are combined with the simplicity and solidity suited to the grain distilleries of the United Kingdom". Dephlegmating and preheating is achieved in a metal vessel immersed in a cold water-bath in which a movable open trough conveys the mash (Fig. 190) which is pre-heated by the vapours passing over it. The apparatus was no succes but it is remarkable for the use of the thermostat invented by URE (British patent of October 20th 1830), probably its first technical application. Two pieces of metal differing in thermal

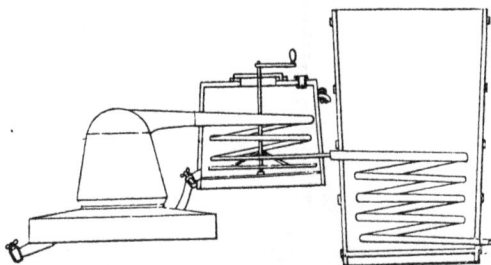

Fig. 191. German still for th'ck washes, about 1800

expansion are joined together and the movement of this joint is here used to regulate the flow of the water in the water-bath by moving a valve.

In 1834 JOSEPH SHEE invented an apparatus (574) in which "the process of redistillation is carried on through a series of stills working simultaneously and by the same heat as is used to work the first still and whereby the product is advanced one stage of the necessary process at each changing of the first still and the products are kept distinct". These stills were simply imperfect Woulfe bottles and it is difficult to see how these bottles (each requiring two secondary vessels and a pump) could have achieved any rectification worth mentioning. The whole apparatus was too complicated to be of much use.

Passing over to *Germany* we find that the development of the new still in that country came much later. This was due to the fact that the German distillers worked with much thicker mashes than the French, mashes which contained a large amount of residue and many solids. It will be clear from this that special stirring apparatus had to be built into the coolers and pre-heaters and that the rectifying part of the apparatus developed sooner than the distilling part. Fig. 191

shows the typical German still of the beginning of the nineteenth century.

However, the German scientist was clever in devising new apparatus. DÖBEREINER mentions a simple extraction apparatus (187) in 1821, the first in which the liquid was constantly drawn off from the extraction residue. It was improved by MOHR and later SOXHLET

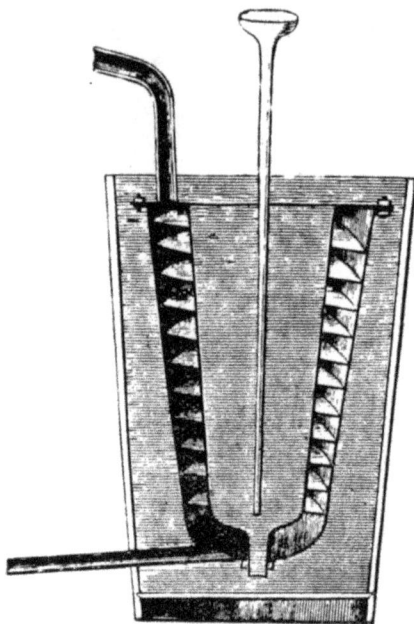

Fig. 192. The later Gedda cooler

gave it its modern form. HAGER built a technical apparatus on the Döbereiner principle in 1862.

There was also some development of the condenser in this period. Apart from the Gedda condenser (Fig. 192) which remained popular in an improved form, we also find the condenser of SCHWARZ, professor at Copenhagen (Fig. 193), a type that shows some similarity to the dephlegmators of PISTORIUS. Possibly the two are connected. A major point was the possibility of cleaning the condenser easily and from this point of view the cooler devised by BABO, a technical Liebig condenser, was the best (Fig. 194). It is the proto-type of our modern double-tube condensers.

The most influential author on distillation in the early nineteenth

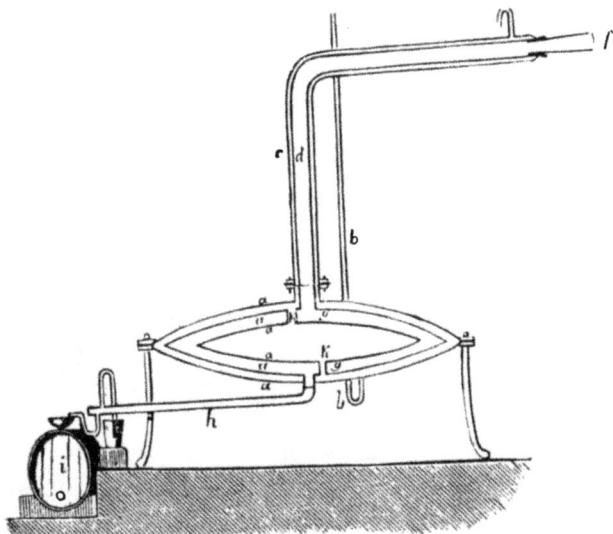

Fig. 193. The Schwarz cooler

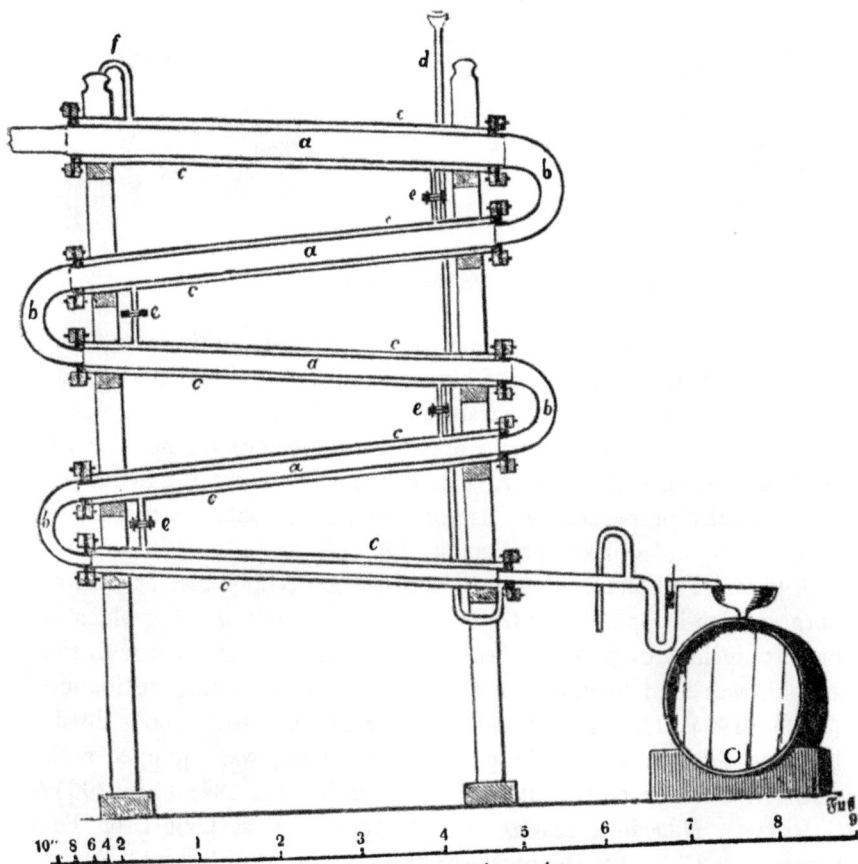

Fig. 194. The Babo cooler

FORBES, Art of distillation

century was SIGISMUND FRIEDERICH HERMBSTÄDT, born at Erfurt on April 14th 1760. He studied pharmacy and medicine in his native town and then worked for several years in an apothecary's shop at Hamburg under ROSE. After travelling and studying factories in Saxony and Thuringia he settled at Berlin and gave private courses until he became professor of chemistry and pharmacy at the Berlin University. Before his death on October 22nd 1833 he promoted the opening of many chemical works in Prussia and tried to spread the

Fig. 195. The Dorn still

new knowledge by giving courses to the general public in the way DAVY did in London. Many of his works deal with distillation (298) (299) (300) and some were translated into Dutch as standard works on the subject (301). In these works he also discusses the new French stills for instance those of ADAM, BÉRARD, DUPORTAL and others. He especially propagated the use of low stills to obtain larger yields of distillate and quicker production.

JOHANN FRIEDERICH DORN, chemist and technologist at Berlin, born at Neu-Ruppin in 1782, was the first to improve the old still by introducing complete pre-heating. A small coil connected with the alembic was used to draw samples of the vapours during distillation (190) (191). This still proved quite useful for discontinuous distillation of thick mashes as the pre-heater/condenser was equipped with a stirrer. The apparatus continued in use until about 1845 (Fig. 195).

JOHANN HEINRICH LEBRECHT PISTORIUS, born at Loburg on February 21st 1777, was the possessor of an estate, a "Rittergut", Weissensee near Berlin, where he died on October 27th 1858. In 1817 he obtained a Prussian patent on an apparatus which he conceived after thorough investigation of the difficulties of the earlier apparatus (484). PISTORIUS introduced a double still, an idea which DEROSNE is said to have taken from him, though this is doubtful. It is certain that PISTORIUS worked without knowing anything of CELLIER's invention which was first described in a French paper in the year 1817, when PISTORIUS took out his patent. It is even possible that POGGENDORF's claim is correct that PISTORIUS had worked on his still since

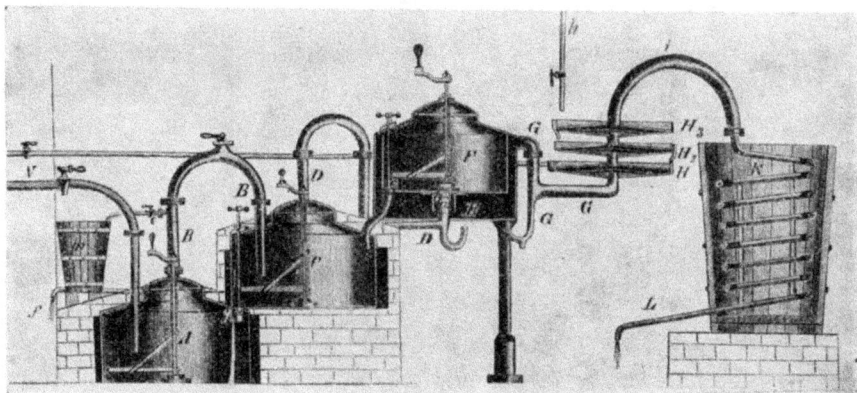

Fig. 196. The Pistorius still (1817)

1811. His rectifying trays are certainly most original (Figs. 196 & 197) though based on ADAM's principle. According to GUMBINNER (273) PISTORIUS especially prevented the blowing off of the alembic by puking of the still contents, a difficulty often encountered with thick mashes in the old stills. The only remarks that GUMBINNER has to make are that the pre-heater might have been somewhat larger (about ¾ of the contents of the still) and that the worm-condensor should have at least 5-7 windings the inlet being for instance 3" and the outlet 2". The most efficient dimensions of the still were a height about half the diameter of the bottom. The Pistorius apparatus was very efficient and continued in use for at least 70 years. The only feature that we miss was an inlet on the upper still to admit a constant stream of mash. In 1840 a still of 550 L hourly capacity cost about 1200 Thaler. Small modifications were introduced in the course of the years and later it was even built in the form of a column and made suitable for steam distillation by the firms of Weigelt at

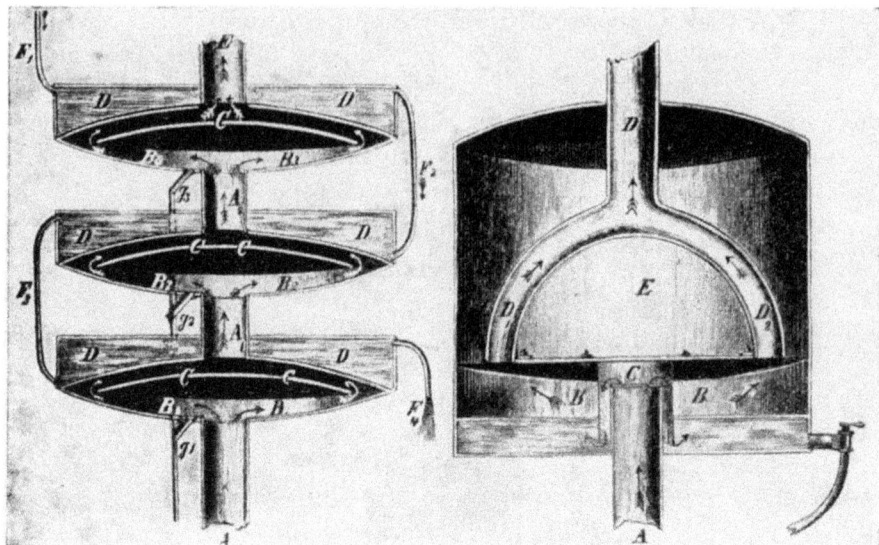

Fig. 197. Details of the Pistorius dephlegmator (1817)

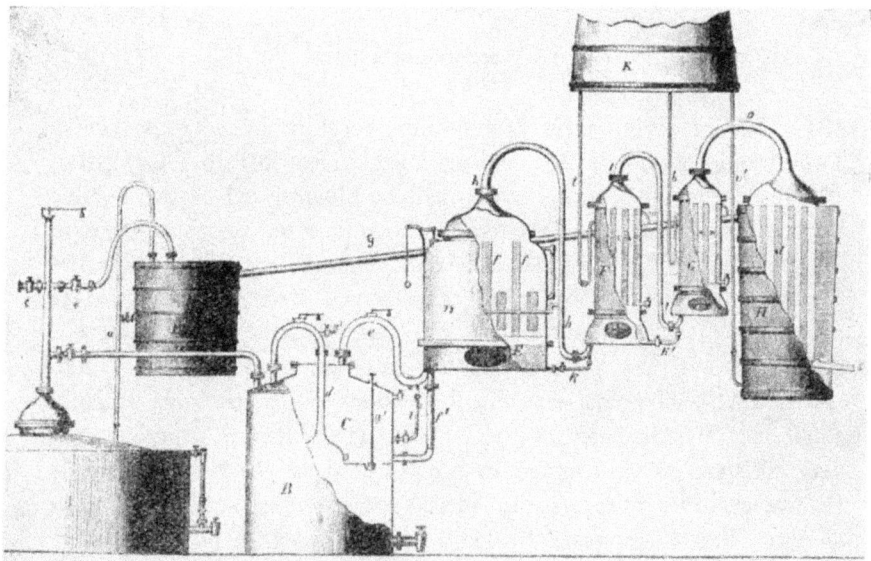

Fig. 198. The Schwarz still

Neuss and Volkmar, Hänig and Co. at Dresden. His rectification colums have rectangular caps, of which PISTORIUS may have been the original inventor. Some of his apparatus are constructed for direct, others for indirect steam heating.

One of the difficulties of the Pistorius dephlegmators ("Becken") was the problem of taking them apart and cleaning them. This point had the special attention of SCHWARZ but though his apparatus was used for some time on south-western Germany (Fig. 198) its rectifying capacity was insufficient, as the original model produced only 70 % alcohol. Also large quantities of heat and water were consumed which made it practically valueless.

ROMERSHAUSEN constructed a still which had a series of shells in the alembic through which the vapours had to pass before leaving it for the cooler. In this construction it was supposed that the rectification could be achieved in the alembic itself. A similar apparatus was published two years later (BUCHNER's *Repertorium für die Pharmacie* XIII, 1822, p. 383) (520).

About 1830, HEINRICH LUDWIG LAMBERT GALL, "Regierungs-sekretär" at Coblenz, born at Aldenhove in 1791, invented an apparatus with a double still (236) (237). Next to the two stills a somewhat higher pre-heater and above this dephlegmator and rectifier were placed (Figs. 199 & 200). Each still is distilled with steam in its turn and in meanwhile the second one is emptied and refilled. The second design had a third still which served as a pre-heater and later the stills were even placed placed together in a steam-bath for indirect heating. In effect this apparatus, that was very popular in southern Germany and Russia, is nothing but a Pistorius apparatus rebuilt for continuous operation. There is less heat lost than in the Pistorius still. GALL had a certain preference for wooden apparatus, like his contemporary SCHICKHAUSEN.

FRANZ ERNST VON SIEMENS who designed apparatus from 1818 (578) was the first to discuss the value of returning the condensate of the dephlegmators to the still. His designs were, however, useless as the distillate obtained was far too weak.

It is clear from these few examples and the perusal of contemporary handbooks such as that of OTTO (461) of 1838, etc. that the new French distillation column for continuous distillation had not yet penetrated to Germany by 1850. The stills of STEIN, GRIMBLE, SAINT MARC and the French models were much admired but not imitated on a large scale. One of the difficulties was the lack of skilled labour

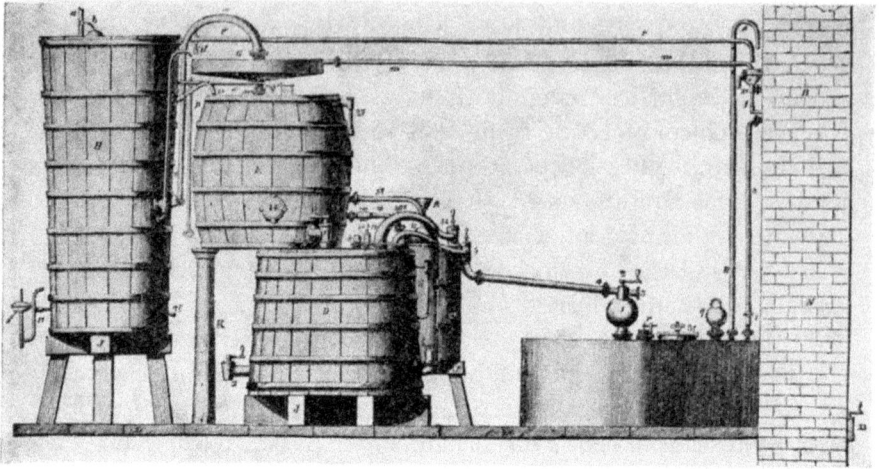

Fig. 199.

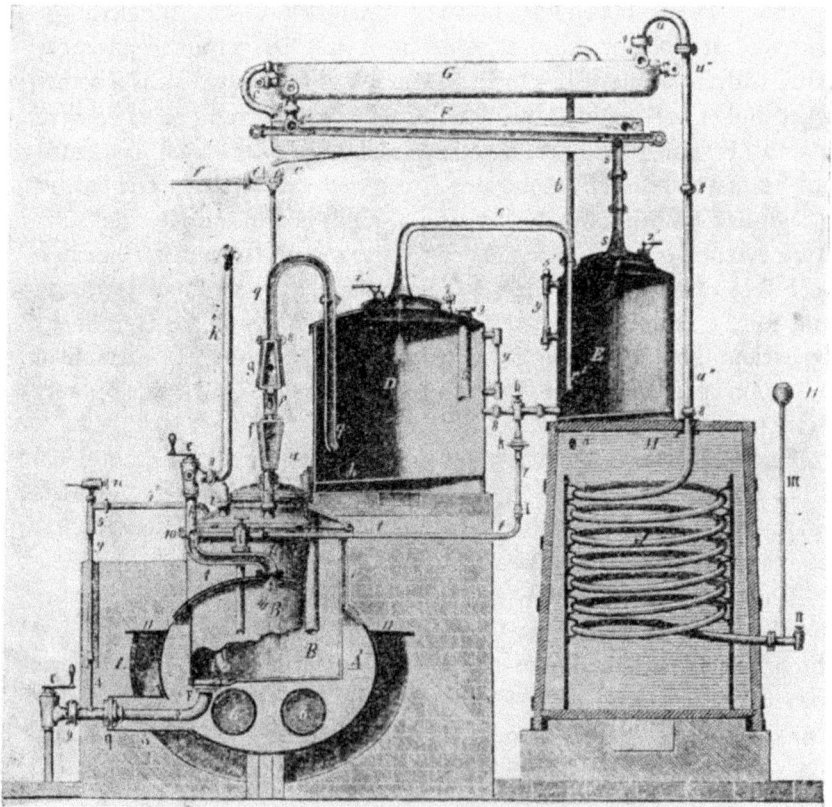

Fig. 199 & 200. The Gall distilling apparatus

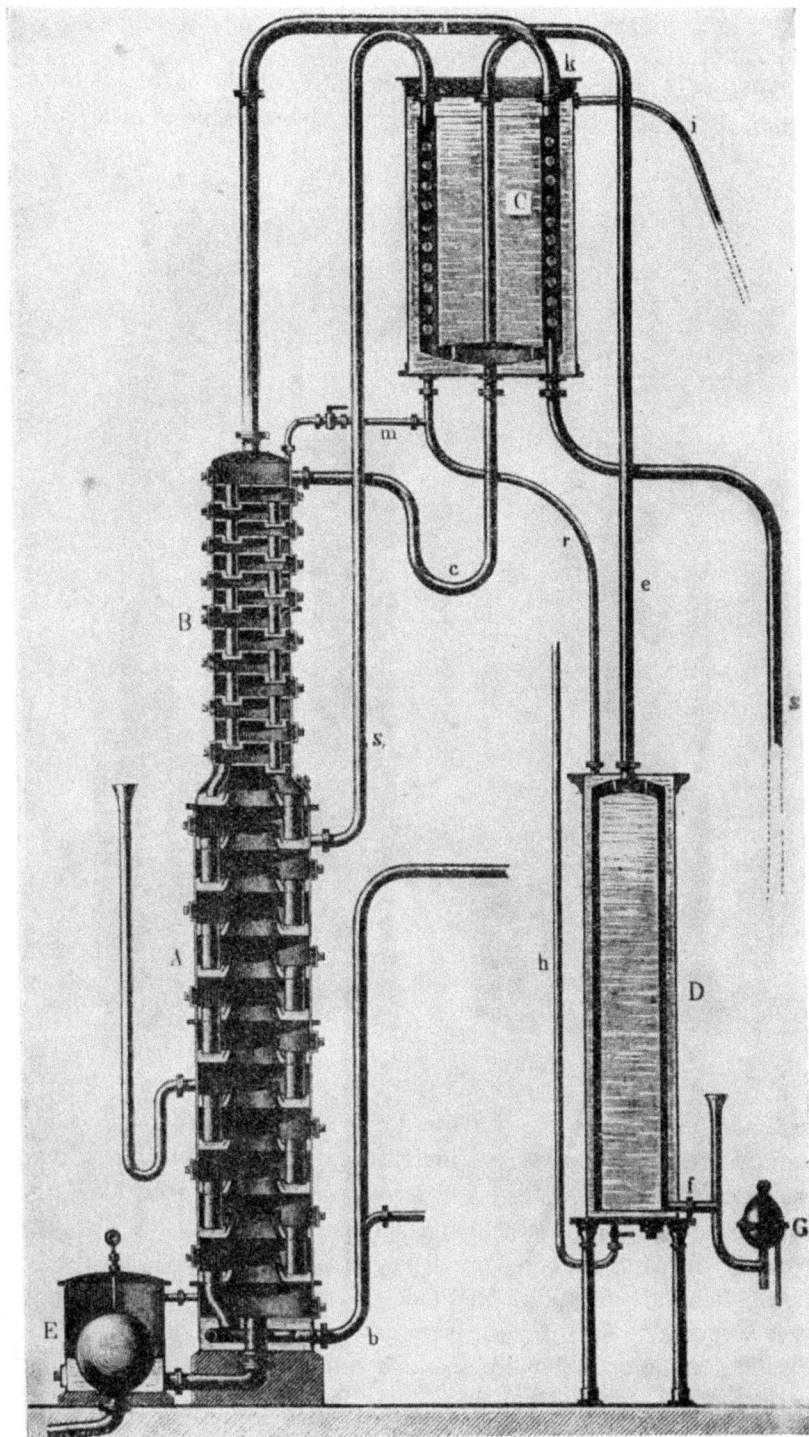
Fig. 201. The Bohm distilling column

in Germany and the undeveloped workshops, which made a German author of the period exclaim: "In England it is possible to make such

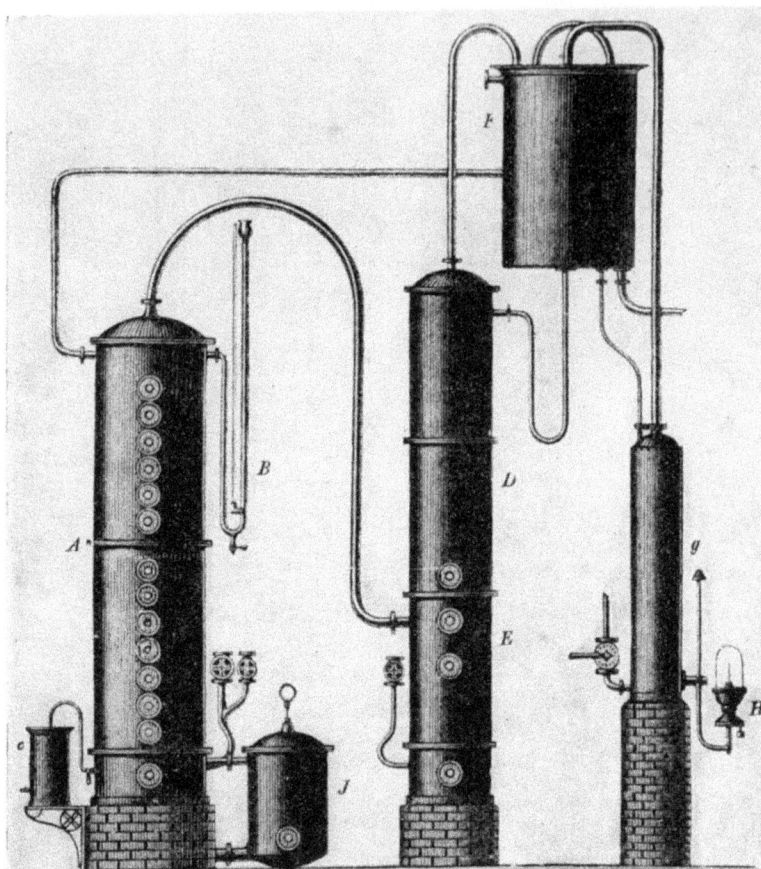

Fig. 202. The Paulmann column

apparatus in a form ready for use but in Germany this is impossible because of the clumsiness and inefficiency of our labour, but it may come one day" (DINGLER's *Polytechnisches Journal* vol. XXV, p. 464). And indeed it came soon enough. For by 1860 the distillers generally began to ask for continuously distilling columns, which were mostly built after the French models. The Germans however built their columns to suit thick mashes and to produce a distillate of a certain minimum quality in one operation (and if possible of all qualities) and they did not imitate the French in their extreme heat economy.

The earlier stills built before 1850 were practically all derived from the Pistorius still. But the later German columns were built along the

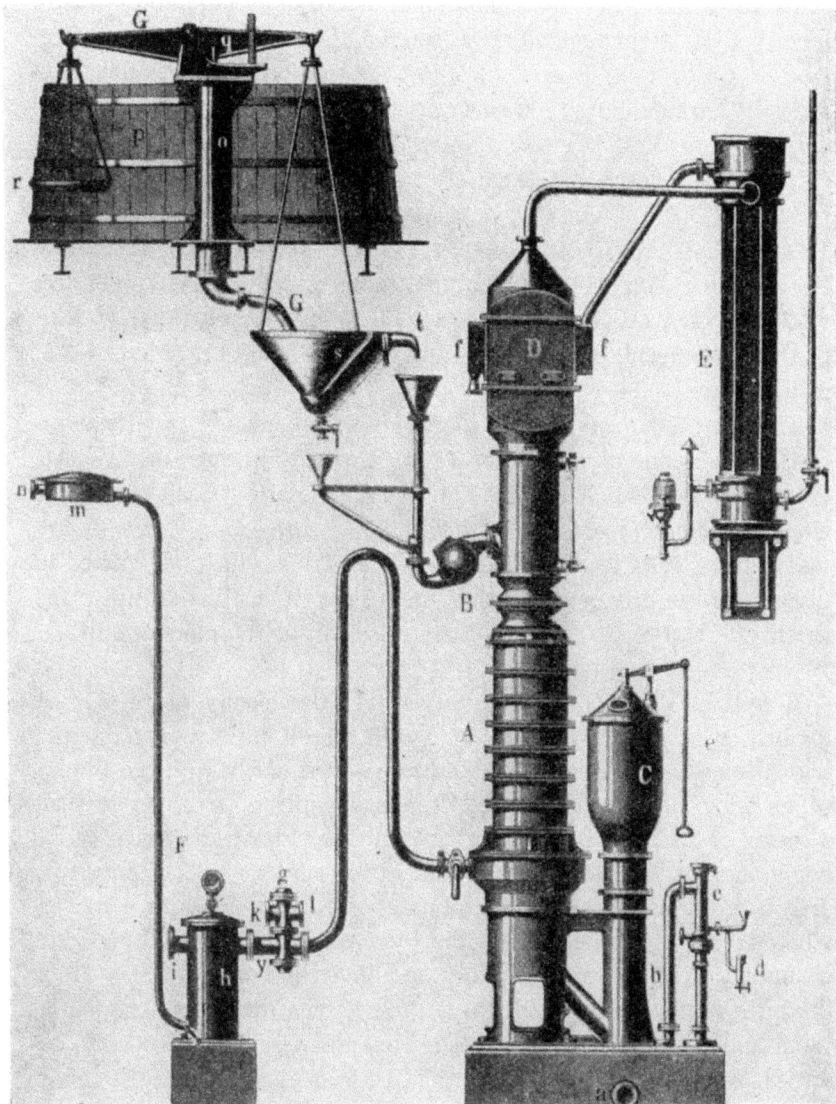

Fig. 203. The Ilges "automatic" distilling column

lines of those of SAVALLE and DUBRUNFAUT, using large caps as introduced by CHAMPONNOIS. Early German columns of this type are those of the Gebr. BOHM (of Fredersdorf) (Fig. 201). The Paulmann apparatus with separate distillation and rectification columns (Fig. 202) was built by F. H. MEYER at Hannover-Heinholz.

In 1873 the first "Ilges einteiliger Automat" (Fig. 203) with a new type of steam regulator was built. Other types follow soon, e.g. that of CRISTOPHE, PAMPE (working under vacuum), PERRIER, SIEMENS (Charlottenburg), WERNICKE (Halle), HECHT-SALZMANN for very thick washes, WEIGEL (Neisse), SCHMIDT (Nauen), NEDWIG, OSTROWSKI (Posen), GALLAND, LUHN (Haspe, D.R.P. 52.440), BONDY (D.R.P. 53.443), FROMMEL (D.R.P. 53.700), FRANÇOIS HAECK (D.R.P. 57.169), BARBET (D.R.P. 58.733) ALFRED BANDHOLZ (D.R.P. 58.741), W. PAALZOW (Reval), BURGHARDT (D.R.P. 58.790), SAVALLE (D.R.P. 64.428), DICK (D.R.P. 68.416), Maschinenfabrik Grävenbroich (D.R.P. 68.567), M. STRAUCH (Neisse) and many others.

The later German rectification columns are mainly modifications of the Champonnois column of 1854. We mention those of SAVALLE, C. HECKMANN (D.R.P. 39.577), PAMPE (Halle a/Saale) and the Braunschweigische Maschinenbauanstalt.

But these stills lead us far beyond the period which we wanted to discuss for we proposed to bring the history of distillation up to the death of CELLIER BLUMENTHAL and the birth of the modern still.

It will be clear from these pages that the history of the art of distillation is a long one and that it has been intimately connected with the history of civilization and more particularly with the history of alchemy and chemistry. Still it was possibly of more value than alchemy of which it is said that "it is an art without art which has its beginnings in falsehood, its middle in toil and its end in poverty". This is certainly not the case with the art of distillation which remained what it was at the beginning, the most easy way of obtaining pure chemicals, both in the laboratory and in technology. Its story from the birth of the modern still to its present state may be more complicated and to some more interesting, but this is a knot worthy to be untied by other hands.

BIBLIOGRAPHY

1. (*Abhandlung*) *Rechtsgegründete* —— *vom Brannteweinbrennen,* etc. (Rostock, 1754)
2. (ABULCASIS) ABû AL-QâSIM HAᵓALAF IBN ᶜABBAS AL-ZAHRâwî, *Kitâb al-taŝrif li-man ᶜağiza ᶜan al-taᵓâlîf* (for translation see No. 3)
3. (——), *Liber servitoris sive Liber XXVIII Bulchasin Beneberacerin* (edit. SIMONE DI GENOVA et ABRAHAM IUDAEUS, Venetia, 1471)
4. (ALKHWARAZMI) ᶜABû ᶜABDALLâH MUHAMMAD BIN AHMAD BIN YûSUF AL-KâTIB AL-KHWâRAZMI, *Mafatîh al-ᶜUlûm* (*Key of the Sciences*) (Edit. G. VAN VLOTEN, Leyden, 1895)
5. (IBN AL BAITAR) ABû MUHAMMAD ᶜABD ALLâH BIN-AHMAD DIYâᵓAL-DîN IBN AL-BAITAR AL-MâLAKI, *Kitâb al-Djamiᵓli-Mufradat al Adwiya waᵓl-Aghdiya* (*Book of the whole of the simples*) (edit. Cairo, 1874/75, 4 vols.)
6. (——), *Le Traité des simples par Ibn al-Baithar,* edit. L. LECLERC (Paris, 3 vols., 1877-1883)
7. (IBN AL AWWAM) ABû ZAKARîyâ YAHYâ BIN MUHAMMAD BIN AHMAD BIN ALᶜAWWâM AL-ISHBîLî, *Kitâb Agriculture*) (Edit. J. J. CLÉMENT-MULLET, *Le livre de l'agriculture,* Paris, 1866, 2 vols.)
8. ADAM, EDUARD, *Procédé pour retirer du vin tout l'alcool qu'il contient* (French patent No. 453 of May 29th 1801)
9. (——), GASPARD Z., *Nouvel appareil de distillation* (French patent of December 2nd, 1817)
10. ADAMS, G., *Natural and Experimental Philosophy* (London, 1794, 5 vols.)
11. AGNIUS, JEAN, *Procédé propre à dégager le genièvre de son goût empyreumatique* (French patent of December 14th, 1813)
12. AGRICOLA, GEORG, *De Re Metallica* (Basel, Froben 1556, 1557; German version Basel, 1561, 1563, 1621, 1657; Italian version Frankfurt, 1580; English version London, 1912)
13. ——, *De Re Metallica* (*Zwölf Bücher vom Berg- und Hüttenwesen*) (Berlin, 1929, Books IX, X, XII)
14. AHMAD, MAQBûL, *A Persian translation of the Eleventh Century Arabic Alchemical Treatise ᶜAin As-sanaᶜh WaᶜAun as-sanaᶜh* (*Mem. Asiat. Soc. Bengal,* vol. VIII, 1922/29, pp. 418-460)
15. ALBERTUS MAGNUS, *De secretis mulierum libellus .... ejusdem de virtutibus herbarum, lapidum et animalum .... item de mirabilibus mundi* (Lugduni, 1566; Amstelodami, 1669)
16. (ALDIMASHKI) ABû ᶜABDALLâH MUHAMMAD B.ABî TALIB AL-ANSAR AL-SûFî, *Kitâb Nukhbat al Dahr fi ᶜadjâᵓib al-Barr wal-Bahr* (French edition see 17)

17. (AL DIMASHKI), *Manuel de la cosmographie du moyen-âge par al-Dimašqî*. Publié par A. F. M. MEHREN (Kopenhagen, 1874)

18. ALEXANDER OF ... APHRODISIAS, *Commentaria in Aristotelem graeca*: *Meteorologica* (edit. M. HAYDECK, Berlin, 1899)

19. (AL DJAWBARI) ᶜABD AL-RAHMAN B. ᶜOMAR ZAIN AL-DÎN AL-DIMASHKÎ, *Kitab al Mukhtâr fî Kashf as-Asrâr wa Hatk al-Astar* (*Chosen Book of the Revelation of Secrets and the Tearing of the Veils*) (Damascus, 1885; Istanbul, 1899; Cairo, 1908)

20. (——), *Buch der Geheimnisse* (Transl. by E. WIEDEMANN, *Mitt. z. Gesch. d. Med.* vol. IX, 1910, pp. 386-390)

21. ALÈGRE, PIERRE, *Appareil distillatoire servant à déflegmer l'esprit de vin et à le porter à son plus haut point de rectification* (French patent of June 21th, 1813, taken under the name of J. B. DUROSELLE, Paris)

22. ——, *Appareil distillatoire dit réducteur propre à réduire le titre de l'esprit de vin* (French patent of February 26th, 1813, taken under the name of J. B. DUROSELLE, Paris)

23. ALTENHOF, FRIED., *Der wohlerfarene Distillateur und Liqueurist* (Altona, 1793)

24. AMANDUS, SANCTUS, (JOHN OF ST. AMAND), *Nicolai Repositi Antodotarium* .... (Venetiae, 1589, apud Juntas ed Costaeus)

25. ANGLICUS, BARTH., *La Propriété des choses* (Paris, 1510)

26. (*Anweisung*) *Gründliche und Nützliche —— zur Verbesserung der Brannteweinbrennery* (Riga, 1794)

27. (*Appeal*) *An —— to the Public concerning the Distilling Trade with a rational Scheme to extirpate it from the Nation* (London, 1757)

28. ARISTOTELES, *Meteorologica* (edit. F. H. FOBES, Cambridge (Mass.), 1919)

29. ARMSTRONG, E. F., *Alcohol through the ages* (*Chemistry and Industry* vol. 52, 1933, pp. 251-257)

30. ARNAUD (DE LYON), *Introduction à la Chimie et à la vraie Physique* (Lyon, 1655, chez Prost)

31. AXTH, JOH. CONR., *Tractatus de arbori coniferis et pice conficienda aliisque ex illis arborus provientibus* (Jena, 1679)

32. BAASCH, E., *Holländische Wirtschaftsgeschichte* (Jena, 1927)

33. BACCIO, ANDREA, *De Naturali Vinorum historia de Vinis d'Italiae et de conviviis antiquorum* libri VII (Roma, 1596)

34. BACKER, H. J., *Oude chemische werktuigen en laboratoria van Zosimos tot Boerhaave* (Groningen, 1918, pp. 13-21)

35. BACON, FRANCIS, *Novum Organum* (London, 1620)

36. ——, *Sylvae sylvarum or a naturall historie in ten centuries* (London, 1626; 5th edit. London, 1639)

37. ——, *De interpretatione naturae sententiae duocem* (London, 1665) (Works edit. ELLIS and SPEDDINGS Vol. III)

38. BADGER, W. L., *Some phases of the history of chemical engineering* (*J. Chem. Education* vol. XII, 1932, pp. 691-707)

39. BAGLIONI, *Appareil propre à la destillation continue des vins et des autres liquides alcooliques* (French patent No. 883 of August 24th, 1813 for a period of 10 years) (Additional patents of January 28th 1814, September 20th 1814, November 7th 1815, March 26th 1816, November 9th 1816)

40. BAILLEUL, *Appareil distillatoire monté sur un fourneau fumivore* (French patent of December 31st, 1812)

41. BARBET, EMILE, *Les appareils de distillation et rectification* (Paris, 1890)

42. BARCHUSEN, JOH. CONR., *Synopsis pharmaceutica* (Frankfort, 1690; Leiden, 1712)

43. ——, *Elementa Chimiae* (Lugdun. Batav., 1718)

44. BARLET, ANNIBAL, *Le Vray et Méthodique Cours de la Physique résolutive vulgairement dite Chymie, réprésenté par figures générales et particulières pour connoistre la Théotechnic ergocosmique, c'est-à-dire l'art de Dieu en l'ouvrage de l'univers* (Paris, 1657; 3rd edit. 1677)

45. BARNE (F. BARRE), *Appareil distillatoire* (French patent No. 108 of January 15th, 1802 for a period of 5 years) (additional patent of November 16th 1804)

46. BARNES, W. H., *The apparatus, preparations and methods of the ancient Chinese alchemists* (*J. Chem. Education* vol. XI, 1934, pp. 655-659)

47. BARRE, A., *Appareil propre à distiller les vins et des marcs de raisin en même temps, ainsi que les produits se mêlent* (French patent No. 323 of December 13th, 1803; additional patents of April 11th 1806 and August 29th 1806) (Patents 699 & 700 of August 26th 1816 for a period of 5 years with additional patent of October 14th 1816)

48. (BARRY) *John Barry's Extract Bereitung* (*Buchners Repertorium für die Pharmacie* vol. XI, 1821, p. 316)

49. BARTHELOMAEUS OF SALERNO, *Practica* (edit. S. DE RENZI, *Collectio salernitana* vol. IV, 1856, p. 321)

50. BASCOU, HONORÉ, *Appareil de distillation au moyen duquel on obtient du 3/6 par une seule distillation ou une seule chauffe* (French patent May 2nd, 1806)

51. BAUDOT, A., *La pharmacie en Bourgogne* (Diss. Paris, Dijon, 1905)

52. BATTUM, CAROLUS, *Secreet-boeck van vele diversche Konsten in veelderley Materien . . . .* (Amsterdam, 1661)

53. BAUMÉ, ANTOINE, *Eléments de pharmacie théorique et pratique* (Paris, chez Samson, 1762, 1769, 1773, 1777, 1784)

54. ——, *Mémoire sur la manière de construire les alembics et les fourneaux propres à la distillation des vins pour en tirer les eaux-de-vie* (*J. de Physique* XII, 2, 1778, pp. 1-37)

55. BECKMANN, JOH., *Dagboek van zijn reis door Nederland in 1762* (edit. G. W. KERNKAMP, *Bijdragen en Mededeelingen van het Historisch Genootschap, vol.* XXXIII, Amsterdam, 1912)

56. BECKMANN, JOH., *Anleitung zur Technologie oder zur Kenntniss der Handwerke, Fabriken und Manufakturen* (Göttingen, 1777)

57. ———, *Beyträge zur Oeconomie, Technologie, Polizey und Kameralwissenschaft* (Göttingen, 1779-1791; I, 143; III, 434; IV, 114)

58. ———, *Beyträge zur Geschichte der Erfindungen* (Leipzig, 1780-1805; I, 2, 179; III, 3, 435; V, 2, 206)

59. *Bemerkungen über die Branntweinblasen, nebst einem Vorschlage die Kühlgeräte auf eine ganz neue Weise vorteilhaft zu benutzen ....* (*Hoffmann's allg. Annalen der Gewerbekunde* vol. I, Leipzig-1803, p. 364)

60. BENTHAM, M. A., *Some seventeenth century views concerning the nature of heat and cold* (*Annals of Science*, vol. II, 1937, pp. 431-450)

61. BENVENUTO OF IMOLA, *Comentum super Dantis Aldigherii comoediam* (Firenze, 1887)

62. BÉRARD, ISAAC, *Appareil distillatoire propre à retirer du vin dans une seule opération de l'eau-de-vie épreuve d'Hollande, de l'esprit trois-cinq, trois-six, à la volonté du fabricant* (French patent No. 406 of August 16th, 1805 for a period of 10 years) (extended by decret of Jan. 17th 1814 to May 1st 1821)

63. ———, *Appareil à rectifier l'esprit de vin et le marc de raisin, appareil applicable aux chaudières de distillation* (French patent No. 1183 of April 23rd, 1816 for a period of 10 years) (Additional patents of December 13th 1805, May 30th 1806, July 18th 1806, October 31st 1806, March 13th 1807, June 5th 1807, December 26th 1811, September 20th 1814)

64. BERNARON, *Application des caloriques des usines à la distillation et à la vapeur* (French patent of January 17th, 1806)

65. BERNOUILLI, DANIEL, *Hydrodynamic* (Paris, 1738, sect 10)

66. BERTHELOT, M. et M. RUELLE, *Collection des anciens alchimistes grecs* (Paris, 1888)

67. BERTHELOT, M., *Introduction à l'étude de la chimie des anciens et du Moyen Age* (Paris, 1938)

68. *Beschreibung und Abbildung einer verbesserten Destillierblase wofür im vorigen Jahre Joseph Corty in der Provinz Middlesex ein englisches Patent erhalten hat* (*Buchner's Repertorium für die Pharmacie* vol. VII, 1819, p. 96)

69. BERZELIUS, J., *Über die verschiedene chemischen Operationen und Gerätschaften* (*Erdmanns Journ. f. techn. und ökon. Chemie* vol. XIII, 1832, 3, p. 320)

70. BESSON, JACOB (JACQUES), *De absoluta ratione extrahendi olea et aquas è medicamentis simplicibus, Liber* (Zürich, 1559) (also published as an appendix to LIBAVIUS' *Praxis Alchymiae*, Francfort, 1604)

71. ———, *Art et moyen parfaict de tirer huyles et eaux de tous les médicaments simples et oleagineux* (Paris, 1571, 1573)

72. BIRINGGUCCIO, V., *De la Pirotechnia* (Venezia, 1540; Napoli 1550; Venezia 1558, 1559, 1678) (German translation edit. Johannsen, Braunschweig, 1925)

73. BIRINGGUCCIO, V., *La Pyrotechnie ou art du feu, contenant dix livres, traduit par le feu maistre Jacques Vincent* (Paris, 1556, 1572, 1627)
74. BLACK, JOSEPH, *Lectures on natural philosophy* (London, 1760-1765, vol. I, pp. 79, 504)
75. ——, *Lectures on the Elements of Chemistry* (Edinburgh, 1803, edit. J. ROBISON, vol. I, pp. 116, 157, 171)
76. (*Bloemlezing*), *Een* —— *uit de Onderwereld* (De Vereenigde Distillateurs en de Nederlandsche Vereeniging van Distillateurs en Likeurstokers, Schiedam, 1935?)
77. BLÜMNER, H., *Römische Privataltertümer* (München, 1911)
78. BOERHAAVE, H., *Elementa chemiae* (Leiden, 1731/1732, 2 vols.)
79. ——, *Eléments de chimie* (Paris, 1754, 6 vols.)
80. BOISSONNADE, P., *Life and work in Medieval Europe* (London, 1927)
81. J. K. B. (BOLS?), *Een uytvoerig en omstandig bericht van de nieuw ontdekte distilleerkonst, waarin niet alleen de nodige instrumenten, regeering van het vuur, materien en stoffen, maar ook de manier van met weinig onkosten, doch echter met overgrote winst, allehande fyne en ruwe wateren te distilleren, alsmede d'inwerking, toezetting, fermentatie, &c, &c, klaar en duidelijk worden beschreven en aangewezen door J. K. B., reets 40 jaren lang ende noch tegenwoordig distillateur* (Amsterdam, Pieter Aldewereldt, 1736)
82. ——, *Verbeterde nieuw ontdekte distilleer- en stookkonst, behelzende een uitvoerig bericht van de maat en gewichten, kunstwoorden, werktuigen, luteren, het vuur, inweeking, moutmaking, stooken en distilleren van ruwe en fyne wateren, oliën en cordialen; het distilleren van gewassen en planten, wyn en azyn, &c. Eerst beschreven door J. B. K. ervaren distillateur. Tweede druk, overzien, verbeterd, vermeerderd en verrykt met vele fraaye geheimen door Polyhistor* (Amsterdam, Steven van Esveldt, 1763)
83. BONTOUX, *Procédés propres à extraire du même marc de raisin l'huile, l'eau-de-vie et le vinaigre qu'il contient* (French patent of October 31st, 1806)
84. BOYLE, ROBERT, *New Experiments Physico-Mechanical touching the spring of air and its effects* (Oxford, 1660, p. 338) (See also *Works*, 1744, vol. , p. 74)
85. ——, *The Sceptical Chymist* (London, 1661) (edit. Everyman's Library, London, 1910, pp. 36, 109)
86. ——, *The mechanicall origine of heat and colde* (London, 1675)
87. ——, *Memoirs for a general History of the Air* (London, 1692)
88. ——, *Experiments and observations upon the producibleness of chymical principles* (*Philosophical Works* III, London, 1738, p. 386) (*Works*, 1744, vol. I, p. 377)
89. (*Brannteweinbrenner*), *Der geschickte* —— (Leipzig, 1754)
90. BRAUMÜLLER, J. G., *Eine vorteilhafte Einrichtung für Brannteweinbrennereyen* (*Annalen der ökon. Ges. zu Potsdam* Band I, Heft 3, p. 105)

368 BIBLIOGRAPHY

91. BREEN, JOH. C., *Aanteekeningen, uit de geschiedenis der Amster-damsche nijverheid.* X. Branderijen en likeurstokerijen (*Neder-landsch Fabrikaat* vol. VII, 1921, 20/10)

92. *Brennereyen, Über die —— zu Appingedam bey Delfzeyl* (*J. f. Fabrik, etc.,* vol. XX, 1801 (Leipzig), p. 411)

93. BROUAT, J., *Traité de l'Eau-de-vie ou Anatomie théorique et pratique du Vin* (Paris, 1646, chez Jean Henault)

94. BROUQUIÈRES, *Appareil distillatoire* (French patent of January 14th, 1814)

95. BRUGNATELLI, L., *Elementi di chimica* (Pavia, 1795, vol. I)

96. ——, *Beschreibung einer Destillations-Anstalt vermöge welcher man zu gleicher Zeit Aquavit und Alkohol erhalten kann* (*Crell's Chem. Annalen,* vol. II, 1798, pp. 267-270)

97. BRUGNIÈRE, *Appareil propre à obtenir des vins et eaux-de-vie par une seule distillation de l'esprit à toutes les titres connues dans la commerce* (French patent No. 643 of March 8th 1805)

98. BRUNET, P. et A. MIELI, *Histoire des sciences. I.* Antiquité (Paris, 1935)

99. BRUNSCHWYGK, HIER, *Liber de arte distillandi de simplicibus oder Buch der rechten Kunst zu Distillieren die eintzigen Dinge* (Strassburg, 1500, Johann Grüninger). (This book was used by Gröninger in composing his *Liber de arte distillandi Simplicia et Composita/Das nue Buch der rechten Kunst zu distillieren.* Auch von Marsilio Ficino und anderen hochberömpten Erzte natür-liche und gute kunst zu behalten den gesunden leib und zu ver-treiben die kranckheiten mit erlengerung des lebens. Strassburg 1509) (the so-called *Small Book of Distillation*)

100. ——, *Liber de arte Distillandi de Compositis; Das Buch der waren Kunst zu distillieren die Composita und simplicia und das Buch thesaurus pauperu, Ein schatz der arme genannt Micarium die brosamlin gefallen von den büchern d'Artzny und durch Expe-riment von mir Jheronimo brunschwick uff geclubt und geoffen-bart zu trost denen die es begehren,* Strassburg 23 Februar 1512. (This is the so-called *Big Book of Distillation.*) (Other editions: Strassburg 1519, 1531; Francfort 1553, 1594, 1598, and several adaptions by ULSTAD, RYFF, UFFENBACH etc.)

101. ——, *The vertuose Boke of Distyllacyon of the Waters of all maner of Herbes, with the fygures of the Syllatoryes .... and now new-ly translated out of Duyche into Englyshe* by L. ANDREW (Lon-don, 1527)

102. (—— ?) *Die distellacien ende virtuyten der wateren* (Brussels, by Tho-mas van der Noot, 1517, 1520) (This may be a reprint of No. 103)

103. (—— ?) *Die rechte conste om alderhande wateren te distileeren* (Brussels, by Willem Vorsterman, 1520)

104. BUCH, MAX, *Die Wotjäken, eine ethnologische Studie* (*Acta Soc. Scient. Fenn.,* Helsingfors, 1882)

105. BULLEYN, WILLIAM, *Book of Simples* (London, 1562, 1579)

106. BURGHART, G. H., *Die zum allgemeinen Gebrauch wohleingerichteten Destillation welche im ersten Theil von Ab- und Einteilungen, Werkzeugen, allgemeinen Arbeiten und allem was diese Kunst überhaupt angehet genugsame Nachricht gibt. In den anderen Theil aber in beynahe zwey hundert processen die Bereitung verschiedener destillierter Wässer, Branteweine, Aqua vitae, Rossolis, flüchtiger säurer, mineralischer Geister, Oele, Essenzen, Extracte und andrer truckner Chemischen Artzneyen deutlich vorträgt und endlich in den dritten Theile in vierzig Processen vom Einmachen mit Zucker und andern dahin gehörigen Nicht nur den Aertzten, Wund-Aertzten und Apotheckern, sondern auch Weinbrennern und Destillatoribus ingleichen Hausvätern und andern Liebhabern dieser Wissenschaft zum besondern Nutz und Gebrauch ausgesetzt* (Breslau 1736, 1754).

107. BUSCH, GABR. CHR. B., *Handbuch der Erfindungen* (4th edit., Eisenach, 1806, Theil III, Abt. 2, p. 69)

108. BIJLSMA, R., *Oud-Rotterdamsche gebrandwijnbranders* (*Rotterdamsch Jaarboekje* 1915, pp. 46-50)

109. CAESALPINUS, ANDREAS, *De metallicis rebus libri tres* (Roma, 1596)

110. CAMERARIUS, JOACHIM, *Hortus medicus philosophicus* (Francfort, 1588)

111. CAMPBELL, D. J., *Arabian medicine* (London, 1926, 2 vols.)

112. CARDANUS, HIERONYMUS, *De subtilitate, de varietate rerum* (Nürnberg, 1550, 1557) (German translation by HEINR. PANTALEON, *Offenbarung der Natur*, Basel, 1559)

113. CAMPY, PLANUS, *Bouquet composé des plus belles fleurs chimiques* (Paris, 1629)

114. CANCRIN, F. L. VON, *Von einer vollkommen eingerichteten Branntweinbrennerey* (*Vermischte, meist ökonomische Schriften*, Riga, 1786)

115. CARNOT, SADI, *Réflexions sur la puissance motrice du feu* (Paris, 1824)

116. CELLIER-BLUMENTHAL, J. B. (together with LAPORTE), *Appareil propre à réduire en pulpe les betteraves, à la mettre en digestion avec l'alcool au bain marie, à extraire la matière sucrée de cette digestion par la distillation et à crystalliser le produit de la distillation* (French patent No. 727 of November 16th, 1811)

117. ——, *Appareil propre à la distillation des vins, grains et pommes de terre* (French patent No. 1886 of November 24th, 1813 for a period of 10 years)

118. ——, *Appareils destinés à la distillation continue et à l'évaporation* (French patent of No. 2266 of January 12th, 1818)

119. ——, *Appareil propre à cuire dans le vide le sucre de betteraves et le sirop de sucre et à concentrer les dissolutions salines* (French patent of December 4th, 1834)

120. CHAMBERS, E., *Cyclopaedia* (*sive Distillation and Still*) (London, 1728-1751)

121. CHAPTAL, J. A. C. (together with ROZIER, PARMENTIER and DUSSIEUX), *Traité théorique et pratique sur la culture de la Vigne*

avec l'art de faire le vin, les aux-de-vie, l'esprit de vin et les vinaigres (Paris, 1801, 1811)

122. CHAPTAL, J. A. C., *L'art de faire, gouverner et perfectionner les vins* (Paris, 1801, 1807, 1839)

123. ——, *La Chimie appliquée aux arts* (Paris, 1807) (German edition 1808)

124. ——, *De l'industrie française* (Paris, 1819, 2 vols.)

125. ——, *La Chimie appliquée à l'agriculture* (Paris, 2 vols. 1823, 1829)

126. CHARAS, MOYSE, *Pharmacopée Royale Galénique et Chymique* (Paris, 1676, 1682, 1717, Lyon 1753)

127. CHASSARY, *Appareil et procédés de distillation des eaux-de vie* (French patent of September 13th, 1805)

128. CHASSINAT, E., *Un papyrus médical copte* (*Mém. Instit. Franc. Archéol. Orient.* vol. 32, 1921, pp. 140, 196, 220)

129. CHAUCER, GEOFFREY, *Complete Works* (edit. W. SKEAT, London 1915, p. 659)

130. CHILD, ERNEST, *The tools of the chemist* (New York, 1940)

131. CHIKASHIGE, M., *Alchemy* (Tokyo, 1936)

132. CHRIST, J. L., *Chemisch-physikalische und praktische Regeln vom Fruchtbranntteweinbrennen nebst* .... (Francfort, 1785)

133. Chu-I-Ching (2544, 5507, 2875) Pei-Shan Chiu Ching (8871, 9663, 2260, 2122) (written about 1120, see A. WYLIE, *Chinese Litterature*, 1902, p. 105)

134. CLEGHORN, W., *Disputatio Physica inauguralis Theoriam Ignis complectens* (Edinburgh, 1779)

135. COFFEY, AENEAS, *Certain Improvements in the Apparatus or Machinery used in the Process of Brewing and Distilling* (Brit. patent No. 5974, August 5th, 1830)

136. *Conste, Die rechte — om alderhande wateren te distilleeren* (Antwerp, 1920) (This is a reprint of a work printed in Brussels, 1520, probably a Dutch edition of the work by BRUNSCHWYGK, see No. 103)

137. COOLHAES, C. JANSZ., *Van seeckere seer costelycke wateren, dien men met recht soude mogen noemen aquae vitae, ende sommige uutgelesene oliën den edelen welruyckenden balsemolie niet sonder reden te vergelycken; der welcker cracht ende menichfoudiche deuchden in dit boecxken cortelyc (uut de schriften sommiger hoochgeleerden ende exoerten doctoren ende professoren der loffelycker medicynen, welcker namen op dandere zyde deses blats verhaelt syn) beschreven worden. Met eener voorrede des distillateurs van den stercken ende beroemden smaeck deser wateren, de welcke met Godts hulpe cunstelyk digereert, circuleert oft substileert ende distileert worden tot Leyden op Rapenburch, al waer men die te coop vindt om een redelycken prys, ten huyse van Caspar Coolhaes* (Amsterdam, 1588)

138. ——, *Waterboexcken, het welcke aanwyst, hoe men zeeckere edele ende seer goede spiritus, aquae vitae compositae, wateren, crachtwateren, ende gedistileerde oliën tot een yeder cranckheyt ende*

gebreken des menschen lichaems, die uyt koude humoren ende catharnen haren oorspronck hebben soo wel uytwendich als inwendich met grooten nut sal moghen gebruijcken (Amsterdam, 1600, 1608, 1622)

139. COOPER, *The Complete Distiller* (2nd edit. London, 1760)
140. CORDIER, V., *Die chemische Zeichensprache, einst und jetzt* (Graz, 1929)
141. CORDUS, VALERIUS, *Dispensatorium sive pharmacorum conficiendorum ratio* (Nürnberg, 1535)
142. ——, *Annotationes in Pedacei Dioscorides de Materia Medica liber quinque. Liber de artificiosis extractionibus. Liber II de destillatione oleorum* (Nürnberg 1540, edit. C. GESNER, *Strassburg*, 1561)
143. *Cordus, Dispensatorium van* ——, *dat is de Maniere van de Medicynen te bereiden . . . met annotatien van den Autheur en corte verklaringen van Peter van Coudenberch, excellent Apotheker van Antwerpen*. Overgeset in Nederduyts deur MARTEN EVERAART (Amsterdam, 1592, by Cornelis Claesz.)
144. CORTY, JOSEPH, *Certain improvements on and Additions to stills, or the Apparatus used for Distilling, and also in the Process of Distilling and Rectifying* (Brit. Patent No. 4203 of January 20th, 1818) (See also *Buchners Repertorium f. d. Pharmacie* vol. 7, 1819, p. 616)
145. COSTAEUS (DE LODI), G., *In mesues simplicia et antidotarii novem posteriores sectiones adnotationes* (Venetiae, 1602)
146. CRAWLEY, A. E., *Drinks (Hastings' Encyclopaedie of Religion and Ethics* vol. V, Edinburgh, 1912)
147. CROLL, OSWALD, *Basilia chymica* (Frankfort 1608, and more than 17 later editions, e.g. London, 1670)
148. CULLEN, WILLIAM, *Essays and Observations* (Edinburgh, 1755, vol. II)
149. CURAUDAU, F. R., *Observations physiques sur les causes de l'imperfection des fourneaux d'évaporation et sur une nouvelle manière de les construire pour y brûler économiquement toute espèce de combustible (Ann. de Chimie* vol. XLVI, 1803, pp. 279-288)
150. ——, *Observations pyrotechniques et leur application aux fourneaux d'évaporation (Ann. de Chimie* vol. L, 1805, pp. 134-139)
147. ——, *De l'influence que la forme des alembics excerce sur la qualité des produits de la distillation (Ann. de Chimie* LXVII, 1808, pp. 198-204)
152. ——, *Sur l'évaporation par l'air chaud (Annales de Chimie* vol. LXXX, 1811, pp. 109-111)
153. DARIOT, CLAUDE, *De praeparatione medicamentorum* (Lyon, 1582)
154. ——, *La grande chirurgie de Paracelse . . . plus un Discours sur la Goutte et trois traités de la préparation des médicaments* (Lyon, 1603; Montbéliard, 1608)
155. ——, *Die güldne Arch, Schatz- und Kunstkammer in drei Theile unterschieden. Im Ersten werden ausführlich behandelt drey Gespräch von spagirischer preparation und zubereitung der Arz-*

neyen. Als warumb die nicht allein von den vegetalibus und animalibus, sondern auch von die den Mineralibus hergenommene eintzele Medicamenta anderst als bishero von den Galenisten geschehen sollen und müsssen praepariert werden und dann auch wie selbige praeparationen recht und wol vollbracht werden solle. Im andern und letzten Theil hat der kunstbegieriger Leser vieler als der fürnembsten auserlesentsten Philosophorum, medicorum und Spagicorum Geschriften und Bücher. Allen den Chemie Liebhabern sonderlich den Jungen angehenden nutzlich zu lesen ins Teutsch mit sonderbaren Fleisz übersetzt durch. I.A.M.D. (Getruckt zu Basel in Verlegung des Authoren, 1614; Nürnberg 1614)

156. DARIOT, CLAUDE, *Vereinigung der galenischen und paracelsischen Artzneykunst* (Basel, 1623)

157. DARMSTAEDTER, E., *Die Alchemie des Geber* (Berlin, 1922)

158. DAVIS, T. L., *Primitive science* (*J. Chem. Educ.* vol. 12, 1935, p. 3-10)

159. ——, *The Problems of the origin of alchemy* (*Scient. Monthly* vol. XLIII, 1936, pp. 551-558)

160. ——, *Pictorial representations of alchemical theory* (*Isis*, vol. 28, 1938, pp. 73-87)

161. DEBRAINE-HELFENBERGER, J., *Manuel du distillateur et du liquoriste* (Paris, 1825)

162. DEGERING, H., *Ein Alkoholrezept aus dem 8. Jahrhundert* (*Sitzber. Preuss. Akad. Wiss.* vol. 36, 1917, pp. 503-515)

163. DEJEAN, *Traité raisonné de la distillation ou la distillation réduite en principes avec un traité des odeurs* (1753, 1759, 1769, 1801 (in 2 vols.))

164. DELBRÜCK, H., *Illustriertes Brennerei Lexikon* (Berlin, 1915, p. 90)

165. DEMACHY, J. F., *L'art du distillateur des eaux-fortes* (Acad. des Sciences, Paris, *Description des Arts et Métiers*, Vol. XIV, 1773) (Paris, 1775, 1780)

166. ——, *Receuil de dissertations physico-chimiques, etc.* (Amsterdam et Paris, 1774)

167. ——, *Traité de l'art du distillateur liquoriste, brûleur d'eau-de-vie, fabriquant de liqueurs, débitant ou cafetier-limonadier* (Paris, 1775)

168. ——, *De sterkstooker, zoutzuur- en vitrioolbereider,* grootendeels uit het Fransch door J. P. KASTELEYN (Dordrecht, 1788)

169. *Demachys Laborant im Grossen* mit Anm. *von Dr. Struve und Abh.* Wieglebs, aus dem Franz. übersetzt durch SAMUEL HAHNEMANN (Leipzig, 1784, 2 vols.; 1801)

170. *Demachys und Dubuisson Liqueurfabrikant,* aus dem Franz. von STRUVE und HAHNEMANN (Leipzig, 1785)

171. DERIVES, *Machine propre à extraire le liquide contenu dans le marc de raisins et autres quelconques* (French patent of November 12th, 1813)

172. DICHAEUS, K., *Beschreibung, welcher Gestalt Theer und Kohlenöfen einzurichten sind*, nach dem Schwedischen (Lüneburg, 1780)
173. DICKINSON, H. W., *History of Vitriol Making in England* (*Trans. Newcomen Soc.* vol. XVIII, 1937/38, pp. 43-60)
174. ——, *Matthew Boulton* (London, 1936, p 38).
175. DIDEROT et D'ALEMBERT, *Encyclopédie ou Dictionnaire raisonné des Arts et des Métiers* (Lausanne, Berne, 1782) (vide alembic, cornue, cucurbite, chapiteau, cohobation, descensum, distillation, filtration, fermentation, lit, liqueurs, récipient, rectification, serpentin, vin)
176. DIELS, H., *Die Entdeckung des Alkohols* (*Abh. Preuss. Akad. Wiss. phil.-hist. Kl.* 1913, No. 3)
177. ——, *Antike Technik* (3. Aufl., Leipzig, 1924, pp. 107-154)
178. DIGBY, KENELM, *Choice experiments and receipts in Physic and Chirurgery as also Cordials and distilled waters and spirits, Perfumes and other Curiosities*, translated by G. HARTMANN (London, 1668) (Later edition under the name of *Hartmann's choice collection of Chymical secrets*, London, 1682)
179. ——, *Philosophische Geheimnisse und Chymische Experimenta* (Hamburg, 1684)
180. ——, *Closet opened whereby is discovered several ways for making excellent metheglin, Sider, Cherrywine &c together with excellent directions for cookery. As also Preserving, Conserving, Candying* (London, 166 .)
181. DINGLER, *Beschreibung meiner Vorrichtung zur Bereitung destillierter Oele* (Trommsdorff *J. d. Pharmacie* vol. XI, 1803, pp. 241-245)
182. DIOSCORIDES, *Greek Herbal* (transl. GOODYER, edit. Gunther, Oxford, 1934)
183. *Distiller, The —— of London, compiled and set forth for the sole use of the Company of Distillers of London* (London, 1639)
184. DOBBELAAR, P. J., *De toestand van de nijverheid in Schiedam in 1816* (*De Economist* vol. 69, 1920, p. 430-435)
185. ——, *Over de opkomst van het Schiedamsche korenwijnbrandersbedrijf* (*De Economist*, vol. 69, 1920, pp. 549-569)
186. ——, *De branderijen in Holland tot het begin der negentiende eeuw* (Rotterdam, 1930)
187. DÖBEREINER, J. W., *Zur mikrochemischen Experimentierkunst* (Jena, 1821)
188. DOMBASLE, MATHIEU DE, *Instructions sur la fabrication des eaux-de-vie de grains et de pommes de terre* (Paris, 1834, 1854)
189. DONOVAN, M., *Domestic Economy* (London, 1830, vol. I, p. 43)
190. DORN, JOH. FRIED., *Zwei neue, sehr zweckmässige Branntweingeräte* (Berlin, 1819)
191. ——, *Praktische Anleitung zum Bierbrauen und Branntweinbrennen und zur künstlichen Anfertigung der Hefe* (Berlin, 1833)
192. DOSSIE, ROBERT, *Elaboratory laid open* (London, 1758, p. 162)
192a. DUBOIS, G., *Origine de la colonne continue* (*Rev. Gén. des Appl. Industr.* 1939, 145)

193. DUBRUNFAUT, A. P., *Traité complet de l'art de la distillation* (2 vols., Paris, 1824; Brussels, 1825)
194. *A complete treatise on the art of distillation.* From the French by J. SHERIDAN; prefixed *The Distiller's practical guide* by P. JONAS (London, 1830)
195. ——, *Notice historique sur la distillation des betteraves* (Paris, 1856)
196. DUBUISSON, F. R. A., *L'art du distillateur* (Paris, 1803)
197. DUJARDIN, J., *Recherches rétrospectives sur l'art de la distillation* (Paris, 1900)
198. DUPLAIS, *Traité de distillation* (Paris, 1855, 1856, 1858)
199. DUPORTAL, A. S., *Recherches sur l'état actuel de la Distillation du vin en France et sur les moyens d'améliorer la distillation des eaux-de-vie* (Paris, 1811)
199a. ——, *Anleitung zur Kenntniss des gegenwärtigen Zustandes der Branntweinbrennereyen in Frankreich, etc.* Aus dem Französischen übersetzt, sowie mit Anmerkungen und Zusätzen den deutschen Branntweinbrennereyen betreffend begleitet von S. F. HERMBSTÄDT (Berlin, 1812, 1817)
200. VAN DYK, *Beschreibung und Abbildung eines verbesserten Dampf-Apparates zur Bereitung von Arztneien* (*Buchners Repertorium f. d. Pharmazie* vol. XXIX, 1828, pp. 94-141)
201. ——, *Beschryving en afbeelding van een verbeterd stoomapparaat ter bereiding van artsenyen* (*Algemeene Konst- en Letterbode*, 1824, Nos. 7 & 8)
202. DYMENT, S. A., *Some Eighteenth Century ideas concerning aqueous vapour and its evaporation* (*Annals of Science* vol. II, 1937, pp. 465-473)
203. EBESSEN, J., *Fragmente aus dem Tagebuche eines Fremden während seines Aufenthaltes in den dänischen Staaten* (Kopenhagen, 1800, p. 264)
204. EBBELL, B., *The Papyrus Ebers* (Kopenhagen, 1937)
205. EGLOFF, G. and LOWRY, C. D. *Distillation Methods, Ancient and Modern* (*Industr. and Eng. Chem.* vol. XXI, 1929, pp. 920-923) (German transl. *Petroleum*, vol. XXV, 1929, pp. 1533-1538) (*Chem. & Metal. Engineering*, April 1935)
206. ——, *Distillation as an alchemical art* (*J. Chem. Education* vol. VII, 1930, p. 2063)
207. ELSHOLTZ, JOH. S'GISM., *Distillatoria curiosa sive ratio ducendi liquores coloratos per alembicum, hactenus si non ignota certe minus observata atque cognita* (Berlin, 1674; Berlin and Francfort, 1704)
208. ELYOT, Sir THOMAS, *The Castell of Helth* (London, 1534)
209. DONATO EREMITA (DI ROCCA DEVANDRO), *Dell'elixir vite* Libri IV (Napoli, 1624)
210. ERMAN, AD. und KREBS, FR., *Aus den Papyrus des Königlichen Museums* (Berlin, 1899, p. 66)
211. ERNSTING, ARTH. CONR., *Nucleus totius medicinae* (Berlin, 1770)
212. ESTIENNE, CH. et LIÉBAUT, J., *L'agriculture et maison rustique* (Rouen, 1647)

213. EULER, L., *Über das Feuer und seine Ausbreitung* (*Recueil des pièces* ... Paris, 1752, vol. IV, p. 13)

214. *Evans, Hrn.* ——*'s Destillir-Apparat* (*Dinglers Polyt. J.* vol. XXVIII, 1828, pp. 116-117) (See also *Mechanic's Magazine* No. 228 of January 5th, 1828)

215. FABRE, PROSPER, *Appareil distillatoire dont la chaudière est en pierre* (French patent of October 4th, 1815)

216. FAIRLEY, T., *Notes on the history of distilled spirits, especially whisky and brandy* (*Analyst*, vol. XXX, 1905, p. 293-306)

217. FELDHAUS, FR. M., *Ein Destillierapparat vom Jahre 1500* (*Chem. Ztg.* vol. 36, 1912, p. 301)

218. FERBER, JOH. JAK., *Nachrichten und Beschreibungen einiger chemischen Fabriken* (Halberstadt, 1793)

219. FESTER, G., *Die Entwicklung der chemischen Technik* (Berlin, 1923)

220. FIEDLER, C. W., *Über die Methode aus Kürbisse und Kartoffeln Branntewein zu brennen nebst Beschreibung einer Quetschmaschine* (Erfurt, 1792)

221. FLICKWIER, J., *Appareil propre à la rectification de l'alcohol que l'auteur nomme par cette raison rectifigène* (French patent No. 380 of May 21st, 1805)

222. FÖRSTER, H., *Praktische Anleitung zur Kenntniss der Gesetzgebung über Besteuerung des Branntweins und des Braumalzes in den Königlich-Preussischen Staaten* (Berlin, 1830)

223. FOURNIER, J. B., *Appareil ambulant propre à la distillation des esprits, eaux-de-vie et principalement des marcs à raisin* (French patent No. 153 of December 27th, 1803) (Additional patents of February 28th 1804 and January 30th 1806) (By decret of January 17th 1814 the main patent was extended to May 1st 1821)

224. FRANCIS OF SIENA, *Liber de venenis* (MSS Bibliothèque Nationale, Paris, No. 6979, date: about 1375)

225. FRANKLIN, A., *Dictionnaire historique des Arts, Métiers et Professions* (Paris, 1906)

226. FRENCH, J., *The Art of Distillation or a Treatise of the Choicest Spagyricall Preparations performed by the way of distillation* (London, 1651, 1664)

227. FUCHSIUS, REMACLUS, *Historia omnium aquarum quae in communi hodie practicantium sunt uses vires et recta eas distillandi ratio. Accessit conditorum (ut vocant) et specierum Aromaticorum quorum usus frequentior apud pharmacopolas tractatus, etc.* (Venetiae, 1542; Paris, 1542, 1552)

228. FUNCK, A. VON, *Beschreibung der Theer- und Kohlenöfen* (*Hannoverische gelehrte Anzeigen*, 1752, 15. Stück)

229. FUNKE, K. PH., *Naturgeschichte und Technologie* (Braunschweig 1790/91, 3 Bde.; Wien, 1812, III. 225, 251)

230. FURNO, VITALIS DE, *Liber selectiorum remediorum pro conservanda sanitate ad totius corporis humani morbos* (Mainz, 1531, p. 12)

231. *Gabelmann, Verfahren des Apothekers* —— *in Barby Branntewein aus rohen Kartoffeln und aus getrockneten Kartoffeln zu ziehen* (*Reichsanzeiger aus dem Jahre 1793* Bd. II, Nos. 46/47)

232. GADOLIN, J., *Vorschlag, die Schlange beym Brannteweinbrennen zu verbessern* (*Schwedische Abhandl. f. d. Jahr 1778*, Leipzig vol. XI, 1783, p. 271)

233. —— (and MACON¹, N.), *Dissertatio chemico-physica de Theoria Calorio Corporum Specifici* (Abö, 1784)

234. ——, *Beschreibung einer verbesserten Abkühlungsanstalt bey Brannteweinbrennereyen* (*Neue Abhandl. königl. schwed. Akad. Wiss.* 1791, p.178) (*Crell's Chem. Ann.* 1792, 4. Stück p. 368)

235. GALE, T., *Antidotarium* (Londen, 1563)

236. GALL, H. L., *Beschreibung und Abbildung eines verbesserten Dampfdestillierapparates* (Trier, 1829)

237. ——, *Das Brannteweinbrennen mittels Wasserdämpfen von A. Kölle geprüft und beleuchtet von ...* (Trier,1830)

238. GALL, HEINR. LURW. LAMBERT, *Ausführliche Beschreibung und Abbildung eines Dampfdestillierapparates nebst Nachtrag* (Trier, 1830)

238a. GALL, L. A., *Ausführliche Beschreibung und Darlegung eines durchaus neuen und eigenthümlichen ... unmittelbar aus der Maische gewährenden patentierten Dampf-Destillierapparat zur Frucht- und Kartoffelbrennerei* (Erstes Heft den praktischen Theil enthaltend, Trier bei F. A. Gall, 1831, 8. VIII, 78 S.)

238b. ——, *Darlegung der Vorzüge des in Preussen, Östreich, Baiern und Würtemberg patentierten rheinländischen Dampfbrennapparates ...* (Trier bei F. A. Gall, 1831, VI, 48 S.)

239. GALL, H. L. L., *Der Rheinländische Dampfbrennapparat in seiner höchsten Vereinfachung* (Trier, 1834)

240. ——, *Beschreibung des Schwarz'schen Brennapparates* (Trier, 1843)

241. GANZEMÜLLER, W., *Die Alchemie im Mittelalter* (Paderborn, 1938).

242. GARDIE, E. DE LA, *Versuch Brodt, Branntewein, Stärke und Puder aus Kartoffeln zu machen* (*Schwed. Abhandl.* 1747, p. 261)

243. *Gedanken über die Schädlichkeit der Brannteweinbrennereyen in einem Lande* (Leipzig, 1790)

244. GEDDA, Le Baron DE, *Un condenseur conique pour les distillateurs* (*Annales des Arts et Manufactures* vol. 19, p. 92)

245. GEOFFROY, *Méthode pour connoitre et déterminer au juste la qualité des liqueurs spiritueuses, qui portent le nom d'eau-de-vie* (*Mém. Acad. des Sciences à Paris*, 1718)

246. GESNER, CONRAD (EUONYMUS PHILIATER), *De Remediis secretis, Liber Physicus, Medicus et partiam Chymicus et Oeconomicus in vinorum diversi apparatu, Medicis & Pharmacopoiis omnibus praecipi necessarius nunc primum in lucem editus* (Zürich, 1552, 1557; second book édited by C. WOLFF, Zürich 1569; Francfort 1578)

247. ——, *Thesaurus Euonymus Philiatri, Ein köstlicher Schatz, usw.* (Zürich, 1555, 1583)

248. ——, *Evonyme Philiâtre Trésor des rémèdes secretz, livre physic, médical, alchymic et dispensatif de toutes substantiales liqueurs et appareils de Vins divers saveurs, nécessaires à tous Gens prin-*

cipalement aux Médicins, Chirügiens et Apothicaires (Traduit par BART. VANEAU, Lyon, 1557, 1572)

249. GESNER, CONRAD (EUONYMUS PHILIATER), *The treasure of Evonymus conteynenge the Wonderfull hid Secretes of Nature touching the most apt Formes to prepare and destyl medicines* (John Daye, London, 1559)

250. —— *A new booke of destyllatyon of waters, called the Treasure of Euonymus.* Translated with great dilligence and labour out of the Latin by PETER MORWYNG (London, John Daye, 1565)

251. ——, *New Jewell of Health* ... translated by GEORGE BAKER (London, 1570, 1576)

252. ——, *The practise of the new and old physicke wherein is contained the most excellent Secrets of Phisicke and Philosophie by* ——, newly corrected and published in English by GEORGE BAKER (London, 1599). (This is the fourth edition of the *New Jewell,* no. 251)

253. GILBERT, O., *Die meteorologischen Theorien des griechischen Altertums* (Leipzig, 1907)

254. GILDEMEISTER, E. and HOFFMANN, F., *Die ätherischen Oele* (Leipzig, 1928)

255. GILG, E. and SCHÜRHOFF, P. N., *Aus dem Reiche der Drogen* (Dresden, 1926, pp. 67-80)

256. GILLY, *Procédé propre à convertir les esprits en eaux-de vie de bon goût et de première qualité* (French patent of April 15th, 1807)

257. GIRARDIN, J. P. L., *Notice biographique sur Edouard Adam* (Rouen, 1836, 1856)

258. ——, *Leçons de Chimie appliquée élémentaire aux Arts industriels* (Paris, 1846; 4th edition 1861)

259. ——, *Handboek der toegepaste scheikunde* (Brussel, 1854)

260. ——, *Considération sur l'usage et l'abus de l'eau-de-vie et des autres liqueurs fortes* (Paris 1864)

261. GLASER, CHRISTOPHE, *Traité de Chimie contenant une méthode claire et facile d'obtenir les préparations de cet art les plus nécessaires à la médecine* (Paris, 1663, 1667, etc.)

262. ——, *Chymischer Wegweiser* (Francfort, 1677)

263. GLAUBER, J. R., *Furni novi philosophici oder Beschreibung der neu erfundenen Destillierkunst* (Amsterdam & Leyden, 1648 and many other editions, e.g. Prague, 1700)

264. ——, *Furni novi philosophici. Descriptio artis destillatoriae novae nec non spiritum, oleorum, florum etc.* (Amsterdam 1651, Francfort 1652, Amsterdam 1658, etc.)

265. ——, *Curieuser Tractat vom Gebrauch und Nutzen des Weins, Korns und Holtzes Gott und dem lieben Vaterlande zu Ehren und allen frommen und getreuen Wein-Händlern, Bier-Brauern, Korn- und Holtz-Händlern zu guter Lehr und Erinnerung wohlmeinend beschrieben und an Tag gegeben* (Amsterdam, 1686)

266. GLOTZ, G., *L'histoire de Délos d'après les prix d'une denrée* (*Revue des Etud. grècques* vol. XXIX, 1916, p. 281)

378 BIBLIOGRAPHY

267. GMELIN, JOH. FR., Geschichte der Chemie (Band I, Göttingen, 1797)
268. GOLDSCHMIDT, G., Von den medizinischen Handschriften zu Basel
(Bericht über die Verwaltung der Öffentl. Bibliothek der Univ.
Basel im Jahre 1940)
269. GRIMBLE, WILLIAM, Construction of apparatus for distilling spirituous
liquors (Brit. Patent No. 5167 of May 14th, 1825)
270. GROTJAN, JOH. AUG., Eines Nordhäusers güldene Kunst Branntewein
zu brennen (Nordhausen, 1754, 1761)
271. GRUPPY, H. B., Samshu-Brewing in North China (J. North China
Branch Royal Asiat. Soc. vol. XVIII, 1884, p. 163)
272. GUAYNERIUS, ANT., Opera (Pavia, 1481; Venezia, 1500)
273. GUMBINNER, J. L., Handbuch der praktischen Branntweinbrennerei
(Berlin, 1840, 1943)
274. GUTMANN, OSCAR, The early manufactur of sulphuric and nitric acid
(J. Soc. Chem. Ind. vol. XX, 1901, p. 5)
275. GUTSMUTH, F. W., Brenn- und Destillierapparat mit Abbild. (Quedlingburg, 1832)
276. ——, Gründliche Anweisung nach einer bis jetzt noch wenig bekann-
ten Methode aus Kartoffeln einen fuselfreien Branntwein zu fa-
bricieren (Quedlingburg, 2. Aufl., 1835)
277. ——, Neuer Brenn- und Destillierapparat, durch welchen Branntwein
sogleich aus der Maische und guter Essig aus dem Niederschlage
der Dämpfe gewonnen wird (Quedlingburg, 1835)
278. GUY, PIERRE, Chaudière et fourneau propres à la destillation des
eaux-de vie (French patent of December 7th, 1804)
279. HAGEN, C. G., Lehrbuch der Apothekerskunst (Königsberg und Leipzig, 1778; 7. Aufl., 1821; 8 Aufl. 1829)
280. HALES, ST., Vegetable Statisticks, (London, 1727)
281. ——, Statistical Essays (London, 1733)
282. ——, Philosophical Experiments (London, 1739)
283. ——, An account of a useful discovery to distill double the usual
quantity of Sea Water, by blowing Showers of Air up through the
Distilling Liquor and also to have the Distilled Water perfectly
fresh and good by means of a little Chalk, etc. (London, 1756)
284. HALL, The distiller (Philadelphia, 1818, 2nd Edit.)
285. HANCOCK, PORTLOCK and EELE, A Way to extract and Make great
Quantities of Pitch, Tarr and Oyle out of a sort of Stone (Brit.
Patent No. 330 of January 29th, 1694)
286. HASKINS, C. H., Studies in the history of medieval science (Cambridge,
1927)
287. HASKINS, HENRY, A new method for extracting a spirit or oil from
tar and from the· same process obtaining a good pitch (Brit.
Patent No. 619 of August 7th, 1764)
288. HAUSBRAND, E., Die Wirkungsweise der Rectificir- und Destillir-
Apparate (Berlin, 1893)
289. HAUTON, An extract of a letter from a learned French gentleman
concerning a way of making Sea-Water sweet (Phil. Trans. Royal
Soc. vol. V, 1670, p. 2048)

290. HEERINGA, K., *Uit de geschiedenis van Schiedam* (*Ber. Vaderlandsche Geschiedenis en Oudheden* vol. IV, 1912, 10, p. 195-240)

291. HEIDE, ANTON DE, *Nieu ligt der Apothekers, aanwysende de onkennis omtrent de kragt der Geneesmiddelen en verbeterende grove misslagen in het voorschryven en bereyden van geneesmiddelen gemeenelyk begaan* (Amsterdam, 1682, 1684, 1742)

292. ——, *Neuen Licht vor die Apotheker, wie solche nach den Grundregeln der heutigen Destillierkunst ihre Arzeneien bereiten sollen* (Leipzig, 1690)

293. HELMONT, J. B. VAN, *Ortus medicinae, id est initia physicae inaudita* .... (Amsterdam, 1648, 1652, 1655, 1656, 1657, 1667)

294. ——, *Aufgang der Arztneykunst* .... (Salzbach, 1683)

295. HENNEBO, R., *De lof der genever* (Amsterdam, 1718, 1723, 1735, 1736, 1800, 1829; Schiedam, 1830; Amsterdam, 1920)

296. HERMANNI, PH., *Een constelijck Distileerboec inhoudende de rechte ende waerachtige conste der distilatiën om alderhande wateren der cruyden, bloemen ende wortelen ende voorts alle andere dinge te leeren distileren opt alder constelijcste, alsoo dat dies gelyke noyt en is gheprint geweest in geen derley sprake. Want de geleerde de selve const verborgen ende noyt geopenbaert en hebben. Nu beschreven van eenen gheleerden meester, gheheeten Philippus Hermanni gelijck als hi tselve veel jaren gheëxperimenteert heeft, so dat hi hier verclaert ende gheleert heeft alle tgene dat ter const der Distilaciën mogelyc is sonder eenigh secreet te verbergen oft achter te laten. Met noch een cleyn tractaat van den selven meester hoe men Ghebranden wyn sal distileren oft maken, al met sonderlinghen consten ende figuren verchiert* (Antwerp, 1552, 1558, 1566) (Amsterdam, 1622)

297. HERMBSTÄDT, S. F., *Sammlung praktischer Erfahrungen für Branntweinbrenner* (Berlin, 1800 . . . . .)

298. ——, *Grundriss der Technologie* (Berlin, 1819)

299. ——, *Chemische Grundsätze der Destillierkunst und Liqueurfabrikation* (Berlin, 1819)

300. ——, *Chemische Grundsätze der Kunst Branntwein zu brennen* (Berlin, 1819)

301. ——, *Algemeene schets der technologie* (Amsterdam, 1825, vol. II, Chapters 21 & 22)

302. HOCH, J. H., *Alchemistical symbols* (*J. Amer. Pharm. Assoc.* vol. 23, 1934, pp. 431-437)

303. HOEFER, F., *Histoire de la chimie* (Paris, 1866)

304. HOFMAN, K., *Finden und Forschen in der älteren Chemie* (*Sitzb. Preuss. Akad. Wiss.* 1931, pp. LVIII-LXVI)

305. HOLMYARD, E. J., *The great chemists* (London, 1928)

306. HOOKE, R., *Micrographia* (London, 1664)

307. HOPKINS, A. J., *A defense of Egyptian alchemy* (*Isis* vol. 28, 1938, pp. 424-432)

308. ——, *A study of the kerotakis process* (*Isis* vol. 29, 1938, pp. 326-355)

309. HOPPE, E., *Geschichte der Physik* (Braunschweig, 1926)

310. HORNBY, TH. and HUNTER, J., *The distillation of ardent spirits from carrots* (*Tillochs Philosophical Magazine* vol. VI, 1800, p. 12)
311. HUMBOLDT, A. VON, *Examen critique de l'histoire de la géographie du nouveau continent* (Paris, 1857, vol. II, p. 300)
312. JACOBI, J. C., *Vom flüssigen Goldschwefel des Spiessglases zur Verbesserung des widrigen Branntweingeschmackes* (*Übersetzungen und deutsche Abhandlungen der Mainzischen Akademie*, vol. I, p. 86)
313. JARS, G., *Voyages Métallurgiques* (Paris, 1781, vol. II, p. 309)
314. JENSEN, I. H., *Deux papyrus à contenu d'ordre chimique* (*Oversigt over det Kgl. Danske vidensk. Selskabs Forhandlinger* 1916, no. 4, pp. 279-302)
315. ——, *Die älteste Alchemie* (Kopenhagen, 1921)
316. JOHANNSEN, O., *Peder Manssons Schriften* (Berlin, 1941)
317. JOHN OF RUPESCISSA, *La vertu et propriété de la quinte essence de toutes choses* (Lyons, 1549, chez Jean de Tournes)
318. ——, *De consideratione quintae essentiae, liber de famulata philosophiae* (Basel, 1561, 1597)
319. JOHNSON, O. S., *A study of Chinese alchemy* (Shanghai, 1928)
320. JOL, G. Z., *Ontwikkeling en organisatie der Nederlandsche brouwindustrie* (Haarlem, 1933)
321. K., G. B., *Aufrichtige Anweisung zum Destillieren der Breslauer und Danziger Liqueure, etc.* (Berlin, no date)
322. KASPEROWSKI, ADAM, *Die Brannteweinbrennerei mit Wasserdämpfen in Holzgefässen* (Leipzig, 1835, 2 vols.)
323. KERSTING, H. H. M., *Freymüthige Gedanken und Vorschläge, in wie weit Brannteweinbrennereyen nötig und nützlich oder schädlich sind, und wie solche einzurichten sind, damit sie zu allen Zeiten im Gange bleiben und beyhalten werden können* (Cassel, 1790)
324. KEULEN, PIETER VAN, *Die nieuw gevonden en geapprobeerde distilleerkonst, waarin verhandelt wordt van verscheydene fyne en andere Wateren om te distileren* (Amsterdam, 1696)
325. KHUNRATH, CONR., *Medulla distillatoria et medica oder Bericht, wie man den spiritus vini zur exaltation bringen soll* (Schleswig, 1594, 2 vols.; Hamburg, 1605, 1623) (German Leipzig 1682)
326. McKIE, D. and HEATHCOTE, A. V., *Discovery of Specific and Latent Heats* (London, 1935)
327. KIRCHER, ATHANASIUS, *Mundus subterraneus in XII libri digestus.* (Amsterdam 1664, 1665, 1668, 1678) (German edit. Augsburg, 1688)
328. KLAR, M., *Technologie der Holzverkohlung und der Fabrikation von Essigsäure, usw.* (Berlin, 1903)
329. KLEMM, G., *Allgemeine Kulturwissenschaft* (Leipzig, 1850, p. 339)
330. KLETLER, P., *Nordwesteuropas Verkehr, Handel und Gewerbe* (Wien, 1924)
331. KÖLLE, A., *Die Branntweinbrennerei mittels Wasserdämpfen, etc.* (Berlin, 1830)
332. KOPP, HERMANN, *Geschichte der Chemie* (Leipzig, 1931, vols. II, III, IV)

333. KRAFFT, G. W., *De Calore ac Frigore Experimenta Varia* (*Comment. Acad. Sci. Imp. Petrop.*, 1744/1746, vol. 14, p. 218)
334. KRAFFT, *The American distiller* (Philadelphia, 1804)
335. KRAMERS, COEN, *De Schiedamsche branderij* (Schiedam, no date)
336. KRAUSS, S., *Sammlung mehrerer wichtigen neuen Angaben und Erfindungen für jeden Bierbrauer und Branntweinbrenner* (Leipzig, 1835)
337. KULISCHER, J., *Allgemeine Wirtschaftsgeschichte* (München, 1928)
338. ———, *La grande industrie aux XVIIe et XVIIIe siècle* (*Ann. d'hist. écon. et soc.*, vol. III, 1931, pp. 11-46)
339. KUNCKEL, JOHANN, *Vollständiges Laboratorium chymicum, worinnen von den wahren principiis in der Nathur, der Erzeugung, den Eigenschaften und der Scheidung der Vegetabilien, Mineralien und Metalle gehandelt wird* (Hamburg, 1716; 4th edit. Berlin, 1767)
340. ———, *Epistola contra spiritum vini sine acido* (Berlin, 1681)
341. LABO, A., *Quintessenz und Destillation bei den alten Aertzten* (*Riv. Ital. Essenze Porfumi* vol. X, 1928, p. 25)
342. LACAMBRE, G., *Traité complet de la fabrication des bières et de la distillation des grains, pommes de terre, vins, betteraves, mélasses, etc.* . . . . . (Brussels, 1851)
343. LAGERCRANTZ, O., *Alchemistische Rezepte des späten Mittelalters* (Berlin, 1925)
344. LAMPADIUS, W. A., *Sammlung praktisch-chemischer Abhandlungen, usw.* (Dresden, 1800, vol. III, p. 127)
345. LAMPERTI, A. VON, *Die allerneuesten Fortschritte der 'Destillierkunst* (Dorpat, 1809, Heft I & II, etc.)
346. ———, *Nachricht von einer sehr vorteilhaften Branntweinbrennerey mittelst Dämpfen in hölzernen Brennkessel* (*Hermbstädt's Bull. des Neuesten und Wissenswürdigsten aus dem Gebiete der Naturwissenschaften*, vol. VI, p. 32)
347. ———, *Der Dampfdestillier-Apparat für Branntweinbrennereyen* (*Hermbstädt's Bulletin des Neuesten* … vol. IX, p. 49 and vol. X, p. 218)
348. LANFRANC, *Chirurgia* (Venetiae 1490, 1519, 1546; Lyons, 1553)
349. LASSWITZ, KURD, *Geschichte der Atomistik vom Mittelalter bis Newton* (Leipzig, 1926, 2 vols.)
350. LAUFER, BERTHOLD, *Sino-Iranica* (Chicago, 1919, p. 238)
351. LAVOISIER, A. L., *Mémoire sur la chaleur* (*Mém. Acad. Royal des Sciences* 1780)
352. ———, *Oeuvres, mémoire sur la fermentation spiritueuse* (Paris 1865, vol. III, p. 777)
353. LEBON, PHILIPPE, *Distillation au moyen du vide et du froid* (French patent of September 11th, 1796 (25 fructidor de l'an IV))
354. LE FÈVRE, NICOLAS, *Traité de Chimie* (Paris, 1649, 1660, 2 vols.)
355. ———, *A Compleat Body of Chemistry* (London, 1664, p. 342)
356. LEJEUNE, FRITZ, *Betrachtungen zur Destillierkunst* (*Die Pharmazeutische Industrie* VIII, 1941, pp. 66-74)

357. LELOUIS, *Appareil propre à extraire du vin par une seule distillation tout l'esprit qu'il contient sans mélange du flegme* (French patent of November 20th, 1807)

358. LEMERY, N., *Cours de chymie contenant la manière de faire les opérations en usage dans la médecine par une méthode facile avec des instructions et raisonnements sur chaque opération pour l'instruction de ceux qui veulent s'appliquer à cette science* (Paris, 1675, 1744 and 21 more editions)

359. LENORMAND, L. S., *Essai sur l'art de la destillation* (Paris, 1811, 2 vols.)

360. ——, *L'art du destillateur des eaux-de-vie et des esprits* (Paris, 1817, 1824)

361. LENTIN, D. A. G., *Etwas über den Prozess der Destillation* (Göttingen, 1799)

362. LIBAVIUS, ANDREAS, *Alchemia e dispersis passim optimorum auctorum . . . . collecta* (Francfort 1595, 1597, 1506 (instead of 1606)

363. ——, *De judicio aquarum mineralium* (forms part II Lib. VI of his *Comment. Alchymiae,* Francfort, 1597)

364. ——, *Praxis Alchymia* (Francfort, 1604)

365. ——, *Syntagma selectorum undiquaque et perspicue traditorum alchymiae arcanorum* (Francfort, 1611, 1613, 1615)

366. *Libellus de distillatione philosophica contra vulgarem modem* (MSS Wolfenbüttel No. 3284 fols. 23r-31v)

367. LIÉBAUT, JEAN, *Thesaurus sanitatis* (Paris, 1577; Francfort 1578)

368. ——, *Quatre livres des secrets de médecine et de la philosophie chimique lesquels sont descrits plusieurs remèdes singuliers pour toutes maladies tant intérieures qu'extérieures du corps humain, traittées bien amplement les manières de distiller eaux huyles et quintessences de toute sorte de matières, etc.* (Paris, 1593; Rouen 1600, 1616, 1628)

369. LIPPMANN, E. O. VON, *Abhandlungen und Vorträge zur Geschichte der Naturwissenschaften* (Leipzig, 1906/1913, 2 vols.)

370. ——, *Zur Geschichte des Wasserbades* (*Beiträge zur Gesch. d. Chemie,* Wien 1908, p. 143)

371. ——, *Einige Bemerkungen zur Geschichte der Destillation und des Alkohols* (*Z. f. angew. Chemie* vol. 25, 1912, p. 1680)

372. ——, *Zur Geschichte des Alkohols und seines Namens.* (*Z. f. angew. Chemie* vol. 25, 1912, p. 2061)

373. ——, *Zur Geschichte der Destillation und des Alkohols* (*Chem. Ztg.* vol. 37, 1913, p. 1)

374. ——, *Beiträge zur Geschichte des Alkohols* (*Chem. Ztg.* vol. 37, 1913, p. 1313)

375. ——, *Thaddäus Florentinus (Taddeo Alderotti) über den Weingeist* (*Arch. Gesch. d. Med.* vol. 7, 1914, p. 379)

376. ——, *Zur Geschichte der ununterbrochenen Kühlung bei der Destillation* (*Chem. Ztg.* vol. 39, 1915, p. 1)

377. ——, *Neue Beiträge zur Geschichte des Alkohols* (*Chem. Ztg.* vol. 41, 1917, p. 865)

378. LIPPMANN, E. O. VON, *Entstehung und Ausbreitung der Alchemie* (Leipzig, 1919-1932, 2 vols.)

379. ——, *Zur Geschichte des Alkohols* (*Chem. Ztg.* vol. 44, 1920, p. 625)

380. ——, *Das Sammelbuch des Vitalis de Furno* (*Chem. Ztg.* vol. 46, 1922, p. 25)

381. ——, *Beiträge zur Geschichte der Naturwissenschaften und der Technik* (Berlin, 1923)

382. ——, *Zur Entstehung von Wasser aus Luft* (*Chem. Ztg.* vol. 55, 1931, p. 681)

383. LISTER, MARTIN, *A Way* (*which seems to be the true method of Nature*) *of distilling Sweet and Fresh water from Sea Water by the Breathing of Sea Plants growing in it* (*Phil. Trans. Royal Soc.* vol. XIV, 1684, p. 493)

384. LONICER, ADAM, *Naturalis historiae opus novum* (Francfort, 1551/1552, 2 vols.)

385. ——, *Herbarium* (Francfort, 1555, 1587)

386. ——, *Kräuterbuch,* (Francfort, 1578, 1737, 1783)

387. ——, *Kreuterbuch, Kunstliche Conterfeytinge der Bäume, Stauden, Hecken, Kräuter, Getreydt, Gewürtze, etc. mit eygentlicher Beschreibung derselben Namen . . . und derselben Gestalt, natürlicher Krafft und Wirckung. Sampt vorher gesetztem und Gantz auszführlich beschriebenem Bericht der schönen und nützlichen Kunst zu Destillieren . . . Nunmehr durch Petrum Uffenbach auf das allerfleissigst übersehen, corrigiert und verbessert* (Francfort, 1630)

388. (——) (PETRUS UFFENBACH) *Adam Loniceri Kräuter Buch und künstliche Conterfeytungen sammt der schönen und nützlichen Kunst zu Destillieren von Petrus Uffenbach in's Teutsche übertragen* (Ulm, 1703)

389. LOWITZ, T., *Anwendung der Schmiedekohlen zur Verbesserung des Branntweins* (*Auswahl ökon. Abhandl. der freyen ökon. Ges. St. Petersburg,* 1793, Theil III, p. 1)

390. ——, *Bemerkungen über die Reinigung des Branntweins durch Kohlen* (*Acta Acad. Elect. Moguntinae* 1794) (Erfurt, 1794)

391. LÜDERSDORFF, F., *Das Wesen der Destillierkunst, nebst einem Repertorium der gebräuchlichsten Gewürze* (Berlin, 1833)

392. ——, *Beschreibung des Pistorius'schen Dampfbrennapparates* (Berlin, 1838)

393. LÜDY, F., *Über die alchemistischen Zeichen* (*Schweiz. Apoth. Ztg.* vol. 64, 1926, p. 25)

394. LULL , RAMOND, *Testamentum novissimum* (Strassburg, 1571)

395. ——, *Arbor scienciae* (Mainz, 1721, 10 vols.)

396. LUYKEN, JAN en KASPAR, *Spiegel van het menschelijk bedrijf* (Amsterdam, 1790)

397. MACQUER, P. J., *Traité de Chimie pratique* (Paris, 1751, 2 vols.)

398. ——, *Elements de Chimie théorique* (Paris, 1751, p. 214)

399. ——, *Chymisches Wörterbuch* (Leipzig, 1781; 1788/91, bearbeitet von J. G. LEONHARDI)

400. MACTEAR, JAMES, *History of the Technology of Sulphuric Acid* (*Proc. Glasgow Phil. Soc.* vol. XIII, 1881, p. 409)
401. MAERCKER, M., *Handbuch der Spiritusfabrikation* (Berlin, 1877; 6th edit. 1894; 7th edit. 1898)
402. ——, *Traité de la fabrication de l'alcool* (Paris 1889)
403. ——, *Anleitung zum Brennereibetrieb* (Berlin, 1898, 2nd edit., 1899)
404. MAGELLAN, *Essai sur la Nouvelle Théorie de Feu Elémentaire et de la Chaleur des Corps* (Paris, 1780)
405. MAGNAN, *Appareil distillatoire ambulant* (French patent of June 7th, 1816)
406. MANGET, J. J., *Bibliotheca chemica curiosa* (Köln, 1702)
407. MANN, H. H., *Analyses of potable spirits used by the native population in India* (*Analyst* vol. XXIX, 1904, pp. 149-152)
408. *Manuel, Le —— du destillateur* (Collection Roret) (Paris, 1856)
409. *Mappae Clavicula or "Little Key to Painting"* (edit. Th. Phillips, *Archaeologie* vol. 32, 1847, p. 183-244)
410. SAINT MARC, *Process of and apparatus for distilling* (British Patent no. 5197 of Juni 28th, 1825)
411. MARGRAF, A. S., *Continuation des preuves fondées sur des expériences qui font voir qu'il se trouve de la terre dans l'eau destillée la plus pure* (*Mém. Berl. Akad. Wiss.* 1756, pp. 20-31)
412. MARTIN, B., *A Sure Guide for Distillers* (London, 1759)
413. MESUE, *Yahya ibn Mâsawaih al-Mârdînî: De medecinis universalibus et particularibus* (Venetiae, 1471)
414. ——, *Collected Works* (Venice, 1549) (and many others; the best being those of GIUNTA and VALGRISI, 1561, 1602)
415. ——, *Simplicia et composita et antidotarii novem posteriores sectiones adnotationes* (Venetiae, 1602)
416. MATTHIOLUS, PETR. ANDR., *Commentarii in VI libros Pedacii Dioscorides Anazarbei de materia medica* (Venezia, 1554, 1558, 1559, 1560, 1563, 1674; French Lyon, 1562; German Prague, 1563) (Contains an appendix called *De ratione distillandi aquas ex omnibus plantis et quomodo genuini odores in ipses aquis conservari possint*)
417. ——, *Opera quae extant omnia* (Supplementum *De ratione distillandi …*) (Francfort, 1586) (Basel, 1565)
418. ——, *Les commentaires sur les six livres de Pedacius Dioscorides anazarbéen, de la matière médicale*, traduit en français par M. ANTOINE DE PINET, …. *comme aussi de distillation, etc. etc.* (Lyon, 1642, 1652)
419. *Maundeville, The Voiage and Travayle of Sir John ——* (edit. J. BRAMONT, Everyman's Library 812, London, 1928)
420. MAUNG, HTIN AUNG, *Alchemy and alchemists in Burma* (*Folklore* vol. XLIV, 1933)
421. MAURIZIO, A., *Geschichte der gegorenen Getränke* (Berlin, 1933)
422. MEGENBURG, KONRAD VON, *Das Buch der Natur* (edit. F. PFEIFFER, Stuttgart, 1861, p. 81)

423. MENEANDER, JUVELIN, *Zurichtung des Theers in Ostbothnien* (*Schrebers Neuen Sammlungen* . . . vol. IV, p. 20)
424. MESCH, A. H. VAN DER BOON, *Verhandeling ter beantwoording van de vrage: Naardien het verschil van onderscheidene geestrijke vochten en dranken uit granen, vruchten of andere planten gestookt, vooral derzelver meer of minder bedwelmend vermogen niets slechts aan de verschillende hoeveelheid Alcohol, maar ook vooral aan de vlugge scherpe olie, de zoogenaamde Foezelolie, die in dezelve bestaat, schijnt te moeten worden toegeschreven, zoo wordt gevraagd: Welke geestryke dranken bevatten de foezelolie in de grootste hoeveelheid? Op welke wyze kan de olie daaruit worden afgescheiden en gezuiverd? Bestaat er eenig verschil in dezelve uit onderscheiden gewassen verkregen? Welke zyn de eigenschappen en de uitwerking op het menschelyk ligchaam? In hoeverre worden door deselve voor de gezondheid schadelyke hoedanigheden aan de gedestilleerde wyngeest houdende vochten medegedeeld en hoe kunnen deze daarvan gezuiverd en min schadelyk gemaakt worden?* (*Nat. Verhand. v. d. Holl. Mij. der Wetensch.*, Haarlem, vol. XXIII, 1836, pp. 118-294)
425. MEYER, KARL, *Geschichte des Nordhäuser Branntweins* (Nordhausen, 1907)
426. MEZ, A., *Die Renaissance des Islâms* (Heidelberg, 1922)
427. MIELI, A., *La science arabe et son rôle dans l'évolution scientifique* (Leiden, 1938)
428. MILLER, JAMES, *Improvements in the Construction of Stills, Furnaces, Chimnies and other Apparatus connected with the Art of Distillation* (British Patent no. 3878 of January 28th, 1815)
429. *Mittel, Leichtes mechanisches —— das Anbrennen der Mösche in der Blase zu verhüten* (*Hoffmann's allgem. Annalen der Gewerbekunde* vol. I, 1803, p. 398)
430. MODEL, J. G., *Kleine Schriften, bestehende in ökonomischen, physischen und chemischen Abhandlungen* (Book III: *Abhandlung vom Brannteweinbrennen*) (St. Petersburg, 1773)
431. ——, *Vom Brannteweinbrennen* (*Abhandl. ökon. Gew. St. Petersburg* vol. III, p. 132)
432. ——, *Beurtheilung des Aufsatzes Ritschkows in Band V* (*Abh. ökon. Ges. St. Petersburg*, vol. IX, p. 41)
433. MÖLLER, J. A. A., *Über Verbesserung der Brannteweinbrennerey, etc.* (Dortmund, 1796)
434. MOLINE, *Alambic et fourneau proposé par M. l'Abbé ——*, prieur chefcier de la commanderie de Saint-Antoine, ordre de Malte (*J. de Physique* vol. XII, 2, 1778, pp. 81-111) (Paris, 1777)
435. MONZERT, *Practical Distiller* (London, 1889)
436. MOREWOOD, J., *An essay on the inventions and customs of both ancients and moderns in the use of inebriating liquors* (London, 1824, 2nd edit. Dublin, 1838)
437. MORIN, J., *Astrologia Gallica* (The Hague, 1661)

438. MÜLLER, NIC., *Freundschaftliche Belehrung an seine Landesleute über die leichteste und sicherste Art aus Kartoffeln einen recht guten Branntwein zu gewinnen* (Nürnberg, 1792)

439. ——, *Über die Gewinnung von vielem und gutem Branntwein, Essig und Likör aus Kartoffeln* (Leipzig, 1797)

440. ——, *Beobachtete Vortheyle beim Brannteweinbrennen* (*Abhandl. ökon. Ges. St. Petersburg*, vol. III, p. 62)

441. MUNIER, ETIENNE, *Mémoire sur la manière de brûler les eaux-de-vie couronné par la Société d'Agriculture de Limoges en 1766* (Paris (?), 1770)

442. MURBERG, JOH., *Historiska anmerkningar om branvinets älder* (*Kongl. Svenska Vitt. Hist. och Antiq. Acad. Handl.* vol. IV, 1797, p. 308)

443. MUSPRATT, K., *Technische Chemie* (Berlin, 1856)

444. (MYREPSOS), *Nicholas Myrepsis Alexandrini medicamentorum opus in sectiones quadraginta octa digestum hactemus in Germania non visum* (edit. L. FUCHS, Basel, 1549)

445. NAUDÉ, W., *Die Getreidepolitik der europäischen Staaten vom XIIten bis zum XVIIIten Jahrhundert* (Berlin, 1896)

446. NAZO, NICHOLAS, *Procédé propre à extraire de l'eau-de-vie de toutes sortes de fruits secs* (French patent of May 17th, 1815)

447. NETTLETON, W., *The manufacture of spirits as conducted in the distilleries of the United Kingdom* (London, 1893)

448. NED. MAATSCHAPPIJ VOOR NIJVERHEID EN HANDEL, *Verhandeling en Teekening nopens de beste en voordeeligste inrichting der ketels, helmen en slangen in de branderijen om in den kortsten tijd de grootst mogelijke hoeveelheid geestrijk vocht over te halen* (Prijsvraag no. 38, blz. 40 van 16 Augustus 1836)

449. ——, *Verhandeling om uit eenig Inlandsch of hier geteeld kunnende worden Buitenlandsch hier te lande niet tot gewoon voedsel voor Menschen dienend Gewas Jenever te stoken* (Prijsvraag no. 61, blz. 48 van 1816)

450. NEUENHAHN, C. C. A., *Die Branntweinbrennerey nach theoretischen und praktischen Grundsätzen* (Erfurt, 2 vols. 1789, 1791, 1803/04, 1810)

451. ——, *Beyträge zur Brennteweinbrennerey in Briefen an Westrumb* (Erfurt, 1793)

452. ——, *Über die Helme der Branntweinblase nebst Beschreibung eines Holzsparenden Blasen-Herdes* (Erfurt, 1795)

453. NEUGEBAUER, A., *Die Technik des Altertums* (Leipzig, 1921)

454. NEUMANN, CASPAR, *Chymia medico-dogmatica-experimentalis* (Züllichau, 4 vols., 1749-1755)

455. ——, *Lehrbuch der Apothekerskunst* (Leipzig, 1786)

456. NEUMANN, C. A., *Die Behandlung der Feuerwärme* (Stück 1 & 2, Altona 1800 & 1802)

457. NEWTON, ISAAC, *Optice or a treatise of the reflexions, inflexion and colours of light* (London, 1704, vol. III, quaest. XVIII, p. 280)

458. NORBERG, J. E., *Beskrifning öfver åtskilliga förbättringar vid bränvins-brännings värktygen* (*K. Svensk. Vetensk. Acad. Nya Handl.* A. 1799, vol. XX, p. 257)

459. ORTOLANUS, *Prcatica vera alchimica* (MSS Bibl. Nationale, Paris, no. 7149) (*Theatrum chemicum* vol. IV, p. 1038) (*Compendium alchimiae*, Basel, 1560)

460. OTTO, F. W., *Über die Brannteweinbrennereien in Flensburg* (Flensburg, 1794)

461. OTTO, JULIUS, *Lehrbuch der rationellen Praxis der landwirtschaftlichen Gewerbe* (Berlin, 1838, 1840, 1850, 1852)

462. PAGANUCCI, J., *Manuel historique, géographique et politique des Négociants* (Paris, 1762, vol. II, p. 280)

463. PAGE, JOHN, *Receipts for preparing and compounding the principal Medicines of the late Mr. Ward* (London, 1763)

464. PARKES, SAMUEL, *Chemical Essays* (London, 1815, vol. II, p. 398)

465. PARMENTIER, A. A., *L'art de faire les eaux-de-vie suivant la méthode Chaptal et les vinaigres simples et composés* (Paris, 1805, 1819)

466. ——,*Notice historique et chronologique de la matière sucrante* (*Ann. de Chim.* vol. LXXX, 1811, pp. 306)

467. ——, *Méthode pratique de la distillation de l'eau-de-vie* (Paris, 1819)

468. PARTINGTON, J. R., *Chinese Alchemy* (*Nature*, vol. 119, 1927, p. 11)

469. ——, *Chemical Arts in the Mount Athos Manual* (*Isis* vol. 22, 1934, pp. 136-150)

470. ——, *Origins and development of applied chemistry* (London, 1935)

471. ——, *The chemistry of Râzi* (*Ambix* vol. 1, 1937/38, pp. 192-196)

472. ——, and McKᴵE, D., *Historical studies on the Phlogiston Theory* (*Annals of Science* vol. II, 1937, pp. 361-404; vol. III, 1938, pp. 1-59, 337-372)

473. PAYEN, A., *Traité de distillation des betteraves* (Paris 1855)

474. ——, *Traité complet de la distillation des diverses substances qui peuvent fournir de l'alcool, vins, grains, tiges, racines, tubercles, etc.* (Paris, 1858, 1861, 5th edit. 1866)

475. ——, *Précis de chimie industrielle* (Paris 1859, vol. II, p. 358)

476. PAYNE, JOHN, *New and more advantageous method of expanding fluids which being convey'd into a proper Ignified Vessel or Vessells contrived for that purpose are immediately rariefied into Elastick Impelling Force, sufficient to give Motion to Hydraulo-Pneumatical and other Engines and Machines for raising Water and other Uses; and also in Brewing and Distilling by a new Form or Make of the Boyler, Still, Evaporating Vessell or Vessells and other contigencies thereunto belonging* (British Patent No. 555 of November 10th, 1736)

477. PERRIER, Sir ANTHONY, *Certain Improvements in the Apparatus for distilling, boilling and concentrating by evaporation various sorts of liquids and fluids* (British Patent no. 4694 of July 27th, 1822)

478. PERTOLD, O., *The liturgical use of Mahuda liquor by Bhils* (*Archiv Orientalni* vol. III, 1931, p. 406)

479. PETERS, HERMANN, *Die Chemie des Markgrafen Friedrich I von Brandenburg* (*Mitt. a. d. german. Nationalmuseum*, 1893, p. 98)
480. ——, *Aus pharmazeutischer Vorzeit in Bild und Wort* (Berlin, 2 vols. 1898, 1910)
481. PETRUS HISPANUS, *A Marvelous Treatise on Waters which master Petrus Hispanus composed with natural industry guided by the intellect* (MSS Addition. No. 32.622) (MSS Egerton No. 2852) (both of the XIVth century)
482. PILCHER, R. B., *Boyle's Laboratory*·(*Ambix* vol. II, 1938/39, pp. 17-20)
483. PIQUE, R., *Histoire de l'alcool et de la distillerie* (*Chimie et Industrie* No. spécial no. 1928, pp. 785-803)
484. PISTORIUS, JOH. HEINR. L., *Praktische Anleitung zum Branttwein-brennen nebst Beschreibung eines Brennapparates* (Berlin, 1821)
485. PLEDGE, H. T., *Science since 1500* (London, 1939, pp. 17, 111)
486. PLESSNER, M., *Arabische Alchemie im lateinischen Abendlande* (*Orient. Lit. Ztg.* vol. 33, 1930, 722)
487. PLOUQUET, W. G., *Warnung an das Publikum vor einem in manchen Branntwein enthaltenden Gifte, sammt dem Mittel es zu entdecken und auszuscheiden* (Tübingen, 1780)
488. POGGENDORFF, J. CH., *Handwörterbuch zur Geschichte der exakten Wissenschaften* (Berlin, 1863)
489. POISONNIER, P. J., *Appareil distillatoire présenté au Ministre de la Marine* (Paris, 1779)
490. PONCELET, POLICARPE, *Le chimie du goût et de l'odorat* (Paris, 1755, 1774, etc.)
491. POPPE, J. H. M. VON, *Versuch einer Kulturgeschichte von den ältesten bis zu den neuesten Zeiten* (Francfort, 1798)
492. ——, *Geschichte der Technologie* (Göttingen, 1807-1811, vol. III, pp. 225, 251)
493. PORTA, G. B. DELLA, *Magia naturalis sive de miraculis rerum naturaliam Libri IV* (Napoli 1558, Antwerp 1560, 1561, 1564, 1570) (French edit. Lyons 1565, 1571, 1650, 1678, 1688; Rouen 1570, 1606, 1626, 1631, 1668) (Italian edition Venetiae, 1560, 1579, 1628)
494. ——, *Magia naturalis Libri XX* (Napoli, 1569, 1588, 1589; Rouen, 1588, 1650; Antwerp, 1576, 1580, 1585; Francfort 1591, 1597, 1607, 1619, 1644; Hanau, 1619, 1644; Leyden 1644, 1650, 1651, 1652; Amsterdam, 1664) (Italian, Napoli 1677) (German edition Nürnberg, 1713)
495. ——, *De distillatione libri IX. Quibus certa methodo, multiplici artificii: penitioribus naturae arcanis detectis cuius libet mixti, in propria elementa resolutio perfectur i docetur* (Roma, Strassburg, 1609)
496. ——, *Natural Magick* (London, 1658, 1669)
497. PRINGSHEIM, O., *Beiträge zur wirtschaftlichen Entwicklungsgeschichte der Vereinigten Niederlände im 17. und 18. Jahrhundert* (Leipzig, 1890)

498. PUFF VAN SCHRICK, MICHAEL, *Hienach volget ein nüczliche materi von manigerley ausgeprañten wasser wie man die nüczen und pruchen sol zu gesuntheyt der menschen Uñ das puchlein hat meyster Michel Schrick doctor der erczney durch lijebe und gepet willen erberen personen ausz den pûchern zu sammen colligiert un beschriben* (Augsburg, 1478, 1479, 1483, etc.)
499. QUERCETANUS (JOSEPH DU CHESNE), *De priscorum philosophorum verae medicinae materia (De dogmaticorum medicorum legitima et restituta medicamentorum praeparatione)* (St. Gervais, 1603)
500. ——, *Pharmacopoea dogmatoricum restituta* (Lipsiae, 1613)
501. RAY, P. C., *A history of Hindu chemistry* (Calcutta, 2 vols. 1903/1927)
502. READ, JOHN, *Alchemy under James IV of Scotland* (*Ambix*, vol. II, 1938/1939, pp. 60-67)
503. READ, T. T., *Chinese Alchemy* (*Nature* vol. 120, 1927, pp. 877-878)
504. REBOUL, *Appareil distillatoire pour les eaux-de-vie 3/5 et 3/6* (French patent of January 17th, 1806)
505. REDGROVE, H. S., *Alchemy, ancient and modern* (London, 1922)
506. REISNER, G., *Aus der Geschichte der Holzdestillation* (*Z. f. Angew. Chemie* vol. 37, 1924, p. 233)
507. REISS, *Abbildung und Beschreibung eines neu erfundenen, einfachen und wenig kostspieligen Brenn- und Destillier-Apparates* (Berlin, 1820)
508. (REPORTS), *Some remarks on the Scotch Distillery and a Description of an improved still which may be charged and run seventy Times in Twenty-Four Hours;* extracted from the Reports of the Committee of the House of Commons July 1799 (*Philos. Magazine* vol, VI, 1800, pp. 70-74)
509. (REPORT), *An account of the improvements introduced by the Scotch Distillers which enable them to charge and run off the same Still upwards of Four Hours;* extracted from the Report of the Committees of the House of Commons om Scotch Distilleries, 1798, 1799 (*Philos. Magazine* vol. VI, 1800, pp. 161-165)
510. RETTBERG, C. F., *Erfahrungen über die Lagerstätte der Steinkohlen, der Braunkohlen und des Torfes ... nebst einem Anhange über das Destilliergeschäft, vorzüglich in Bezug auf das Branntweinbrennen* (Hannover, 1801)
511. RICHMANN, G. W., *De Quantitate Caloris* (*Nov. Comment. Acad. Sci. Imp. Petrop.* vol. I, 1747/48, p. 152)
512. RIEM, J., *Auserlesene Sammlung ökonomischer Schriften* (Dresden, 1790, Abth. II, Heft 2)
513. RIEMSDIJK, J. VAN, *Het brandersbedrijf te Schiedam in de 17de en 18de eeuw* (Schiedam, 1916)
514. RHENANI, JOANNIS, *Aureus tractatus, Latine datus Solis e putep emergentis ...* (Francofurti, 1613, Pars I)
515. RIPLEY, JOHN, *The Compound of Alchemy* (edit. ELIAS ASHMOLE, *Theatrum chemicum britannicum*, London 1652, p. 140)
516. RITSCHKOW, P., *Versuche das Brannteweinbrennen mit möglichster Ersparung des Holzes und besonders des Getreides zu treiben* (*Abh. ökon. Ges. St. Petersburg*, vol. V, p. 34)

517. RIVAZ, *Appareil propre à obtenir par la distillation toutes les sub-stances volatilisées par la chaleur et spécialement les acides et l'ammoniaque* (French patent of June 23rd, 1810)

518. ROBISON, JOHN, *A system of mechanical philosophy* (Edit. Brewster, Edinburgh, 1822, vol. II, p. 1-45)

519. ROBINSON, C. A., *The elements of fractional distillation* (New York, 2nd edition, 1930)

520. ROMERSHAUSEN, *Destillier- und Abdunstungsapparat* (Zerbst bei Füchsel, 1820)

521. ROUX, J. P., *La fabrication de l'alcool et la rectification* (Paris, 1883)

522. (ROZ'ER), *Receuil des Mémoires qui ont concouru pour le prix pro-posé en 1766, par la Société d'Agriculture de Limoges, pour l'année 1767 sur cette question: Quelle est la manière de brûler ou de distiller les vins la plus avantageuse, relativement à la quantité et à la qualité de l'eau-de-vie et à l'épargne des frais, imprimés par l'ordre de la Société* (Paris/Lyons, 1770) (*J. de Physique* vol. I, 1771, septembre, pp. 184-189) (*Crell's Chemisches Journal* vol. VI, 1781, pp. 197-207)

523. ROZIER, M. l'abbé, *Cours complet d'agriculture, théorique, pratique, économique et de médécine rurale et vétérinaire suivi d'une méthode pour étudier l'agriculture par principe ou Dictionnaire universelle d'Agriculture* (Paris, 1781)

524. RUBEUS, HIERON, (GERONIMO ROSSI) *De destillatione Liber in quo stillatitiorum liquorum qui ad medicinam facuint, mithidus ac vites explicantur. Et chemice artis veritas ratione et experimento comprobatur* (Ravenna, 1582; Basel, 1586)

525. RUMFORD, Sir B. THOMPSON, Count, *Recherches sur la chaleur* (Paris, 1813)

526. ——, *On the use of steam as a vehicle for transporting heat* (*Essays political, economical and philosophical*, London, 1802, vol. III, pp. 475-498) (*Complete works* of Count Rumford, Boston, 1873, vol. II, p. 324)

527. ——, *Note on the use of steam as a source of heat* (*Complete Works*, Boston, 1875, vol. IV, p. 789)

528. RUSCELLO, ALEXIUS (PEDEMONTANUS), *De secretis Libri VII mira-quandum reremi varietate utilitatque referte, longe castigitatiores et ampliores quam priore editione* (Venetiae, 1557, 1558, 1560, 1562, 1563, 1603; Basel, 1559, 1563, 1568; Lyons, 1561; French editions Antwerp, 1557; Rouen, 1564, 1600, 1614; English edit. Warde, London, 1562, 1615; Dutch edition Amsterdam, 1614; German edition Basel 1560, 1570, 1571, 1573, 1580, 1581, 1593, 1611, 1615, 1616)

529. RUSKA, J., *Das Steinbuch des Aristoteles* (Heidelberg, 1912, p. 158)

530. ——, *Ein neuer Beitrag zur Geschichte des Alkohols* (*Der Islam*, vol. IV, 1913, pp. 320-324)

531. ——, *Chemische Apparatur bei den Arabern und Persern und im Abendlande am Ausgang des Mittelalters* (*Chem. Apparatur*, vol. X, 1923, p. 137)

532. RUSKA, J., *Probleme der Gâbir Forschung* (*der Islam* vol. XIV 1924, p. 101)
533. ——, *Arabische Alchemisten. II. Ga'far al Sâdiq, der sechste Imâm* (Heidelberg, 1924, p. 73)
534. ——, *Tabula Smaragdina* (Heidelberg, 1926, p. 29)
535. ——, *Chemie in Iraq und Persien im Xten Jahrhundert* (*Der Islam* vol. XVII, 1928, p. 280)
536. ——, *Turba Philosophorum* (Berlin, 1931)
537. ——, *Das Buch der Salze und Alaune* (Berlin, 1935)
538. ——, *Al Razi's Buch der Geheimnisse* (Berlin 1937)
539. ——, *Über die von Abulqâsim az-Zuhrâwi beschriebene Apparatur zur Destillation des Rosenwassers* (*Chem. Apparatur* vol. XXIV, 1937, pp. 313-315)
540. RUSSELL, RICHARD, *The Works of Geber englished by ...* (London, 1678, 1686) (edit. E. J. HOLMYARD, London, 1928)
541. RYFF, W. H., *Teutsche Apotheck* (Francfort 1545, 1561, 1569, 1580, Strassburg, 1602, Francfort 1603)
542. ——, *Neu gross Destillierbuch wohl gegründeter künstlicher Destillation* (Francfort, 1545, 1556, 1567, etc.)
543. SACHS, PH. JAC., *Ampelographia, Vitis Viniferae ejusque partium consideratis, etc.* (Leipzig, 1661)
544. SALA, A., *Hydrelaeologia* (Rostock, 1639)
545. ——, *Opera omnia medico-chymica quae exstant omnia* (Francfort, 1647; Rouen, 1650; Francfort, 1680, 1712)
546. SALADIN OF ASCOLA, *Lumen Apothecariorum* (Bonon., 1468, etc.)
547. ——, *Compendium aromatariorum* (Bononiae, 1488)
548. (SALERNUS), *Catholica Magistri Salerni* (edit. PIERO GIACOSA; *Magistri salernitani mondum editi*, 1901, pp. 69-166)
549. SANDYS, Sir EDWYN, *Europae Speculum* (The Hague, 1629; London, 1632)
550. SARTON, GEORGE, *Introduction to the History of Science* (Washington, 1931)
551. SAVALLE, DÉS. FRANC., *Progrès récents de la destillation* (Paris, 1873)
552. ——, *Appareils et procédés nouveaux de destillation* (Paris, 1875)
553. SAVARY, *Dictionnaire universel du Commerce* (Paris, 1760)
554. SAVONAROLA, MICHAEL, *Ad divum Leonellum Marchionem Estensem libellus de aqua ardenti Michaellis Savonarole phisici sui feliciter incipit. Cum gravissimus Antonius Rosellus ...* (Gregor. de Gentis Pisa, 1484)
555. ——, *De balneis et thermis naturalibus omnibus Italiae* (Ferrara, 1485; Venetiae, 1553)
556. ——, *De arte confectionis vitae simplicem et compositam et de eiusdem admirabili virtute ad conservandam sanitatem et ad diversas humani corporis segritudinis curandas* (Hagenau, 1532; Basel, 1561, 1597)
557. SCARISBRICK, *Spirit* (Stoke on Trent, 1891)
558. SCHELENZ, H., *Geschichte der Pharmazie* (Berlin, 1904)
559. ——, *Zur Geschichte der pharmazeutisch-chemischen Destillierapparate* (Leipzig, 1911)

560. SCHLICHTE, H., *Das Branntweingewerbe in Steinhagen* (Hamburg, 1924)

561. SCHMIDT, C. W., *Das Ganze der Destillierkunst* (Königsberg, 1823)

562. SCHMUCK, MARTIN, *Secretorum naturaliam chymicorum et medicorum thesaurolium* (Schleusingen 2 vols., 1637; Nürnberg, 1652-1653)

563. SCHOFIELD, M., *The rise and decline of charcoal burning* (*Science Progress* vol. XXVI, 1932, pp. 654-661)

564. SCHREGER, TH., *Beschreibung der chemischen Gerätschaft* (Fürth, 1802)

565. SCHREIBER, W. L., *Die Kräuterbücher des XV. und XVI. Jahrhunderts* (München, 1924)

566. SCHREINER, OSWALD, *History of the art of distillation and of distillation apparatus* (Monograph No. 6 of the Pharmaceutical Science Series, Milwaukee, 1901)

567. SCHRIEB, MICHAEL, *Von dem geprannten Weyn und Apothek für den gemeinen Mann* (Augsburg, 1484, 1494; Nürnberg, 1529)

568. SCHRÖDER, JOH., *Pharmacopoea medico- chymica, thesaurus pharmacologus* (Ulm, 1655) (Francfort, 1640, 1669, 1677) (Lyons, 1649, 1656, 1665, 1681) (Ulm, 1641, 1649, 1662, 1705) (Leyden, 1672) (Genève, 1689) (Nürnberg, 1746)

569. SCHROHE, A., *Aus der Vergangenheit der Gärungstechnik und verwandter Gebiete* (Berlin, 1917, Teil II)

570. SEDLACEK, FRANZ, *Thyrsenblut (Bl. f. d. Gesch. d. Technik,* vol. I, 1932, p. 73)

571. SERRES, OLIVIER DE, *Théatre d'Agriculture et Message des Champs* (Paris, 1600, 1620)

572. SHAND, WILLIAM, *Distillation* (British Patent No. 5828 of August 10th, 1829)

573. SHEARS, DANIEL TOWERS, *Distilling* (British Patent No. 5925 of March 31st, 1830)

574. SHEE, JOSEPH, *Distilling* (British Patent No. 6598 of April 22nd, 1834)

575. SHERLEY, T., *Curious Distillatory or the Art of Distilling Coloured Spirits, Liquors, Oyls etc. from Vegetables* (London, 1677)

576. SHIRLEY, J., *Closet of Rarities* (5th Edit. London, 1666)

577. SIEMENS, F. E. VON, *Beschreibung eines neuen Betriebes des Kartoffelnbrennes und einer neuen Dampfdestillation* (Hamburg, 1829)

578. ——, *Beschreibung einer neuen Vorrichtung zum Zerkleinern und Einmeischen von Kartoffeln* (Stuttgart, 1840)

579. SIMON, J. C., *Vollständiger Unterricht vom Branntweinbrennen* (Dresden, 1765, 1795)

580. SINGER, *Die Destillierkunst der Neuzeit* (4. Aufl., Berlin, 1900)

581. SIZAIRE, *Appareil à distiller le vin et à fabriquer de l'eau-de-vie et de l'esprit de vin* (French patent of 1806, February 21st.)

582. SKYTTE, C., *Versuche aus den Kartoffeln Branntewein zu brennen* (*Schwedische Abhandlungen* 1747, p. 252)

583. SMITH, G., *Distilling* (London, 1725)

584. SMITH, E. F., *Old chemistries* (New York, 1927)

585. SNELLER, Z. W., *Rotterdams bedrijfsleven in het verleden* (Rotterdam, 1940)

586. SOLIMANI, LAURENT, *Appareil propre à la distillation du vin et à la formation des eaux-de-vie et esprits* (French patent No. 401 of June 6th, 1801 (le 17 prairial, an IX)) (Additional patent of September 9th, 1801)

587. ——, *Appareil distillatoire propre à la fabrication des eaux-de-vie* (French patent No. 697 of December 20th, 1803)

588. SOMBART, W., *Der moderne Kapitalismus* (6 vols., München, 1919)

589. SOREL, E., *La rectification de l'alcool* (Paris 1894)

590. ——, *La distillation* (Paris, 1895)

591. ——, *Distillation et rectification industrielles* (Paris, 1899)

592. SPEIGHT, W. L., *African Folk Chemistry* (*Pharm. J.* vol. 76, 1933, p. 26)

593. SPETER, M., *Geschichte der Erfindung des Liebigschen Kühlapparates* (*Chem. Ztg.* vol. 32, 1908, pp. 3-5)

594. ——, *Nachtrag zur Geschichte des Gegenstrom-Kühlers* (*Chem. Apparatur* vol. XVI, 1929, pp. 174-177)

595. ——, *Die erst-nachweislichen Anfänge der stetigen Kühlung bei der Destillation* (*Chem. Apparatur* vol. XVI, 1929, pp. 221-224)

596. ——, *Zur Geschichte der Wasserbaddestillation* (*Pharm. Acta Helvetica* vol. V, 1930, p. 116)

597. ——, *Ein Destillierofen um 1500* (*Chem. Ztg.* (Schweiz.) 1932, p. 23)

598. STAMMER, *Manuel complet de la destillation* (Paris, 1865)

599. STAPLETON, H. E. and AZO, R. F., *Alchemical equipment of the eleventh century A. D.* (*Mem. Asiat. Soc. Bengal* vol. I, 1907, p. 47)

600. ——, *An alchemical compilation of the thirteenth century A. D.* (*Mem. Asiat. Soc. Bengal* vol. III, 1910, pp. 57-94)

601. —— and HUSAIN, M. HIDâYAT, *Chemistry in ʿIrâq and Iran in the tenth century A. D.* (*Mem. Asiat. Soc. Bengal* vol. VIII, 1922/ 29, pp. 317-417)

602. STEIN, R., *Distillation* (British Patent No. 5721, of December 1828)

603. STEPHANIDES, M., *Petites contributions à l'histoire des sciences* (*Rev. des Etud. grecques* vol. 28, 1915, p. 39)

604. ——, *Le feu grégeois ou le feu liquide des Byzantins* (*Comptes Rendus Acad. Inscript.* vol. 167, 1918, p. 165-167)

605. ——, *La terminologie des anciens* (*Isis* vol. 7, 1925, p. 468)

606. STILLMANN, J. M., *Chemistry and Medicine in the Fifteenth Century* (*Scientific Monthly* vol. VI, 1918, pp. 167-175)

607. ——, *The Story of early Chemistry* (New York, 1924)

608. SUDHOFF, K., *Deutsche medizinische Inkunabeln* (Leipzig, 1908)

609. ——, *Weiteres zur Geschichte der Destillationstechnik* (*Arch. Gesch. Naturw. Techn.* vol. 5, 1915, pp. 282-288)

610. ——, *Alkoholrezept aus dem 8. Jahrhundert* (*Naturwiss. Wochenschrift* vol. 16, 1917, pp. 681-683)

611. SÜSSENGUTH, A., *Zur Geschichte der chemischen Apparatur* (*Chem. Apparatur* vol. XVII, 1930, pp. 133-135, 159-161)
612. TACHENIUS, OTTO, *Hippocrates chemicus, qui novissimi salis antiquissima fundamenta ostendit* (Venetiae, 1666, Braunschweig 1666, Leyden 1671; Paris 1669, 1673; Brussels 1690) (English transl. London, 1677)
613. TARAJANZ, SEDRAK, *Das Gewerbe bei den Armeniern* (Diss. Leipzig, 1897, p. 20)
614. TATE, P. G. H., *Spirits* (*Encyclopaedia Brittanica* 14th Ed.)
615. TAYLOR, F. SHERWOOD, *A Survey of Greek Alchemy* (*J. Hellenic. Soc.* vol. 50, 1930, p. 109-139)
616. ——, *The beginnings of Alchemy* (*Chemistry and Industry* 1937, pp. 38-42) (*Ambix* vol. I, 1937, pp. 30-47)
617. ——, *Alchemical works of Stephanos of Alexandria* (*Ambix*, vol. II, 1938/39, p. 47)
618. ——, *A Short history of science* (London, 1939)
619. TENNANT, SMITHSON, *On the means of producing a double distillation by the same heat* (*Phil. Trans. Royal Soc.* vol. 102, 1814, p. 587) (*Buchners Repert. f. d. Pharmacie* vol. VI, 1819, p. 142)
620. THADDEUS OF FLORENCE (TADDEO ALDEROTTI), *De virtutibus aquae vitae* (MSS CLM 363; CLM 666)
621. THEOPHRASTUS, *Enquiry into plants* (edit. A. HORT, Loeb Classical Library, London, 1916)
622. THOMPSON, R. C., *An Assyrian chemist's Vademecum* (*J. Royal Asiat. Soc.* 1934, pp. 771-785)
623. THORNDYKE, LYNN, *History of Magic and Experimental Science* (London, 1923, vols. I and II; London, 1934, vols. III and IV)
624. THURNEISSER ZUM THURN, LEONHART, *Quinta Essentia, das ist Die höchste Subtilitet, Krafft und Wirckung* ... (Münster, 1570; Leipzig, 1574)
625. ——, *Historia und Beschreibung influentischer, elementischer und natürlicher Wirckungen aller Erdgewechsen* (Berlin, 1578)
626. ——, *Historia sive descriptio plantarum omnium* .. (Berlin, 1578)
627. TIMMER, E. A. M., *De generale brouwers van Holland* (Haarlem, 1918)
628. TOU P'ING, *Chiu P'u* (see WYLIE, *Chinese literature*, 1867, p. 150)
629. TRITTON, HENRY, *Patent still* (*Buchners Repert. f. d. Pharmacie* vol. VI, 1819, p. 99)
630. TSAO, Y. Y., *The equipment and methods of the ancient Chinese alchemists* (*Science* (*Sci. Soc. of China, Shanghai*) vol. XVII, 1933, pp. 31-54)
631. TSCHIRCH, *Handbuch der Pharmakognosie* (Leipzig, 1910, Band I, Abt. 2, p. 872) (Leipzig, 1930, 2. Auflage, Band I, 1, p. 229)
632. ULSTADT, PHILLIP (VON), *Coelum philosophorum seu De Secretis naturae liber* (Strassburg, 1526, 1630; Francfort, 1551, 1600; Leyden, 1572; Paris, 1544; Lyons, 1553, 1557)
633. ——, *Büchlein von den Heimlichkeiten der Natur* (Francfort, 1551)
634. ——, *Le ciel des philosophes* (Paris, 1547)

635. UNDERWOOD, A. J. V., *The historical development of distilling plant* (*Trans. Instit. Chem. Engineers* vol. XIII, 1935, pp. 34-63)
636. ——, *Alcohol through the ages* (*Chem. Trade J.* vol. 96, 1935, pp. 146-148)
637. URE, ANDREW, *Apparatus for distilling* (British Patent No. 6101 of March 31st, 1831)
638. ——, *A dictionnary of arts, manufactures and mines* (New York, 1843)
639. VIGENÈRE, BLAISE DE, *Traité du feu et du sel, excellent et rare opuscule, trouvé parmy ses papiers après son deceds* (Paris, 1608; Rouen, 1642, 1651)
640. VILLANOVA, ARNALD DE, *De aqua vitae simplici et composita* (Venetiae, 1477, 1478)
641. ——, *Liber de Vinis* (Paris, 1500) (Latin and German edit. Eslingen, 1478; Augsburg, 1479, 1481, 1482, 1483, 1484; Strassburg, 1483, 1484; Reutlingen, 1485; Ulm, 1499, 1500)
642. ——, *Traité des vins suivant l'art de médicine* (Lyons, 1585)
643. ——, *Opera omnia* (Lyons, 1504; Basel, 1585, etc.)
644. ——, *Art und Eigenschaft aller Wein und Bereitung derselben*, übersetzt durch V. HIRNKOFEN genannt KENWART (Wien, 1542)
645. (VILLANOVA), *Tractatus de vino* (Leyden, 1586)
646. WALEY, A., *Notes on Chinese Alchemy* (*Bull. School Orient. Stud.* vol. VI, 1930, i, pp. 1-24)
647. WALTER OF ODDINGTON, *Correctorium alchimiae* (*or Icocedron*) (MSS Brit. Museum No. addit. 15549)
648. WATT, JAMES, *Steam* (article in Robisons' *Mechanical Philosophy*, edt. Brewster, vol. II, Edinburgh, 1822)
649. WEDEL, JOH. WOLFGANG, *De Remora* (Jena, 1730)
650. WEIGEL, CHR. ERENFRIED v., *Observationes chemicae et mineralogicae* Pars I (Göttingen, 1771)
651. WEISS, J. J. G., *Systematische theoretisch-praktische Anweisung zum Fruchtbranntweinbrennen* (Leipzig, 1801, 2 vols.)
652. (WEIN) (HANS FOLZ?), *Wem der geprante Wein nutz sey oder schad, und wie 'er gerscht oder fälschlich gemacht sey* (Bamberg, bey Marx Ayrer und Hans Pernecker, 1493)
653. WESTRUMB, J. F., *Bemerkungen und Vorschläge für Branntweinbrenner* (Hannover, 1793, 3. Aufl. 1803)
654. ——, *Handbuch der Apothekerkunst* (Hannover, 1795, 1798, 1800, 1815)
655. WEULE, K., *Chemische Technologie der Naturvölker* (Stuttgart, 1922, vol. II, p. 55)
656. WIEDEMANN, E., *Über die Naturwissenschaften im Islamischen Mittelalter* (*Der neue Orient* vol. V, 1919, pp. 52-56)
657. ——, *Beiträge zur Geschichte des Zuckers* (*Deutsche Zuckerindustrie* 1921, p. 302)
658. WIESENHAVERN, L. H. I., *Abhandlung über das Theer- und Pechbrennen* (Breslau, 1793)

659. WILCKE, J. C., *Om Snöns kyla vid Smältningen* (*K. Svenska Vet. Akad. Handl.* vol. 33, 1772, p. 97)

660. ——, *Rón om Eldens specifika myckenhet uti fasta Kroppar och dess afmätande* (*K. Svenska Vet. Akad. Nya Handl.* vol. II, 1781, p. 49)

661. WILLIAM OF MARRA, *Sertum papale de venenis* (MSS Vatic. Barb. No. 306)

662. WINKLER, L., *Thyrsenblut und Thyrsenöl* (*Pharm. Monatshefte* vol. IV, 1923, p. 105)

663. WINTER VON ANDERNACH, JOHANNES (GUINTHERUS ANDERNACENSIS), *De medicina veteri et nova faciunda comment. secund.* (Basel, 1571)

664. WOLF, A., *History of science, technology and philosophy in the 16th and 17th centuries* (London, 1936)

665. ——, *History of science, technology and philosophy in the 18th century* (London, 1938)

666. (WOLLENBERG, J. G. V. D.), *De ed'le Hollandsche, Fransche en Duitsche Disteleerkunst met en zonder distelatie, waarin voorkomen de voornaamste likeuren, Elixers, Bitters, fijne Spiritussen benevens het maken van Keulsche Wijn-Azijn* ('s-Hertogenbosch, no date)

667. WORTH, Y., *The whole art of destillation* (London, 1692)

668. ——, *The compleat distiller of the whole art of distillation practically stated* (London, 1705)

669. WRESZINSKI, W., *Der grosse medizinische Papyrus des Berliner Museums* (Pap. Berlin 3038) (Leipzig, 1909 pp. 9, 58)

670. WYATT, CHARLES, *Certain improvements in the apparatus for and mode of distilling, drying coffee and sugar* (British Patent No. 2639 of August 2nd, 1802)

671. ZEISE, HEINR., *Praktische Anleitung zur vorteilhaften und sicheren Benutzung der Wasserdämpfe von einfacher und mehrfacher Spannung zumeist zu pharmazeutischem Gebrauch* (Altona, 1826)

672. ZILLESEN, CORN., *Antwoord op de vraage des oeconomischen taks van de Hollandsche Maatschappye der Wetenschappen te Haarlem over de vuurwerken in de branderyen* (Haarlem, 1784)

673. ZOSIMOS OF PANOPOLIS, *Works* (edit. BERTHELOT et RUELLE, *Collection des anciens alchemistes grecs*, sec. livraison, Paris, 1888)

INDEX: PERSONAL

INDEX : TOPICAL

FORBES, Art of distillation   26

www.ingramcontent.com/pod-product-compliance
Lightning Source LLC
Chambersburg PA
CBHW030757150426
42813CB00068B/3202/J